CONGENITAL AND ACQUIRED BONE MARROW FAILURE

——

CONGENITAL AND ACQUIRED BONE MARROW FAILURE

Edited by

M.D. ALJURF
Adult Hematology and Bone Marrow Transplantation
Oncology Center, King Faisal Specialist Hospital
and Research Center, Riyadh, Saudi Arabia

E. GLUCKMAN
Eurocord, Saint Louis Hospital, Paris, France

C. DUFOUR
Hematology Unit
G. Gaslini Children's Hospital, Genova, Italy

ACADEMIC PRESS

An imprint of Elsevier
elsevier.com

Academic Press is an imprint of Elsevier
125 London Wall, London EC2Y 5AS, United Kingdom
525 B Street, Suite 1800, San Diego, CA 92101-4495, United States
50 Hampshire Street, 5th Floor, Cambridge, MA 02139, United States
The Boulevard, Langford Lane, Kidlington, Oxford OX5 1GB, United Kingdom

Notices
Knowledge and best practice in this field are constantly changing. As new research and experience broaden our understanding, changes in research methods, professional practices, or medical treatment may become necessary.

Practitioners and researchers must always rely on their own experience and knowledge in evaluating and using any information, methods, compounds, or experiments described herein. In using such information or methods they should be mindful of their own safety and the safety of others, including parties for whom they have a professional responsibility.

To the fullest extent of the law, neither the Publisher nor the authors, contributors, or editors, assume any liability for any injury and/or damage to persons or property as a matter of products liability, negligence or otherwise, or from any use or operation of any methods, products, instructions, or ideas contained in the material herein.

Library of Congress Cataloging-in-Publication Data
A catalog record for this book is available from the Library of Congress

British Library Cataloguing-in-Publication Data
A catalogue record for this book is available from the British Library

ISBN: 978-0-12-804152-9

For information on all Academic Press publications
visit our website at https://www.elsevier.com/

Working together
to grow libraries in
developing countries

www.elsevier.com • www.bookaid.org

Publisher: Mica Haley
Acquisition Editor: Tari Broderick
Editorial Project Manager: Kathy Padilla
Production Project Manager: Karen East and Kirsty Halterman
Designer: Victoria Pearson

Typeset by Thomson Digital

Contents

21. Bone Marrow Failure Syndromes in Children

S. ELMAHDI, S. KOJIMA

List of Contributors

S.O. Ahmed Adult Hematology and Bone Marrow Transplantation, Oncology Center, King Faisal Specialist Hospital and Research Center, Riyadh, Saudi Arabia

G. Aldawsari Adult Hematology and Bone Marrow Transplantation, Oncology Center, King Faisal Specialist Hospital and Research Center, Riyadh, Saudi Arabia

M.D. Aljurf Adult Hematology and Bone Marrow Transplantation, Oncology Center, King Faisal Specialist Hospital and Research Center, Riyadh, Saudi Arabia

H. Alzahrani Adult Hematology and Bone Marrow Transplantation, Oncology Center, King Faisal Specialist Hospital and Research Center, Riyadh, Saudi Arabia

M. Ayas Pediatric Hematology–Oncology and Stem Cell Transplantation, King Faisal Specialist Hospital and Research Center, Riyadh, Saudi Arabia

A. Bacigalupo Istituto di Ematologia, Policlinico Universitario A. Gemelli, Universita' Cattolica del Sacro Cuore, Roma, Italy

F. Ciceri Hematology and BMT Unit, IRCCS San Raffaele Scientific Institute; University Vita-Salute San Raffaele, IRCCS San Raffaele Scientific Institute, Milano, Italy

J.N. Cooper National Institutes of Health, National Heart, Lung, and Blood Institute, Bethesda, MD, United States

J.H. Dalle Hematology–Immunology Pediatric Department, Robert–Debre Hospital, Paris, France

C. Dufour Hematology Unit, G. Gaslini Children's Hospital; Unità di Ematologia Istituto Giannina Gaslini, Genova, Italy

S. Elmahdi Department of Pediatrics, Graduate School of Medicine, Nagoya University, Nagoya, Japan

P. Farruggia Pediatric Hematology and Oncology Unit, A.R.N.A.S. Ospedale Civico, Palermo, Italy

F. Fioredda Hematology Unit, G. Gaslini Children's Hospital; Unità di Ematologia Istituto Giannina Gaslini, Genova, Italy

S. Gandhi Department of Haematological Medicine, King's College Hospital/King's College London, London, United Kingdom

S. Giraudier Hematology Laboratory, Henri–Mondor Hospital, Paris–Est–Creteil University, Paris, France

E. Gluckman Eurocord, Hospital Saint-Louis, Paris, France

F. Grimaldi Hematology, Department of Clinical Medicine and Surgery, Federico II University, Naples, Italy

B. Höchsmann Institute of Transfusion Medicine, University of Ulm; Institute of Clinical Transfusion Medicine and Immunogenetics, German Red Cross Blood Transfusion Service Baden–Württemberg–Hessia, and University Hospital Ulm, Ulm, Germany

K. Hosokawa National Institutes of Health, National Heart, Lung, and Blood Institute, Bethesda, MD, United States

S. Kojima Department of Pediatrics, Graduate School of Medicine, Nagoya University, Nagoya, Japan

T. Leblanc Hematology–Immunology Pediatric Department, Robert–Debre Hospital, Paris, France

J.C.W. Marsh Department of Haematological Medicine, King's College Hospital/King's College London, London, United Kingdom

D. Meyran Hematology–Immunology Pediatric Department, Robert–Debre Hospital, Paris, France

M. Miano Hematology Unit, G. Gaslini Children's Hospital; Unità di Ematologia Istituto Giannina Gaslini, Genova, Italy

G.J. Mufti Department of Haematological Medicine, King's College Hospital/King's College London, London, United Kingdom

J.R. Passweg Division of Hematology, University Hospital Basel, Basel, Switzerland

R. Peffault de Latour BMT Unit, French Reference Center for Aplastic Anemia and PNH, Saint-Louis Hospital, Paris, France

A.M. Risitano Hematology, Department of Clinical Medicine and Surgery, Federico II University, Naples, Italy

A. Rovó Hematology, University Hospital of Bern, Bern, Switzerland

A. Ruggeri Eurocord, Hospital Saint-Louis; Hematology Department, Hospital Saint Antoine, Paris, France

S. Samarasinghe Department of Haematology, Great Hormond Street Hospital, London, United Kingdom

P. Scheinberg Oncology Center, Hospital São José National Institutes of Health, National Heart, Lung, and Blood Institute, Bethesda, MD, United States

H. Schrezenmeier Institute of Transfusion Medicine, University of Ulm; Institute of Clinical Transfusion Medicine and Immunogenetics, German Red Cross Blood Transfusion Service Baden–Württemberg–Hessia, and University Hospital Ulm, Ulm, Germany

S. Sica Istituto di Ematologia, Policlinico Universitario A. Gemelli, Universita' Cattolica del Sacro Cuore, Roma, Italy

M.T.L. Stanghellini Hematology and BMT Unit, IRCCS San Raffaele Scientific Institute, Milano, Italy

A. Tichelli Hematology, University Hospital of Basel, Basel, Switzerland

M.T. Van Lint Divisione di Ematologia e Trapianto di Midollo Osseo, IRCCS AOU San Martino IST, Genova, Italy

A.J. Warren Cambridge Institute for Medical Research; The Department of Haematology, University of Cambridge; Wellcome Trust–Medical Research Council Stem Cell Institute, University of Cambridge, Cambridge, United Kingdom

N.S. Young National Institutes of Health, National Heart, Lung, and Blood Institute, Bethesda, MD, United States

Introduction

From the first descriptions by Ehrlich over a century ago to where we are today, the understanding of the pathogenesis and biology and of the diseases that are encompassed under the umbrella of the "bone marrow failure syndromes" have witnessed tremendous progress. This has been coupled with significant improvements in the clinical care and outcomes of both congenital and acquired forms of bone marrow failure.

When considering acquired bone marrow failure, in the last 4 decades, aplastic anemia has progressed from a disease that was almost universally fatal to one in which the majority of patients can now expect to be cured and lead normal lives. We have gained noteworthy insights in understanding the immune mechanism of hematopoietic stem cell destruction, stem cell niche, and T-cell dysregulation. This knowledge has placed immunosuppressive therapy at the center stage in the treatment of aplastic anemia, and has facilitated the initiation of clinical trials for the use of new potential therapies to treat acquired aplastic anemia.

Substantial progress in the knowledge about the role of somatic mutations and their relation to clonal hematopoiesis in acquired bone marrow failure has been facilitated by the use of the next generation sequencing technologies, and has allowed us to understand their relationship with myelodysplastic syndromes, paroxysmal nocturnal hemoglobinuria, and the risk of disease progression.

Advances in HLA-typing technology, progress in the field of alternate donor availability, haploidentical and cord blood transplants, and in supportive care of hematopoietic stem cell transplantation have lead not only to the expansion of the potential donor pool, but also to remarkable improvements in the outcome of hematopoietic stem cell transplantation for bone marrow failure from related and unrelated stem cell sources with excellent rates of cure, particularly in the younger patients, when a fully matched unrelated donor is used.

Telomere shortening and telomeropathies have been increasingly recognized as contributing factors for bone marrow failure and stem cell exhaustion in acquired states and causative factors for the increasing number of entities of congenital bone marrow failure.

Several new treatment modalities have emerged during the last decade, most notable of which is the introduction of thrombopoietin mimetic agents. Possible interventions to rescue telomere length are being actively investigated.

Given the exciting progress in the field, this book is a timely educational initiative produced jointly by the Working Party of Severe Aplastic Anemia of the European Society for Blood and Marrow Transplantation and the European School of Haematology.

The contents are a collection of up-to-date contributions from world-renowned experts in the field of congenital and acquired bone marrow failure. The chapters in this book provide extensive coverage for all aspects of congenital and acquired bone marrow failure, including biology, pathology, and treatments involving hematopoietic stem cell transplantation.

We hope this book will be an important resource for scientists, clinicians, nurses, and all other health-related professionals involved in research and patient care with the bone marrow failure syndromes.

With the endorsement of the European Society for Blood and Marrow Transplantation and the European School of Haematology

Mahmoud Aljurf, Eliane Gluckman, Carlo Dufour

Epidemiology of Acquired Bone Marrow Failure

F. Grimaldi*, S. Gandhi**, A.M. Risitano*

*Hematology, Department of Clinical Medicine and Surgery, Federico II University, Naples, Italy; **Department of Haematological Medicine, King's College Hospital/ King's College London, London, United Kingdom

INTRODUCTION

Paul Ehrlich in 1888 gave the first seminal description of aplastic anemia (AA) in a pregnant woman, where the normal hemopoietic tissue was replaced by a fatty marrow and empty spaces, the "hypocellular" marrow that resulted in pancytopenia. Idiopathic AA is a rare form of acquired bone marrow failure, where improved supportive care, early institution of immunosuppressive treatment, and hematopoietic stem cell transplantation have led to better treatment outcomes [1]. Several retrospective studies from Europe, United States, South America, and Asia suggest that the incidence is 0.6–6.1 cases per million population. In addition, the incidence of AA shows geographical variability, with lower rates reported in Europe, North America, and Brazil and higher rates in Asia. However, the rarity of disease means there have been few prospective studies to gather data on the epidemiology of AA. Based on the two epidemiological studies carried out in Europe and Asia that used the same methodology, the incidence of the disease is two- to threefold higher in Asia than in the West [2–12]. This variability in incidence rates may reflect differences in exposure to environmental factors including viruses, drugs and chemicals, genetic background, diagnostic criteria, and study designs.

In the following sections, a short review of studies performed and available in literature, including case series and reports to determine the epidemiology and demographics of AA across different centers of the world is presented. Discussion is restricted to acquired cases of AA, and hence the inherited bone marrow failure syndromes, the inevitable cases of marrow aplasia that follow intentional chemoirradiation treatments and cytopenias of nutritional deficiencies or other causes have been excluded.

Congenital and Acquired Bone Marrow Failure
http://dx.doi.org/10.1016/B978-0-12-804152-9.00001-4

INCIDENCE OF AA IN DIFFERENT GEOGRAPHICAL REGIONS AND RACE

The annual incidence of AA varies from 0.6 to 6 per million populations per annum across centers in different continents. Most findings are from retrospective studies, and even retrospective reviews of death registries. However, this incidence masks the variability that is seen across continents and different ethnic groups; for example, reports from the Barcelona group (2008), which was a detailed prospective study by Montané et al. [2] had an incidence of 2.34 per million population. This incidence rate is similar to the 2.0 per million reported by the International Agranulocytosis and Aplastic Anemia Study (IAAAS) [3], which was conducted in Europe and Israel from 1980 to 1984, and to rates reported in smaller national studies in Europe that included United Kingdom [4], France [5], Scandinavia [6], and in South Americas and Brazil [7]. The incidence was accurately determined to be 4 cases per million population in Bangkok [8], but based on prospective studies, it may actually be closer to 5.6 cases per million population in the rural areas of Thailand (Khonkaen region) [9]. In the prospective Chinese Epidemiologic Study Group of Leukemia and Aplastic Anemia survey, 7.4 per million was reported as a national incidence, which clearly is much on the higher side, but may have been overestimated, as stringent criteria for the diagnosis of AA, as bone marrow study, were not strictly applied [10]. Increased incidence in Eastern countries may be related to environmental factors, such as increased exposure to toxic chemicals and pesticides on agricultural farms, practiced in the Far East and South Asia. However, the incidence of AA in children of immigrants from the East Asia in a pediatric population from the ages 0 to 14 years, in a study from British Columbia (Canada) was significantly higher at 6.9 per million, as compared to children of white/mixed ethnic descent at 1.7 per million [11]. Children of immigrants from

South Asia, in this study were found to have an incidence even higher than their counterparts from East Asia, at 7.3 per million [13]. This study suggested that Asian children have an increased incidence of AA, possibly as a result of a genetic predisposition. Indeed, in a hospital based case control study from Lucknow (India), the annual incidence of childhood AA was determined to be 6.8 per million [14].

Benzene has been found to be toxic to the hemopoietic progenitor cells. In large collaborative studies between the National Cancer Institute and American and Chinese institutions [15], hematologic susceptibility to benzene has been correlated to nucleotide polymorphism in key drug metabolic patways [16]. Thus, hematotoxicity from exposure to benzene may be particularly evident among genetically susceptible populations, and it is plausible that the increased incidence of AA seen in Asia may be related not only to increased exposure but also to key polymorphism in genes regulating the metabolism and cytokine expression.

Similarly, although AA incidence reported in a multicenter Latin American study remains very low at 1.6 per million; this study corroborated the association between risk of exposure to benzene, chloramphenicol, and also azithromycin and predisposition to AA in this area [17,18].

AGE AND GENDER RELATED DEMOGRAPHICS OF AA

In nearly all modern studies of AA, the sex ratio has been close to 1:1, which is unusual for immune-mediated diseases [2,19]. An exception to this has been a study from the Sabah province in Malaysia [12], where an unusually high male to female ratio was noted at 3:4. Similarly a male preponderance was also noted in studies from Thailand [9], India [20], and Pakistan [21]. This may reflect the underreporting of cases of AA among females and access to adequate healthcare services in Asia. However, it remains unclear why a female preponderance is not seen in a quintessential autoimmune disorder, such as AA among studies in Europe and United States.

In all the largest studies available, including a series of 300 patients reported by Clinical Center at NIH by Young et al. [22], the Barcelona report [2] and epidemiologic study from Thailand [9], 2 patient age peaks of incidence are constantly observed, one among young adults and the second in the elderly. This characteristic biphasic distribution shows two peaks, one from 10 to 25 years and the second above 60 years. Within the younger age group, a small peak in the incidence is observed in childhood, probably due to overlap with inherited marrow-failure syndromes featured by a less penetrating phenotype, where classical physicalanomalies of the inherited marrow-failure syndromes are not obvious. On the other hand, the second incidence peak of AA seen above 60 years may reflect the smaller pool of hematopoietic stem cell reserve left, with age-related telomeric attrition and its capacity to maintain normal hemopoiesis, in the face of an immune insult against the hematopoietic precursor cells [23].

POSTHEPATITIS AA AND AA OCCURRING AFTER VIRAL INFECTIONS

Posthepatitis AA is a stereotypical syndrome, where pancytopenia often presents 2–3 months after an acute attack of seronegative severe but self-limited liver inflammation. This distinct variant has been commonly seen in 5–10% of "classical" AA cases, typically occurring in adolescent boys and young men [24]. Severe imbalance of the T-cell immune system as seen with "classical" AA, human leukocyte antigen (HLA) association, and effective response to immunosuppressive therapy strongly suggest an immune-mediated mechanism. As for other form of AA, higher incidence is noted in East Asia (4–10%), when compared to the Western coun-

tries and Europe, where recently an incidence of 5.4%, stable across the years, has been reported in a large registry study from EBMT [25].

Cases of AA associated with HAV, HBV, HGV, parvovirus B19 [26–29], Epstein–Barr virus (EBV) [30], transfusion-transmitted virus (TTV) [31] or echovirus [32] infections have been reported. In a German–Austrian study of 213 pediatric cases aged 17 years or less of AA, 80% of cases were idiopathic, 9% followed posthepatitis AA, 7% were following viral infections, and 4% were associated with drugs/toxins [33].

Parvovirus B19 is the causative agent for fifth disease, usually in the immunocompetent host. However, transient aplastic crises have been reported in chronic hemolytic anemias, such as sickle cell disease owing to reticulocytopenia, and cases of severe AA have also been reported in normal individuals during an acute episode of infection [34]. The actual incidence of acute parvovirus B19 infection at presentation in AA is not known, but in single case series of 27 patients with AA from India, parvovirus B19 IgM and viral DNA was detected in nearly 40% of the cases [35]. This maybe an important etiological factor, especially in immunocompromised host or patients with chronic hemolytic anemias.

Very recently retrospective observations about eight AA cases occurring during HIV infection have been reported [36]. Even if AA appears to be a late rare complication in HIV patients, report from Pagliuca et al. [36] highlights that immunosuppressive therapy is a feasible strategy in AA patients and that better outcome is observed in patients eligible to transplant, while death for infection remain the principle cause of mortality in undertreated patients.

AA AND ASSOCIATION WITH TOXINS/DRUGS

There is no discernible difference in the demographics or clinical behavior, including response to immunosuppressive therapy, between patients classified as having "drug or toxin induced" versus "idiopathic" AA [37].

Benzene is the most widely studied and implicated amongst toxins causing AA. The relationship was initially brought to light by a case where a series of workers exposed through their specific occupations [15] and has since been detected in some, but not all population-based case-control studies. An association has been seen in some case-control studies but even when present, the proportion of cases that can be attributed to this chemical has been small. Studies on American workers earlier in this century suggested that the risk of AA was about 3% in men exposed to concentration higher than 300 ppm, and in the more recent IAAAS study [3], benzene was accounted for about 1–3% of AA recorded cases. Similarly in Thailand population, benzene carried a relative risk of 3.5 but accounted for an etiologic fraction of only 1% [9].

Pesticides have been associated with AA in a large number of medical records. In the Indian cohort of pediatric AA, although significantly higher blood levels of organochlorine compounds were detected suggesting an association, they were not entirely supported by statistical methods [14]. Anecdotal case reports of AA following use and pesticides, such as dichlorodiethyly-trichlorthane (DDT), chlordane or lindane, or following exposure to organic solvents, such as toluene and other molecules resembling benzene, or containing benzene ring, again point merely to an association. Unfortunately, systematic population case-control studies correlating level and duration of exposure of these identified toxins and AA onset are lacking.

Initially suggested by accumulation of case reports, specific drug associations have been established in different population based study and have changed across time, mainly due to changes in drugs diffusion and utilization. Chloramphenicol, for example, that gained notoriety for its prominent association with AA in the 1950s and for decades was considered the commonest cause of the disease, has progressively declined to the point that it has not been reported as significant risk factor in any recent systematic epidemiologic study of AA in Western countries. Even in Thailand where the need for such effective and inexpensive antibiotic is substantial, and usage is reported 100 times greater than in the West, association with AA is infrequent, probably due to lower-doses prescription [9].

In the IAAAS study [3] approximately 25% of the identified AA cases were related to drug use. Major drug associations were with gold salt (relative risk, RR of 29), antithyroid drugs (RR of 11), and nonsteroidal antiinflammatory agents (RR of 8.2 for Indomethacin). Similarly in 235 patients with AA prospectively followed by the Barcelona group [2], 67 cases (28.5%) had a history of exposure to drugs or toxic agents that have been associated with AA, sometime in the preceding 6 months. Forty nine (20.8%) cases had been exposed to the following drugs (Table 1.1). In addition, 21 (8.9%) cases had been exposed to toxic agents: insecticides ($n = 8$), benzene ($n = 6$), and other solvents ($n = 10$).

TABLE 1.1 Exposures to Drugs Reported to be Associated With Aplastic Anemia in the 2–6 Months Prior to Hospital Admission Among 235 Cases (Montané et al. [2])

Drug	N (%)
Allopurinol	9 (3.8)
Indomethacin	9 (3.8)
Gold salts	9 (3.8)
Sulfonamides	9 (3.8)
Carbamazepine	5 (2.1)
Ticlopidine	4 (1.7)
Chloramphenicol	3 (1.2)
Oxyphenbutazone	3 (1.2)
Phenylbutazone	3 (1.2)
Penicillamine	3 (1.2)
Clopidogrel	2 (0.8)
Methimazole	2 (0.8)

Finally, incidence of drug-associated AA appears to be lower in East Asia [8,9].

AA AND ASSOCIATION WITH HLA GENES

The (HLA) system is a crucial group of genes responsible for starting and regulating immune response. Since antigen presentation and T-cell activation through HLA may represent the early step that precede global hematopoietic stem cell destruction in AA, HLA polymorphism may contribute to pathogenesis of the disease by: (1) increasing or decreasing susceptibility to AA, particularly for specific ethnic or age groups, (2) facilitating activity of drugs or viral antigen to break immune tolerance and starting the disease, (3) influencing response to immunosuppressive therapy.

As for other autoimmune disease, a large number of studies demonstrated an association between polymorphism of HLA class II genes and AA susceptibility (Table 1.2, a).

An increased frequency of HLA-DR2 was first described in several studies for European and American Caucasian patients [38], and finally confirmed in a multiethnic cohort by

Nimer et al. [39]. Subsequently, positive association between HLA-DR2 (precisely DRB1*15 allele) and AA was confirmed in Chinese [40], Japanese [41], Turkish [42], Pakistani [43], and Malaysian [44] series. These data depict a clear role for HLA-DR2 gene as risk factor for AA, and different distribution of this gene across human population may probably account for different incidence seen for AA inspecific geographical areas and ethnic groups. Interestingly, when Maciejewski et al. [45] analyzed distribution of HLA-DR2 (mainly DRB1*15 allele) across disease subgroups (i.e., AA, hemolytic PNH and AA/PNH), they found an increased frequency in the ones where bone marrow failure was associated with a PNH clone. Other DR2 alleles have been correlated with AA; in a single case report, Nakao et al. [46] showed a specific T-cell cytotoxic response associated with the DRB1*0405 allele. The DRB1*07 allele has been reported in a single cohort of Iranian subjects [47], and DRB1*0901 in Chinese children [48]. Finally, evidences available correlate AA even with other HLA-II class genes, such as Dpw3 [49] and DR4 [22].

Less data are available for HLA class I genes. Early reports suggested a correlation with HLA-A2 gene [50,51] and with HLA-B7 and HLA-B14

TABLE 1.2 Summary of AA and Incidence of Specific HLA Genes or Alleles

a. HLA and increased risk of AA (Class II)	b. HLA and increased risk of AA (Class I)	c. HLA and reduced risk of AA	d. HLA and onset age of AA	e. HLA and drug/viral related AA	f. HLA and AA outcome
HLA-DRB1*1501	HLA-A2	HLA-DRB1*13	HLA-DRB*09	Haplotype: -B38, -DR4, -DQ3	HLA-DR2
HLA-DRB1*0405	HLA-A*0206	HLA-DRB1*03:01	HLA-A26	Haplotype: -DRB1*0402, -DQB1*0302	HLA-DRB1*1501
HLA-DRB1*07	HLA-B7	HLA-DRB1*11:01	HLA-B14	Haplotype: -DRB1*1601, -DQB1*0502	HLA-DR4-Ala74
HLA-DRB1*0901	HLA-B14	HLA-DRB1*51:01	HLA-B48	HLA-DRB1*08	
HLA-Dpw3	HLA-B*4002	HLA-DRB1*03			
HLA-DR4		HLA-DRB1*13:02			

genes at least in European patients [52,53]. Moreover, association between HLA-B14 alleles, HLA-Cw7, and AA has been confirmed by Maciejewski et al. [45] in a multiethnic cohort of 212 American individuals. Finally Shichishima et al. [54] reported higher incidence of HLA-B*4002 and HLA-A*0206 in 78 Japanese AA patients, suggesting a potential role of HLA class I genes even in Far East countries.

HLA genes revealing a protective role in developing AA have been identified too (Table 1.2, b). Even if data refer mainly to small series of patients, and may reflect natural polymorphism in HLA system, again a key role for HLA class II genes and its alleles is suggested. HLA-DRB1*13 appeared to be protective in a cohort of patients of Turkish origin [55], as well as HLA-DRB1*03:01, HLA-DRB1*11:01, and HLA-B1*51:01 alleles appear to be protective in cohort of Chinese children [48]. In a small cohort of Pakistani patients [43], HLA-DRB1*03 had an higher frequency in healthy control, suggesting a putative protective role, and a similar observation is available for allele DRB1*1302 in Korean population [56].

Even if AA has a typical bimodal age of incidence, majority of studies did not look specifically into HLA frequency and age differences, mostly due to their retrospective nature and rarity of disease. Therefore, a good assumption could be that data on HLA allele frequency could be inferred to the pediatric setting. However Fuhrer et al. [57] in a recent study involving 181 Caucasian children observed a positive association between HLA-A26 and HLA-B14 alleles, but not a higher incidence of HLA-DR2. Similarly Chen et al. [48] in a retrospective survey conducted on 80 Chinese children showed an higher than expected incidence of class I HLA-B48 alleles and class II HLA-DRB*09 allele. Finally, a frequency of HLA-DR2 B15 allele not different from normal population has been reported by Kook et al. [58] in North Korean AA children and by Yoshida et al. [59] in Japanese AA children. These limited data suggest a po-

tentially different HLA landscape in children with AA, even if further investigation involving a higher number of patient are needed to specifically address this question (Table 1.2, c).

Association between agranulocytosis and HLA genes has been reported for drugs exposure in specific ethnic groups, suggesting that certain alleles may facilitate the initiation of AA through direct presentation of drug-derived antigens to T-cells [37]. HLA-class II haplotypes have been reported to be associated with clozapine-induced agranulocytosis in Askenazi Jewish [60,61] individuals, and association between HLA-DRB1*08 alleles and thionamide-induced AA has been described for Japanese patients [62]. A similar mechanism of disease can be suggested for posthepatitis AA too, where association between HLA class I, specifically HLA-B8, and hematological disease have been documented [24] (Table 1.2, d).

Finally, some groups have shown that response to immunosuppressive therapy can be related to specific HLA genes. HLA-DRB1*1501 is the most clearly allele associated with a better response to Cyclosporine [41], and the presence of HLA-DR2 and a PNH clone independently predict response to therapy in patients under immunosuppression [45]. On the other hand HLA-DRB1*1502 allele [63], that represents a different allele variant of HLA-DR2, doesn't seem to have same influence on treatment, even if its incidence is increased among older Japanese patients with AA. High-resolution genotyping of HLA-DRB1 showed that the HLA-DRB1*04 allele coding for alanine position 74 (HLA-DR4-Ala74) [64] predispose to AA independently from HLA-DRB1*15 and that HLA-DRB1*04 alleles are associated with worse response to cyclosporine and poorest prognosis (Table 1.2, e).

Association between AA and HLA remains an intriguing field of interest. However studies with larger patients populations from different ethnic backgrounds are needed to better address complex relations between HLA, environmental factors, ethnicity, and AA incidence.

AA AND AUTOIMMUNE DISORDERS

Although AA is idiopathic in most cases, associations with other autoimmune disease (AID) have been shown in numerous single-case reports. In rheumatic diseases, such as systemic lupus erythematosus (SLE) and rheumatoid arthritis (RA), there is a recognized associations with AA [65], with a true incidence not known, but presumed to be very low.

Peripheral autoimmune cytopenias are commonly seen in SLE, and they are listed among diagnostic criteria for the disease in the revised American College of Rheumatology classification. Similarly, anemia of chronic disease, and cytopenias in relation to the use of cytotoxic drugs are commonly seen in SLE patients [66]. Less frequently, case reports and small series documented bone marrow abnormalities consistent with AA, suggesting that in rare cases bone marrow may also be a target organ of the disease, showing a full picture of acquired bone marrow failure [67]. Very recently Chalayer et al. [68] reported a MEDLINE methanalysis of 25 patients with SLE and AA. In 12 of these 25 patients, diagnosis of AA was associated with long story of SLE, extensive disease, and drug exposure, suggesting that drug-induced AA may represent a consistent type of AA in these patients.

Similarly the association of RA with AA is better recognized because of the use of therapeutic agents in RA, such as Methotrexate, gold salts, penicillamine, and Eternacept [69] causing AA, likely drug induced.

Correlation with immune disorders is also highlighted by the strong link observed with eosinophilic fasciitis (Shulman disease), a rare sclerodermiform syndrome that, in most cases, resolves spontaneously or after corticosteroid therapy, and that is frequently reported in association with hematological diseases, especially AA [70].

Two large different studies have specifically looked into incidence of autoimmune conditions among AA patients. Stalder et al. in a single center series [71] of 253 individuals showed that 5.3% of patients with AA had a previous diagnosis of AIDS, and 4.5% developed an AIDS after AA was diagnosed. Similarly 4% of patients recorded in EBMT database [72] for AA have been found diagnosed with a previous AIDS in a retrospective analysis. Most frequently diagnosed AIDS were autoimmune gastritis and thyroiditis in Stalder series, and RA in EBMT analysis, respectively. Interestingly, in both studies patient's age at diagnosis of AA seems to be higher (>50 years) than normal. Impact of immunosuppressive treatment for AA on concomitant AIDS, and effects on its natural history, remains at the moment a controversial topic.

AA DURING PREGNANCY

There are no prospective studies about the incidence of AA during pregnancy. Curiously AA was first described in a pregnant women by Ehrlich in 1888. Since then, there have been a few published reports of AA during pregnancy. Tichelli et al. [73] described a case series of 36 pregnancies in women with AA. Their study showed that successful pregnancy with normal outcome is possible in women with AA who have been previously treated with immunosuppression. Nineteen percent of women were found to have relapse of AA and a further 14% needed transfusion support at the time of delivery. Complications appear to be more likely in patients with low platelet counts and paroxysmal nocturnal hemoglobinuria-associated AA. Similarly, Choudhry et al. [74] described 10 cases of AA and concomitant pregnancies, reporting successful delivery in 10 of 11 cases, with adverse outcome related to fatal bleeding in only 2 patients.

Although pregnancy remains an immunomodulatory state, with an higher incidence of autoimmune disease reported, causative relationship with AA remains controversial.

AA POSTVACCINATION

There have been three case reports of AA, following hepatitis B vaccination [75]. Systematic case control studies for the incidence of AA following vaccination are not known. Patients with AA who have been treated with immunosuppressive treatment, should better avoid vaccinations, if possible, including influenza, as there remains a theoretical risk of disease relapse [1].

PROBLEMS WITH EPIDEMIOLOGICAL STUDIES IN AA AND FUTURE STRATEGIES

AA remains a rare disease with an annual incidence between 0.6 and 6 per million per annum. It is innately difficult to conduct a population based study in a rare disease, such as AA. Reporting of cases with AA is likely to be only from areas and centers that have a high coverage of health services. It's likely that there had been underreporting of cases with AA in regions with poor access to tertiary health care services or facilities for diagnostic tests, including special diagnostic tests for genomic and molecular analysis.

Drug recording and reporting to the medicines regulatory body for association studies is not uniformly practiced, leading to under or over estimation of association and causality (Chloramphenicol remains a case in point for an example of the latter).

The distinction between acquired and inherited disease may present a clinical challenge, especially among pediatric cases.

Multicenter clinical trials for newer therapeutic agents or treatment strategies in AA may provide on different countries a uniform panel of recording data, including family history for inherited bone marrow failure syndromes, and special tests, including T-cell subset repertoire analysis and next-generation technique molecular analysis.

Ready availability of nucleotide variants from the human genome project have defined new polymorphisms that are definitely pathogeneic and disease causing as opposed to variants that are disease predisposing and require the interplay of other external factors for disease causation. Furthermore, genetic tests have also defined how environmental influences, such as benzene play a role in the pathogenesis of AA because of nucleotide polymorphisms that affect key drug metabolism pathways or cytokine signaling molecules.

Following this theme, future epidemiological studies for AA, are likely to better explain the association to causality in AA, precise diagnosis and staging prognostic information and also unravel key molecular pathways for therapeutic exploitation.

References

[1] Killick S, Brown N, Cavenagh J, et al. Guidelines for the diagnosis and management of adult aplastic anemia. Br J Haematol 2016;172:187–207.

[2] Montané E, Ibanez L, Vidal X, et al. Epidemiology of aplastic anaemia: a prospective multicenter study. Haematologica 2008;93:518–23.

[3] Kaufman DW, Kelly JP, Levy M, et al. The drug etiology of agranulocytosis and aplastic anemia. New York: Oxford University Press; 1991.

[4] Cartwright RA, McKinney PA, Williams L, et al. Aplastic anaemia incidence in parts of the United Kingdom in 1985. Leuk Res 1988;12:459–63.

[5] Mary JY, Baumelou M, Guiguet M. The French Cooperative Group for Epidemiological Study of Aplastic Anemia. Epidemiology of aplastic anemia in France: a prospective multi-centric study. Blood 1990;75:1646–53.

[6] Clausen N, Kreuger A, Salmi T, et al. Severe aplastic anaemia in the Nordic countries: a population based study of incidence, presentation, course, and outcome. Arch Dis Child 1996;74:319–22.

[7] Maluf EM, Pasquini R, Eluf JN, et al. Aplastic anemia in Brazil: incidence and risk factors. Am J Hematol 2002;71:268–74.

[8] Issaragrisil S, Sriratanasatavorn C, Piankijagum A, et al. Incidence of aplastic anemia in Bangkok. The Aplastic Anemia Study Group. Blood 1991;77(10):2166–8.

[9] Issaragrisil S, Kaufman DW, Anderson T, et al. The epidemiology of aplastic anemia in Thailand. Blood 2006;107:1299–307.

[10] Yang C, Zhang X. Incidence survey of aplastic anemia in China. Chin Med Sci J 1991;6:203–7.

[11] Szklo M, Sesenbrenner L, Markowitz J, et al. Incidence of aplastic anemia in Metropolitan Baltimore. A population based study. Blood 1985;66:115–9.

[12] Yong AS, Goh M, Rahman J, et al. Epidemiology of aplastic anemia in the state of Sabah, Malaysia. Cell Immunol 1996;1284:S75.

[13] McCahon E, Tang K, Rogers PC, McBride ML, et al. The impact of Asian descent on the incidence of acquired severe aplastic anaemia in children. Br J Haematol 2003;121(1):170–2.

[14] Ahamed M, Anand M, Kumar A, Siddiqui MK. Childhood aplastic anaemia in Lucknow, India: incidence, organochlorines in the blood and review of case reports following exposure to pesticides. Clin Biochem 2006;39(7):762–6.

[15] Qing L, Luoping Z, Guilan L, et al. Hematotoxicity in workers exposed to low levels of benzene. Science 2004;306(5702):1774–6.

[16] Qing L, Luoping Z, Min S, et al. Polymorphisms in cytokine and cellular adhesion molecule genes and susceptibility to hematotoxicity among workers exposed to benzene. Cancer Res 2005;65:9574–81.

[17] Maluf E, Hamerschlak N, Cavalcanti AB, et al. Incidence and risk factors of aplastic anemia in Latin American countries: the LATIN case-control study. Haematologica 2009;94(9):1220–6.

[18] Young N, Kaufman D. The epidemiology Of acquired aplastic anemia. Haematologica 2008;93:489–92.

[19] Heimpel H. Epidemiology and aetiology of aplastic anaemia. In: Schrezenmeier H, Bacigalupo A, editors. Aplastic anaemia: pathophysiology and treatment. Cambridge, UK: Cambridge University Press; 2000. p. 97–116.

[20] Mahapatra M, Singh PK, Agarwal M, et al. Epidemiology, clinico-haematological profile and management of aplastic anaemia: AIIMS experience. J Assoc Physicians India 2015;63(3 Suppl):30–5.

[21] Adil SN, Kakepoto GN, Khurshi M. Epidemiological features of aplastic anaemia in Pakistan. J Pak Med Assoc 2001;51:443.

[22] Young NS. Bone marrow failure syndromes. Philadelphia: W.B. Saunders; 2000.

[23] Hao LY, Armanios M, Strong MA, et al. Short telomeres, even in the presence of telomerase, limit tissue renewal capacity. Cell 2005;123:1121–31.

[24] Brown KE, Tisdale J, Barrett AJ, et al. Hepatitis-associated aplastic anemia. N Engl J Med 1997;336:1059–64.

[25] Locasciulli A, Bacigalupo A, Bruno B. Hepatitis-associated aplastic anaemia: epidemiology and treatment results obtained in Europe. A report of The EBMT aplastic anaemia working party. Br J Haematol 2010;6:890–5.

[26] Gonzalez-Casas R, Garcia-Buey L, Jones EA, et al. Systematic review: hepatitis-associated aplastic anaemia—a syndrome associated with abnormal immunological function. Aliment Pharmacol Ther 2009;30:436–43.

[27] Adachi Y, Yasui H, Yuasa H, Ishi Y, Imai K, Kato Y. Hepatitis B virus-associated aplastic anemia followed by myelodysplastic syndrome. Am J Med 2002;112:330–2.

[28] Byrnes JJ, Banks AT, Piatack M Jr, Kim JP. Hepatitis G-associated aplastic anaemia. Lancet 1996;348:472.

[29] Pardi DS, Romero Y, Mertz LE, Douglas DD. Hepatitis-associated aplastic anemia and acute parvovirus B19 infection: a report of two cases and a review of the literature. Am J Gastroenterol 1998;93:468–70.

[30] Lau YL, Srivastava G, Lee CW, et al. Epstein–Barr virus associated aplastic anaemia and hepatitis. J Paediatr Child Health 1994;30:74–6.

[31] Poovorawan Y, Tangkijvanich P, Theamboonlers A, et al. Transfusion transmissible virus (TTV) and its putative role in the etiology of liver disease. Hepatogastroenterology 2001;48:256–60.

[32] Imai T, Itoh S, Okada H, et al. Aplastic anemia following hepatitis associated with echovirus 3. Pediatr Int 2002;44:522–4.

[33] Führer M, Rampf U, Baumann I, et al. Immunosuppressive therapy for aplastic anemia in children: a more severe disease predicts better survival. Blood 2005;106:2102–4.

[34] Young NS, Brown KE. Mechanisms of disease: parvovirus B19. N Engl J Med 2004;350:586–97.

[35] Mishra B, Malhotra P, Ratho RK, et al. Human parvovirus B19 in patients with aplastic anemia. Am J Hematol 2005;79:166–7.

[36] Pagliuca S, Gérard L, Kulasekararaj A, et al. Characteristics and outcomes of aplastic anemia in HIV patients: a brief report from the severe aplastic anemia working party of the European Society of Blood and Bone Marrow Transplantation. Bone Marrow Transplant 2016;51:313–5.

[37] Young NS, Alter BP, Young NS. . In: Young NS, Alter BP, editors. Aplastic anemia, acquired and inherited, drugs and chemicals. Philadelphia: W.B. Saunders; 1994. p. 100–32.

[38] Chapuis B, Von Fliedner VE, Jeannet M, et al. Increased frequency of DR2 in patients with aplastic anaemia and increased DR sharing in their parents. Br J Haematol. 1986;63:51–7.

[39] Nimer SD, Ireland P, Meshkinpour A, et al. An increased HLA DR2 frequency is seen in aplastic anemia patients. Blood 1994;84:923–7.

[40] Shao W, Tian D, Congyan L, et al. Aplastic anemia is associated with HLA-DRB1*1501 in northern Han Chinese. Int J Hematol 2000;71:350–2.

[41] Nakao S, Takamatsu H, Chuhjo T, et al. Identification of a specific HLA class II haplotype strongly associated

with susceptibility to cyclosporine-dependent aplastic anemia. Blood 1995;84:4257–61.

[42] Ilhan O, Beksac M, Arslan O, Koc H, et al. HLA-DR frequency in Turkish aplastic anemia patients and the impact of HLA-DR2 positivity in response rate in patients receiving immunosuppressive therapy. Blood 1995;86:2055.

[43] Rehman S, Saba N, Khalilullah, et al. The frequency of HLA class I and II alleles in Pakistani patients with aplastic anemia. Immunol Investig 2009;38:251–4.

[44] Dhaliwal JS, Wong L, Kamaluddin MA. Susceptibility to aplastic anemia is associated with HLA-DRB1* 1501 in an aboriginalpopulation in Sabah, Malaysia. Hum Immunol 2011;72:889–92.

[45] Maciejewski JP, Follmann D, Nakamura R. Increased frequency of HLA-DR2 in patients with paroxysmal nocturnal hemoglobinuria and the PNH/aplastic anemia syndrome. Blood 2001;98:3513–9.

[46] Nakao S, Takami A, Takamatsu H. Isolation of a T-cell clone showing HLA-DRB1*0405-restricted cytotoxicity for hematopoietic cells in a patient with aplastic anemia. Blood 1997;89:3691.

[47] Yari F, Sobhani M, Vaziri MZ, et al. Association of aplastic anemia and Fanconi's disease with HLA-DRB1 alleles. Int J Immunogenet 2008;35:453–6.

[48] Chen C, Lu S, Luo M, et al. Correlations between HLA-A, HLA-B and HLA-DRB1 allele polymorphism to acquired aplastic anemia. Acta Haematol 2012;128:23–7.

[49] Odum N, Platz P, Morling N, et al. Increased frequency of HLA-DPw3 in severe aplastic anemia. Tissue Antigens 1987;29:184–5.

[50] Albert E, et al. HLA antigens and haplotypes in 200 patients with aplastic anemia. Transplantation 1976;22:528–31.

[51] Dausset J, Gluckman E, et al. Excess of HLA-A2 and HLA-A2 homozygotes in patients with aplastic anemia Fanconi's anemias. Nouv Rev Fr Hematol Blood Cells 1977;18:315–24.

[52] Gluckman E. HLA markers in patients suffering from aplastic anaemia. Haematologia 1981;14:165–72.

[53] D'Amaro J, et al. HLA associations in Italian and non-Italian Caucasoid aplastic anemia patients. Tissue Antigens 1983;21:184–91.

[54] Shichishima T, Noji H, Ikeda K, et al. The frequency of HLA class I alleles in Japanese patients with bone marrow failure. Haematologica 2006;91:857.

[55] Oguz FS, Yalman N, Diler S, et al. HLA-DR15 and pediatric aplastic anemia patients. Haematologica 2002;87:772–4.

[56] Song EY, Park S, Lee DS, et al. Association of human leukocyte antigen-DRB1 alleles with disease susceptibility and severity of aplastic anemia in Korean patients. Hum Immunol 2008;69:354–9.

[57] Fuhrer M, Durner J, Brunnler G, et al. HLA association is different in children and adults with severe acquired aplastic anemia. Pediatr Blood Cancer 2007;48:186–91.

[58] Kook H, Hwang TJ, Seo JJ, et al. The frequency of HLA alleles in Korean children with aplastic anemia and the correlation with the response to immunosuppressive treatment. Korean J Pediatr Haematol Oncol 2003;10:177–88.

[59] Yoshida N, Yagasaki H, Takahashi Y, et al. Clinical impact of HLA-DR15, a minor population of paroxysmal nocturnal haemoglobinuria-type cells, and an aplastic anaemia-associated autoantibody in children with acquired aplastic anemia. Br J Haematol 2008;142:427–35.

[60] Yunis JJ, Corzo D, Salazar M, et al. HLA associations with clozapine-induced agranulocytosis. Blood 1995;86:1777.

[61] Corzo D, Yunis JJ, Salazar M, et al. The major histocompatibility complex region marked by HSP70-1 and HSP70-2 varisnts is associated with clozapine induced agranulocytosis in two different ethnic groups. Blood 1995;86:3835–40.

[62] Tamai H, Sudo T, Kimura A, et al. Association between the DRB1*08032 histocompatibility antigen and methimazole-induced agranulocytosis in Japanese patients with Graves disease. Ann Intern Med 1996;124:490–4.

[63] Sugimori C, Yamazaki H, Feng X, et al. Roles of DRB1*1501 and DRB1*1502 in the pathogenesis of aplastic anemia. Exp Hematol 2007;35:13–20.

[64] Kapustin SI, Popova TI, Lyshchov AA, et al. HLA-DR4-Ala74 beta is associated with high risk and poor outcome of severe aplastic anemia. Ann Haematol 2001;80:66–71.

[65] Bacigalupo A. Aetiology of severe aplastic anaemia and outcome after allogeneic bone marrow transplantation or immunosuppression. Eur J Haematol 1996;57(Suppl):16–9.

[66] Bhatt AS, Berliner N. Hematologic Manifestations of SLE. In: Schur P, Massarotti E, editors. Lupus erythematosus: clinical evaluation and treatment. New York: Springer; 2012. p. 127–40.

[67] Alishiri GH, Saburi A, Bayat N, et al. The initial presentation of systemic lupus erythematosis with aplastic anemia successfully treated with rituximab. Clin Rheumatol 2012;31(2):381.

[68] Chalayer É, Ffrench M, Cathébras P. Aplastic anemia as a feature of systemic lupus erythematosus: a case report and literature review. Rheumatol Int 2015;35(6):1073–82.

[69] Kozak N, Friedman J, Schattner A. Etanercept-associated transient bone marrow aplasia: a review of the literature and pathogenetic mechanisms. Drugs 2014;14(2):155–8.

[70] De Masson. Severe aplastic anemia associated with eosinophilic fasciitis: report of 4 cases and review of the literature. Medicine (Baltimore) 2013;2:69–81.

[71] Stalder MP, Rovó A, Halter J, et al. Aplastic anemia and concomitant autoimmune disease. Ann Hematol 2009;88:659–65.

[72] Cesaro S, Marsh JC, Tridelli G, et al. Retrospective survey on the prevalence and outcome of prior autoimmune diseases in patients with aplastic anemia reported to the registry of the European group for blood and marrow transplantation. Acta Haematol 2010;124:19–22.

[73] Tichelli A, Socié G, Marsh J, et al. Outcome of pregnancy and disease course among women with aplastic anemia treated with immunosuppression. Ann Intern Med 2002;137(3):164.

[74] Choudhry VP, Gupta S, Gupta M, et al. Pregnancy associated aplastic anemia—a series of 10 cases with review of the literature. Haematology 2002;7:233–8.

[75] Shah C, Lemke S, Singh V, et al. Case reports of aplastic anemia after vaccine administration. Am J Hematol 2004;77(2):204.

2

Pathophysiology of Acquired Bone Marrow Failure

K. Hosokawa*, P. Scheinberg**, N.S. Young*

*National Institutes of Health, National Heart, Lung, and Blood Institute, Bethesda, MD, United States; **Oncology Center, Hospital São José National Institutes of Health, National Heart, Lung, and Blood Institute, Bethesda, MD, United States

INTRODUCTION: EVIDENCE AND INFERENCES FROM THE CLINIC

Acquired aplastic anemia (AA) is the prototypical bone marrow (BM) failure syndrome. AA is characterized by pancytopenia of peripheral blood and BM hypoplasia. Low blood counts and "empty" marrow pathology implies the absence of stem and progenitor cells, which is consistent with, the success of BM transplantation (BMT), in which replacement of hematopoietic stem cells (HSCs) (and immune cells) is adequate to cure disease. There is little evidence that AA is mediated by BM stromal cells. If stem cell absence were the only defect, twin donor or syngeneic transplants should be successful with infusion of BM, but a large proportion suffer graft failure; that conditioning eliminates graft failure suggests an immune pathophysiology [1]. Immunosuppressive regimens were developed in the context of graft failure and now are widely employed when stem- cell transplant is not feasible. The efficacy of immunosuppressive therapy (IST) alone is strong evidence of an immune mechanism in most patients with AA: 60–70% respond to one course of horse hATG/cyclosporine (CsA) and an additional 30% of primary nonresponders will respond to a second course of IST. Similarly, patients' blood counts often are dependent on continued administration of CsA, as would be expected of a T-cell driven disease. Eltrombopag, a thrombopoietin (TPO) mimetic, has activity in refractory SAA as single agent and in increasing the response rate and completeness of the response when combined with IST as first therapy. These results suggest that there are residual HSCs that can repopulate BM, even if they are not detectable. Recent genomic data applying whole exome sequencing in AA show that hematopoiesis can be sustained from a single or very few HSC clones. Clonal hematopoiesis has also been inferred from the frequent and unique association of paroxysmal nocturnal hemoglobinuria (PNH), the origin of which is a somatic mutation in an X-linked gene, with acquired AA.

In vitro experiments with patients' cells are consistent with clinical observations in supporting an immune mechanism of AA. The presence of PNH or acquired copy number-neutral loss of heterozygosity of the 6p arms (6pLOH) clones support clonal escape from immune-mediated BM destruction. Immune-mediated destruction of BM can be modeled in animals and animal experiments employing myelotoxic drugs or with myeloablative transplants show that very limited numbers of HSCs can support hematopoiesis for very prolonged time periods.

Clonal evolution, the development of myelodysplastic syndrome (MDS) and acute myeloid leukemia (AML) in a patient with typical AA, often after successful IST, demonstrates genomic instability in the setting of immune or inflammatory disease environment (and occurs in other immune diseases, such as inflammatory bowel disease and chronic hepatitis). Short telomeres appear to predispose to genomic instability, with tissue culture and animal model experiments providing a mechanism of chromosome derangement. In some cases of clonal evolution, there is evidence of origin from a tiny clone of cells harboring a recurrent mutation in an MDS/AML candidate gene.

PATHOPHYSIOLOGY

Hematopoiesis in AA

AA is a BM failure syndrome characterized by pancytopenia of the peripheral blood and BM hypoplasia [2–4]. Profound reduction in hematopoietic stem and progenitor cells (HSPCs) is a consistent finding [5–9]. Stem cells/early progenitor cells can be assayed by long-term culture-initiating cell assays (LTC-ICs) [7,9], or cobblestone area-forming cells [10], these cells are also markedly deficient in AA. The scant numbers of LTC-ICs per mononuclear cell suggest that only a small percentage of residual early hematopoietic cells remain in severely af-

fected patients at presentation. There are few or no CD34+ cells on flow cytometry [8].

Recent HSC fate-mapping analyses in mice have suggested that progenitors and not HSCs are fundamental for hematopoiesis under homeostatic conditions [11,12]. Physiological studies to experimentally test this hypothesis are not possible in normal human subjects. However, certain disease states may indicate the consequences of HSC loss on progenitors and the role of HSCs in human blood production under nontransplant conditions. Using a flow cytometric gating scheme to define MPPs (CD34+CD38−Thy1−CD45RA−CD49f−), CMPs (CD34+CD38+CD10−FLT3+CD45RA−), and MEPs (CD34+CD38+CD10−FLT3− CD45RA−), the progenitor hierarchy was examined in a few cases of AA [13]. Consistent with previous reports, the proportion of CD34+ cells within the overall mononuclear cell pool was much lower in AA compared with normal BM [6,9]. The CD34+CD38− stem-cell compartment in AA was more significantly depleted compared with the CD34+CD38+ progenitor compartment. HSCs and MPPs were virtually undetectable in the residual CD34+CD38− compartment, confirming that HSCs are lost in AA, as determined by phenotype. Despite the loss of phenotypic HSCs, the CD34+CD38+ compartment was detectable in all cases. The percentage of myeloid progenitors was equivalent compared with normal BM. In contrast, erythroid progenitors were low, like HSCs, in all patients. These results suggest that ongoing erythropoiesis is more reliant on HSC input compared with myelopoiesis. Consistent with these findings, clinical data suggested that the baseline absolute reticulocyte count (ARC) and absolute lymphocyte count (ALC) together serve as a simple predictor of response to IST and greater rate of 5-year survival [14].

In AA, BM is not truly empty but replaced by fat cells [15]. BM adipocytes were reported to be possible negative regulators in the hematopoietic microenvironment [16]. To examine the role of adipocytes in BM failure, recent study investigated peroxisomal proliferator-activated receptor gamma (PPARγ), a key transcription factor in adipogenesis, utilizing an antagonist of this factor [17]. While PPARγ antagonists inhibited adipogenesis as expected, it also suppressed T-cell infiltration of BM, reduced plasma inflammatory cytokines, decreased expression of multiple inflammasome genes, and ameliorated marrow failure. These results suggested that PPARγ antagonists acted as negative regulators of T cells in addition to their inhibition of BM adipogenesis.

AA is strongly associated with PNH [18]. PNH is a rare acquired disorder of HSCs, characterized by hemolytic anemia, BM failure, and venous thrombosis. The etiology of PNH is a somatic mutation in the X-linked phosphatidylinositol glycan class A gene (*PIG-A*), resulting in global deficiency of glycosyl phosphatidylinositol–anchored proteins (GPI–APs) [19]. Clinically, AA may coexist or appear to evolve to other hematologic diseases that are characterized by proliferation of distinctive cell clones, as in PNH or MDS [2]. Nearly half of AA patients have clonal populations of cells lacking GPI–APs because of somatic mutations in the *PIG-A* gene; these are called PNH clones [20,21]. Most clones are small and do not lead to clinical manifestations of hemolysis or thrombosis [22], but classic PNH can be dominated by marrow failure (the "AA/PNH syndrome"). *PIG-A* mutant cells can support hematopoiesis for long term, despite accumulation of somatic mutations associated with clonal evolution to MDS/AML [23–25].

Recent studies have demonstrated clonal hematopoiesis in the majority of AA [26,27]. To clarify the origin, importance, and dynamics over time of clonal hematopoiesis in AA, and its relationship to the development of MDS, AML, or both, targeted deep-sequencing, SNP array karyotyping, and whole-exome sequencing, were performed to identify genetic alterations in AA and described their dynamics over long clinical courses [26]. Somatic mutations in myeloid cancer candidate genes were present in one-third of the AA patients, in a limited number of genes and at low initial variant allele frequency [26]. Clonal hematopoiesis was detected in 47% of the AA, most frequently as acquired mutations. The prevalence of the mutations increased with age. *DNMT3A*-mutated and *ASXL1*-mutated clones tended to increase in size over time; the size of *BCOR*- and *BCORL1*-mutated and *PIGA*-mutated clones decreased or remained stable. However, clonal dynamics were highly variable and did not determine the response to therapy and long-term survival among individual patients. These results were consistent with those of recent studies in which candidate-gene targeted sequencing also showed recurrent mutations in a similar spectrum of genes [28,29]. Two hundred and nineteen genes were screened in 39 patients, and found somatic mutations in 9 (23%), comprising *ASXL1*, *DNMT3A*, and *BCOR*. The median allele burden was <10% in 7 patients [28]. In 2 of 38 patients (*SLIT1*, and *SETBP1* with *ASXL1*) using a smaller panel of 42 genes, 3 mutations were found. However, the patient with *SETBP1* and *ASXL1* was tested at time of progression to MDS.

These results show parallels between BM failure and normal aging of the hematopoietic compartment. The characteristic mutation signature and correlation of mutations with patient age suggested age-related, spontaneous conversion of methylated cytosine to thymidine at CpG sites as major source of nucleotide alternations in AA [30]. Similar C-to-T conversion mutations accumulate in hematopoietic progenitors in healthy persons [31–33]. Mutations generally appeared at a low variant allele frequency and involved common mutational targets in myeloid cancers, which suggests that the origin and clonal selections of these mutations are similar to those in AA.

However, the exact mechanism of selection of mutated cells in AA is unclear. *DNMT3A* is essential for hematopoietic stem-cell differentiation [34]. *DNMT3A* loss predisposes murine HSCs to malignant transformation [35]. Deletion of *ASXL1* results in myelodysplasia in vivo [36,37]. Cells containing *DNMT3A* or *ASXL1* mutations may preferentially self-renew rather than differentiate in response to extrinsic signals. In contrast, striking overrepresentation of *BCOR* and *BCORL1* and *PIGA* mutations as well as frequent 6pUPD involving the specific HLA classes suggest a mechanism of protection of mutated cells from immune-mediated destruction by pathogenic T cells [38,39].

Immune Mechanisms in AA

Clinical Data

An immune mechanism was inferred decades ago from the recovery of hematopoiesis in patients who failed to engraft after stem-cell transplantation, when renewal of autologous blood-cell production was credited to the conditioning regimen. Also, the majority of syngeneic transplantations in which BM was infused without conditioning failed due to rejection implying a disease immune mechanism [40]. The responsiveness of AA to IST in most patients is the best evidence of an underlying immune pathophysiology: the majority of patients show hematologic improvement after only transient T-cell depletion by ATGs; relapse also usually responds to ATG; and dependence of adequate blood counts on administration of very low doses of CsA is not infrequent [2].

T Cells and Cytokines

In early laboratory experiments, removal of lymphocytes from aplastic BMs improved colony numbers in tissue culture, and their addition to normal marrow inhibited hematopoiesis in vitro [41]. Immunity to HSCs by activated T cells has been considered to be responsible for the pathogenesis of AA [42,43]. Laboratory in vitro data has further reinforced the immune pathogenesis in AA with the principal findings including:

1. An increased cytokine (IFN-γ) of activated T cells identified both in the blood and

BM [44–50]; CD8$^+$ CTLs are expanded in AA, leading to the production of proinflammatory cytokines (e.g., IFN-γ) that induces apoptosis of CD34$^+$ cells [47,50].

2. Oligoclonal skewing of the T-cell repertoire indicating expansion of pathogenic CD8$^+$ T cells [51–53]. In general, patients at presentation demonstrate oligoclonal expansions of a few Vβ subfamilies, which diminish or disappear with successful IST; original clones reemerge with relapse, sometimes accompanied by new clones, consistent with spreading of the immune response. Very occasionally, a large clone persists in remission, perhaps evidence of T-cell tolerance.

3. A reduction of regulatory T cells (Treg) and an increase in Th17 related T cells resulting in a high Th17/Treg ratio at diagnosis which tends to normalize in responding patients to IST [54,55]; reduction in Treg numbers correlates with disease severity, and the defect is most prominent in severe and very severe AA [56].

4. Increased transcription of Th1-related genes in activated T cells of AA patients [57,58]; recent study in BM failure mouse models identified NOTCH signaling as a primary driver of Th1-mediated pathogenesis in AA and may represent a novel target for therapeutic intervention.

5. Confirmation in murine models on the role of Th1 and Th17 cells and related cytokines to producing destruction of marrow progenitor cells and the positive effects of Th17 blocking antibodies and infusion of regulatory T cells in reversing BM failure in these models [55,59–63].

The impact of T-cell attack on BM can be modeled in vitro and in vivo. IFN-γ (and tumor necrosis factor-α) in increasing doses reduce numbers of human hematopoietic progenitors assayed in vitro; the cytokines efficiently induce apoptosis in CD34$^+$ target cells, at least partially through the Fas-dependent pathway of cell death [43,64]. In long-term culture of human BM, in which stromal cells were engineered to constitutively express IFN-γ, the output of long-term-culture-initiating cells (LTCI-ICs) was markedly diminished, despite low concentrations of the cytokine in the media, consistent with local amplification of toxicity in the marrow milieu [49].

Measurements of soluble circulating mediating factors in BM failure were limited largely to one or two cytokines in AA [65–68]. High TPO levels have been observed in patients with AA, and these abnormal levels correlate with disease severity [66,68]. Comprehensive analysis of 31 cytokines by an immuno-bead-based multiplex assay (Luminex) identified that high levels of TPO and granulocyte colony-stimulating factor, with low levels of CD40 ligand, CXCL5, CCL5, CXCL11, epidermal growth factor, vascular endothelial growth factor, and CCL11 were a signature profile for AA [69]. An increase in IL-17-producing Th17 cells in the peripheral blood and BM of patients with AA has also been reported [55,70,71]. Recent study showed that TPO and IL-17 levels are useful for differentiating hypocellular refractory cytopenia of childhood (RCC) from pediatric AA [72].

HLA and Cytokine Gene Polymorphisms

There have been a number of studies on the human leukocyte antigens (HLA) and their association with AA. HLA-DR2 is overrepresented among patients [73], suggesting a role for antigen recognition, and its presence is predictive of a better response to CsA [74]. Further research showed HLA-DRB1*1501 was associated with a good response to IST in Japanese cohorts [75,76].

Polymorphisms in cytokine genes, associated with an increased immune response, also may be more prevalent: a nucleotide polymorphism in the TNF-α (TNF2) promoter at −308 [77], homozygosity for a variable number of dinucleotide repeats in the gene encoding IFN-γ [78], and polymorphisms in the CTLA4 [79].

Immune Escape Clones (PNH, 6pLOH)

Certain clones may escape the immune attack within the BM environment and proliferate and attain a survival advantage over normal HSCs. The global absence of a large number of cell-surface proteins in PNH has been hypothesized to allow escape and survival of a preexisting mutant clone. Association of present PNH clone with a predictor of responsiveness to IST suggests that the escape is from immune attack [20,80–83]. Small PNH clones present at diagnosis usually remain stable over time, but may expand sufficiently to produce symptomatic hemolysis [84]. Comparison by microarray shows that residual cells of normal phenotype in the PNH BM upregulate the same apoptosis and cell-death genes as do CD34$^+$ cells in aplastic marrow, while the PIG-A mutant clone appears transcriptionally similar to CD34$^+$ cells from healthy donors [85]. Alternatively, NK cell mediated cytotoxicity may play a role; immunoglobulin-like receptors (KIR) may be differentially expressed in PNH compared to normal, resulting in cytotoxicity of normal HSCs [86]. Recent work has suggested expansion of autoreactive, CD1d-restricted, GPI-specific T cells in PNH [38]. These data suggested important roles for cell-extrinsic factors in clonal expansion of PNH cells.

It was recently reported that AA patients possessing clonal/oligoclonal hematopoiesis who had specifically lost either HLA haplotype containing the HLA-B*40:02, HLA-A*31:01, HLA-A*02:01, and HLA-A*02:06 [87]. Approximately 10% of AA have acquired copy number neutral loss of heterozygosity (CN-LOH) in chromosome arm 6p (6pLOH), postulated to emerge by immune selection against specific HLA alleles [87–89]. This clonal hematopoiesis may represent a signature of an escape from cytotoxic T-cells autoimmunity targeting autoantigens and strengthen the hypothesis of the immune-mediated pathogenesis of AA, although the exact mechanism is still unclear. Recent report showed the PNH patient that had acquired CN-6pLOH in GPI–AP$^+$ granulocytes, but not in GPI–AP$^-$

granulocytes, supporting the hypothesis that a hostile immune environment drives selection of resistant hematopoietic cell clones [90]. Recent study reported successful isolation of HLA-B*40:02-restricted CTLs specific for HSCs that were present in AA patient peripheral blood or BM [91].

STAT3 Mutant Clones

Large granular lymphocyte leukemia (LGL) is often associated with immune cytopenias and can occur in BM failure, such as AA and MDS [92,93]. STAT3 mutations in LGL clonal expansions are detected [94,95]. STAT3 clones can be not only in known LGL concomitant cases, but in a small population of other BM failure cases (7% AA and 2.5% MDS) [72]. In STAT3-mutated AA patients, trend toward better responses of IST and an association with the presence of HLA-DR 15 were found. STAT3-mutant clones may facilitate a persistently dysregulated autoimmune activation, responsible for the primary induction of BM failure in a subset of AA and MDS.

Innate Immunity

Transcriptional analysis of T cells from AA has implicated some components of innate immunity in AA, including toll-like receptors and natural killer cells [96].

There are some experimental results that support the natural killer cells involvement in AA [97,98]. KIR and KIR ligand (KIR-L) genotype study showed that AA and PNH showed decreased frequency of KIR-2DS1 and KIR-2DS5 genes [98]. The reduced frequency of these KIRs in AA and PNH may indicate an immunogenetic relationship between these diseases.

microRNAs

There is emerging evidence that microRNA (miRNA) play crucial roles in controlling and modulating immunity. Dysregulation of miRNA can lead to autoimmune diseases, such as rheumatoid arthritis, multiple sclerosis, and inflam-

matory bowel disease. Recent work described previously unknown potential regulatory roles of the miR-145-5p and miR-126-3p in T-cell activation in AA, in which MYC and PIK3R2 are the respective targets of these miRNA [99]. Dysregulated miR-145-5p and miR-126-3p promote T-cell proliferation and increase GZMB and IFN-γ production. Targeting or employing miRNA mimics might be novel molecular therapeutic approaches in AA.

Autoantibodies

Autoantibodies are frequently detected in patients with AA [100–103]. Antimoesin [100], diazepam-binding inhibitor-related protein 1 [101], kinectin [102], postmeiotic segregation increased 1 [102], and HNRNPK antibodies [103] were reported to be expressed in AA. Recent study using SEREX identified autoantibodies that are expressed in AA accompanied by immune abnormality [104]. Eight candidates were identified: CLIC1, SLIRP, HSPB11, NHP2L1, SLC50A1, RPL41, RPS27, and SNRPF.

Immune-Mediated BM Failure Mouse Models

Mouse models of AA, produced by the destruction of BM cells using radiation, cytotoxic drugs, and immune cells, have been useful in defining the hematopoietic stem cell and illustrating the potency of small numbers of lymphocytes in specifically inducing apoptosis of BM targets and their cytokines (e.g., IFN-γ-) as negative effector molecules [105–108]. Murine models mimicking AA have used exposure to agents that result in marrow destruction through a direct toxic effect, but models that explore antigenic disparities between strains have resulted in immune-mediated destruction of the marrow, more closely modeling human AA [109]. Infusion of parental lymph node cells into F1 hybrid donors caused pancytopenia, profound marrow aplasia, and death [63]. Not only a murine version of ATG and CsA but also monoclonal antibodies to IFN-γ and tumor necrosis factor abrogated hematologic disease, rescuing animals. A powerful "innocent bystander" effect, in which activated cytotoxic T cells kill genetically identical targets, was present in secondary transplantation experiments [62]. In a minor histocompatibility antigen-discordant model, marrow destruction resulted from activity of an expanded H60 antigen-specific T-cell clone [61]. Treatment with the CsA abolished H60-specific T-cell expansion and rescued animals from fatal pancytopenia. The development of BM failure was associated with a significant increase in activated CD4$^+$CD25$^+$ T cells that did not express intracellular FoxP3, whereas inclusion of normal CD4$^+$CD25$^+$ regulatory T cells in combination with C57BL/6 LN cells aborted H60-specific T-cell expansion and prevented BM destruction. Trafficking of T cells to the marrow has also been shown to be important in AA pathogenesis in murine models [110].

Mouse Models of Chemical and Drug Hematopoietic Toxicity

In a few instances, mouse models have been utilized to examine chemical and drug toxicity for hematopoiesis. Industrial exposure to benzene has numerous deleterious hematologic effects in human workers. When benzene was subcutaneously injected into CD1 mouse, they showed lethargy, irritability, and weight loss, with decreased hemoglobin, erythrocytes, leukocytes, and BM cells indicative of BM failure [111].

Chronic, delayed hematotoxicity of the chemotherapy drug busulfan was recapitulated in a mouse model: following a course of therapy, animals maintained normal blood and BM cell counts for 1 year before developing pancytopenia and frank marrow aplasia, with significant decline in splenic colony-forming units (CFU) [112]. BALB/c mice were treated 8 times with busulfan over 23 days and found reductions in nucleated marrow cells, granulocyte-macrophage (CFU-GM), CFU-erythroid, erythrocytes, leukocytes, platelets, and reticulocytes on day 1

and continued until days 91 and 112 posttreatment [113].

Genetic Risk Factors in AA

Telomeres are DNA sequences with a structure that protects chromosomes from erosion and that a specific enzyme, telomerase, is involved in their repair after mitosis [114,115]. Telomeres are repeated nucleotide sequences that cap the ends of chromosomes and protect them from damage. Acquired and congenital AA have been linked molecularly and pathophysiologically by abnormal telomere maintenance [116,117]. Telomeres are eroded with cell division, but in HSCs, maintenance of their length is mediated by telomerase. Accelerated telomere shortening is virtually universal in dyskeratosis congenita, caused by mutations in genes encoding components of telomerase or telomere-binding protein (TERT, TERC, DKC1, NOP10, or TINF2) [118–122]. Short telomeres were found in leukocytes from approximately one-third of patients with AA, especially those who do not have a response to IST [123,124]. Systematic screening of patients with apparently AA showed a few patients with TERT or TERC mutations [125,126].

Mutation of SBDS underlie Shwachman–Diamond syndrome (SDS), inherited syndrome featuring BM failure [127]. Heterozygosis for 258 + 2 T > C SBDS gene was associated with AA (4 of 91 AA) and telomere shortening of leukocytes [128]. These mutations cause low telomerase activity, accelerated telomere shortening, and diminished proliferative capacity of hematopoietic progenitors. Sex hormones increase telomerase activity by upregulating the TERT gene [129]. Blood count improvement can be obtained with androgen therapy in patients with mutation in telomere repair complex genes [130]. For more information about telomeres, please refer Chapter 15.

Germ-line GATA2 gene mutations, leading to haploinsufficiency, have been identified in patients with familial MDS/AML [131], monocytopenia and mycobacterial infections [132,133], Emberger syndrome [134], and dendritic cell, monocyte, B-, and NK-cell deficiency [135,136]. *GATA2* mutations have also been identified in a subset of patients presenting with chronic neutropenia [137], and young adults with AA [138,139], highlighting the clinical heterogeneity and variable hematologic phenotypes associated with a single genetic defect. The BM from patients with *GATA2* deficiency is typically hypocellular, with varying degrees of dysplasia. The marrow had severely reduced monocytes, B cells, and NK cells; absent hematogones; and inverted CD4:CD8 ratios. Atypical megakaryocytes and abnormal cytogenetics were more common in *GATA2* marrows. Routine BM flow cytometry, morphology, and cytogenetics in patients who present with cytopenias can identify patients for whom *GATA2* sequencing is indicated [139]. If *GATA2* mutations are identified, it is important to screen family members who may be potential donors, as BM transplantation is the only definitive therapy for *GATA2* deficiency [140,141].

Familial AA is an extremely rare inherited subtype affecting multiple individuals in a family. Patients typically only have features of AA; the absence of any somatic features making it distinct from other inherited AA. By exome sequencing, the causative homozygous *MPL* mutation in a family with familial AA is reported [142]. Biallelic constitutional mutations in *MPL* have been described in congenital amegakaryocytic thrombocytopenia (CAMT) [143]. *MPL* mutations can be found in children with familial AA in whom CAMT was not diagnosed or suspected. Additional studies will be needed to further clarify the relationship between CAMT, AA, and MPL.

For screening, genomic panels of large numbers of genes are now available and routinely used in the clinic to molecularly characterize patients with suspected inherited bone marrow failure syndromes (IBMFS); multiple genes from pathways involved in, Fanconi anemia (FA),

dyskeratosis congenita (DC), and Diamond–Blackfan anemia (DBA) are sequenced by this method [144].

Clonal Evolution in AA

With improved survival, the late development of MDS, AML, or both has been noted in about 15% of AA patients and termed "clonal evolution" [145]. AA patients with clonal cytogenetic patterns are heterogenous; unlike in primary MDS, aberrancies of chromosome 5 and 20 were infrequent [146]. The clinical course depended on the specific abnormal cytogenetic pattern. Most deaths related to leukemic transformation occur in patients with abnormalities of chromosome 7 or complex cytogenetic alterations or both [146]. In contrast, +8 and del13q, may appear in AA that is responsive to IST and associated with good prognosis [146–149]. In AA patients with PNH clone, cytogenetic abnormalities usually occur in hematopoietic cells that are normal phenotype (GPI–AP positive), suggesting these cells have different origin [25,148].

Telomere dynamics play a role in the development of myeloid cancers in patients with AA not associated with a telomeropathy. In adult patients with severe AA undergoing IST (without a known genetic telomeropathy), pretreatment telomere length in the bottom quartile for age was a significant risk factor for evolution to MDS [150]. Patients with the shortest telomeres had more uncapped telomere-free chromosome ends as compared to the patients with the longest telomeres [151]. Analysis of patients, without telomeropathies and with normal telomere length at the time of diagnosis, who developed to monosomy 7 found these patients had dramatically accelerated telomere attrition before developing MDS [152]. Rapid telomere loss led to an accumulation of individual chromosomes bearing extremely short telomeres prior to the development of monosomy 7 as detected by STELA. Dependence on a limited stem-cell pool to support hematopoiesis would require an increased rate of cell division and accelerate telomere attrition.

In recent years, a number of studies have reported the presence of acquired somatic mutations in AA, often associated with low level clones [28,29,153,154]. These mutations were recurrent somatic mutations found in MDS/AML. In a small cohort of predominantly pediatric patients, somatic mutations were detected in 72%, most frequently involved in immune escape (PIGA, 6pLOH) and signal transduction (STAT5B, CAMK2G), and MDS-associated SM were found in only 9% of patients [27]. A large cohort of 150 AA patients with no morphological evidence of MDS identified a subgroup (19%) with relevant somatic mutations in a small number of genes (ASXL1 in 12 patients, DNMT3A in 8, BCOR in 6, and

1 each for SRSF2, U2AF1, TET2, MPL, IKZF1, and ERBB2) [155]. Somatic mutations, when examined together, predicted for risk of later evolution to MDS; the risk was 38% compared to 6% in the absence of a somatic mutation, and if the disease duration of the AA was >6 months, the risk of MDS was even more significant at 40% compared to 4% without a somatic mutation. ASXL1 and DNMT3A mutations were associated with evolution to monosomy 7 in 4 AA patients. They also showed that presence of a somatic mutation was associated with shorter telomere length compared to patients who lacked a somatic mutation. In a combined USA and Japanese study targeted sequencing of 106 genes in 439 AA patients [26], various sets of mutations show distinct behavior and clinical effect; BCOR-mutant and BCORL1-mutant clones and PIGA-mutant clones tended to disappear or remain small, were associated with a better response to IST, and predicted favorable outcomes. These data are compatible with previous studies focusing on the significance of PIG-A mutant clones. The vast majority of patients tended to equilibrate at a clone size, even after IST [23,84]. Recent study reported that 6p CN-LOH clones were present in 11.3% of AA, remained stable over time, and

were not associated with the development of MDS-defining cytogenetic abnormalities [156]. In contrast, clones carrying mutations in DNMT3A, ASXL1, and a few other genes were more likely to increase in size over time, and these mutations (dominated by DNMT3A and ASXL1) as a group were associated with a poorer response to IST, inferior overall survival, and progression to MDS, AML, or both. Computational strategy identified patients with better overall survival (those with PIGA, BCOR, and BCORL1) and patients with worse overall survival (those with ASXL1, DNMT3A, TP53, RUNX1, and CSMD1) than overall survival in the unmutated group, and patients with better progression-free survival (those with PIGA, BCOR, and BCORL1) and patients with worse progression-free survival (those with ASXL1, DNMT3A, RUNX1, JAK2, and JAK3) than progression-free survival in the unmutated group.

A more detailed analysis of clonal evolution was obtained by the sequencing of serial samples from 35 AA patients that spanned years [26]. In most patients, clonal hematopoiesis originated from a minor clone that was already present at the time of diagnosis. However, the subsequent temporal course of these clones was highly variable. Close monitoring of clonal hematopoiesis by means of both deep sequencing and SNP array karyotyping will need to be combined with clinical evaluation to estimate prognosis and to guide treatment of patients with AA.

TREATMENTS FOR AA

BMT With Matched Sibling Donors

HSCT is the treatment of choice in newly diagnosed patients up to 40 years of age eligible for HSCT with a histocompatible donor [157]. For all other patients IST with horse ATG/CsA is the preferred treatment modality. The long-term survival for either therapy is about 80% for patients of all ages, with younger patients (<20) faring better in general [158]. Standard treatment for patients who have a matched sibling donor is HSCT which provides a cure in about 80–90% of patients [159]. Cyclophosphamide (CY) with ATG as a conditioning regimen and the CsA plus methotrexate (MTX) as GVHD prophylaxis represent an effective treatment, with a rate of engraftment of 90–95% and overall survival near 80–90% [159]. These rates tend to be worst in older patients (>40 years of age) approximating 50–60% [160].

BM source, conditioning with CY + ATG, and GVHD prophylaxis by CsA + MTX represent the gold standard in transplantation for AA in patients receiving transplantations from an HLA-identical sibling donors. Given these excellent results, survival is thus no more the sole concern then prevention and early detection of late complications after BMT is the main objective [161]. After transplantation from an HLA-identical sibling donor or from an unrelated donor, the use of peripheral blood stem cells must be strongly discouraged because they have been systematically associated with an increased incidence of chronic GVHD compared with the use of BM as a stem cell source, leading to an unacceptably higher risk of treatment-related mortality in this setting.

BMT From Alternative Donors

For patients who lack an HLA identical sibling donor, IST remains the first treatment of choice. However, 30–40% of the patients will be refractory or relapse to IST and will thus be considered for transplantation using an alternative donor. HSCT is indicated if refractory AA patients are fit and have a suitably matched donor. Transplantation from an unrelated donor (UD) is usually considered after failure of at least one course of IST [162]. This strategy is based on a relatively high risk of complications for UD transplant recipients, such as graft rejection, GvHD, and infections [163,164]. However, the outcome of unrelated donor transplants has

significantly improved [165,166]. Recent study reported that the outcome of UD transplants for AA is currently not statistically inferior when compared to sibling transplants, although patients are at greater risk of acute and chronic GVHD [167]. The risk of death of UD grafts was higher, but not significantly higher, compared to a sibling donor. The strongest negative predictor of survival was the use of peripheral blood as a stem cell source, followed by an interval of diagnosis to transplant of 180 days or more, patient age 20 years or over, no ATG n in the conditioning, and donor/recipient cytomegalovirus serostatus, other than negative/negative.

Experimental HSCT

Other alternative source of stem cells includes umbilical cord blood and a haploidentical family donor [168]. The difficult situation encountered is when a patient with refractory AA has no suitably matched UD, and this is not uncommon. The options for these individuals include a second course of IST, an alternative immunosuppressive drug or novel agent, or an experimental form of transplantation using an alternative donor source, namely cord blood or a haploidentical family donor. HSCT offers possibilities of cure, but risks of alternative donor HSCT remain graft rejection and GVHD, especially chronic GVHD, which affects mortality and quality of life.

The key features of haploidentical graft are availability for most patients, the time to procure the graft is short, and the cost is low. This is not a new approach to HSCT for SAA patients lacking a matched sibling donor or a suitably matched UD. Historically, haploidentical HSCT was invariably unsuccessful, with high rates of graft rejection and GVHD. A recent review on 73 patients receiving HSCT between 1976 and 2011 and mostly using nonmyeloablative regimens showed a 3-year OS of only 37% [169]. A novel approach is nonmyeloablative conditioning with high-dose CY given on days +3 and +4 after transplantation (PTCy) to prevent GVHD by depleting dividing donor-alloreactive T cells but sparing quiescent, nonalloreactive T cells. Such an approach has been reported anecdotally in PNH and is under study in protocols in AA [170]. Recent studies from Johns Hopkins and others showed that nonmyeloablative allo-BMT with PTCy from haploidentical donor appears promising in patients with refractory acquired and inherited AA with acceptable rates of engraftment, eradication of preexisting clonal diseases, low risk of GVHD, and expansion of the donor pool [171,172].

The largest retrospective study of unrelated cord blood transplantation comprised 71 AA patients (9 with PNH) [173]. The main problem was engraftment failure, with a cumulative incidence of neutrophil recovery of only 51% at 2 months and a 3-year OS of 38%. All those patients receiving total body irradiation 12 Gy as part of the conditioning regimen died, indicating that a RIC rather than myeloablative regimen is preferable. Significantly improved OS was seen in recipients of $>3.9 \times 10^7$ TNCs/kg prefreezing. Recent study showed that haplocord HSCT is an effective treatment option for severe AA patients who lack an HLA-matched donor [174].

IST

The possibility that the pathophysiology of AA could be immune mediated was initially proposed by Barnes and Mole and later reinforced by autologous hematologic recovery after ATG exposure [109,175]. The combination of ATG + CsA was shown superior to ATG alone and became the standard IST option in patients not eligible for HSCT [176]. Extensive experience with this regimen showed that 60–70% of patients responded to the combination of horse ATG/CsA becoming the standard IST regimen. Across these studies hematologic response correlated with excellent long-term survival [157,176–179].

Amounting evidence that AA had an immune mediated pathogenesis led to the exploration of more immunosuppressive agents in clinical protocols. This development was conducted in two manners: (1) adding a third immunosuppressive agent to horse ATG/CsA and (2) using more potent lymphocyte depleting agents than horse ATG. Neither strategy improved the outcomes. Adding mycophenolate mofetil or sirolimus to horse ATG/CsA did not improve outcomes in SAA and the use of more lymphocyte depleting agents (cyclophosphamide, alemtuzumab, rabbit ATG) was equally disappointing [180,181]. Cyclophosphamide, despite being an active agent in SAA, is associated with prohibitive toxicity impeding its use [182,183]. Toxicity was considerable, mainly due to prolonged absolute neutropenia, that occurred regardless of pretherapy blood counts, and persisted an average of 2 months [183]. Alemtuzumab associated with a low response rate (about 20%) when given as first therapy [184]. Noteworthy are the disappointing results with rabbit ATG/CsA when compared directly to horse ATG/CsA in a randomized study [185]. The anticipation was for a higher response rate with rabbit ATG/CsA given its more lymphocyte depleting properties, Treg inducing property, activity in horse ATG/CsA refractory cases and superiority to horse ATG in solid organ transplant [186–189]. However, results showed a lower response rate of about 35–40% compared to the expected 60–70% seen with horse ATG/CsA [185]. This difference resulted in superior survival outcomes with horse ATG/CsA compared to rabbit ATG/CsA [185]. Other observations from prospective studies confirmed this observation [190]. Therefore, horse ATG/CsA remains the optimal immunosuppressive regimen as first line therapy in SAA. The options for patients who fail initial horse ATG/CsA include HSCT from a related (in older patients) or unrelated (in younger patients) histocompatible donor or a repeat course of IST [191,192]. Rabbit ATG/CsA or alemtuzumab in this refractory setting can salvage about 30–40% of patients who can achieve a hematologic response [184,188]. The ability to salvage a proportion of patients with transplant and nontransplant modalities along with advances in supportive care have resulted in improved survival outcomes in IST refractory patients over the years [193]. Even among very refractory patients to IST who do not undergo HSCT long-term survival can be achieved with supportive care measure, such as transfusions, antimicrobials, growth factor, androgens and iron chelation. Neutrophil count in this setting are a principal determinant of survival [194]. When neutrophils are adequate in numbers that prevent recurrent life-threatening infections long-term survival with supportive care measures can be achieved [194]. A longer, tapering course of CsA following horse ATG up to 24 months as initial therapy delayed but did not ultimately prevent relapses [195]. Therefore, efforts to improve beyond the outcomes achieved with horse ATG plus CsA were ineffective until recently.

Eltrombopag

TPO mimetics, such as romiplostim and eltrombopag, were developed to treat patients with refractory immune thrombocytopenia but have been investigated for the treatment of BM failure syndromes [196]. TPO is the main regulator for platelet production and its receptor (c-Mpl) is present on megakaryocytes and HSCs [197,198]. Historically the use of G-CSF and erythropoietin have not been effective in systematic AA studies and did not change the natural history of the disease [199]. The use of a TPO agonist was of particular interest given the expression of the TPO agonist receptor in marrow progenitor cells and the reduced numbers of HSCs in murine TPO receptor knockout models [200]. However, the very high endogenous TPO levels in SAA could render this approach ineffective [201].

In a pilot study (n = 26) eltrombopag was investigated at an initial dose of 50 mg/day titrated up to 150 mg in a IST refractory SAA cohort [202]. The overall response rate was 44% with some bi- and trilineage hematologic improvements observed. The degree and quality of improvement in blood counts with eltrombopag of patients with refractory AA was unanticipated. Responses were striking in several respects: they were not restricted to platelets and were robust, resulting in transfusion independence. In an extended experience (n = 43) the response rate was confirmed at 40% and multilineage increment in blood counts were continued to be observed [203]. Discontinuation of eltrombopag was possible in a few patients who had achieved a robust hematologic recovery without worsening of the counts. Eight of 43 patients developed new cytogenetic abnormalities. The most common chromosomal changes were chromosome 7 abnormalities, which developed in five of the eight evolvers. This clonal transformation is associated with a poor outcome, but all patients who developed chromosome 7 changes were successfully transplanted. Increase in bone marrow cellularity was observed in some responding patients suggesting that eltrombopag was stimulating a more primitive progenitor population leading to improvement in marrow function and increase in blood counts. The protection and/or stimulation of residual HSCs by eltrombopag is a strategy with many applications.

Multiple clinical trials of thrombopoietin mimetics, eltrombopag, and romiplostim are in progress. Eltrombopag is being combined with standard horse ATG and CsA in treatment-naive severe AA [204]. In these previously untreated patients, the combination of eltrombopag with horse ATG and CsA yielded overall response rate at 3 and 6 months of 80 and 85%, respectively, which are 20–30% higher than historical rates with the same regimen without Tpo agonist in AA. Clonal cytogenetic evolution occurred in 7 of 88 patients, similar to prior rates observed with standard IST. Longer and more mature data regarding this novel combination is awaited to better define the durability of response and the impact on long-term outcomes of relapse and clonal evolution.

Supportive Care

Management of AA has been dramatically improved since late 1970s with the introduction of allogeneic stem-cell transplantation (allo-SCT) and IST as well as optimized supportive care [194]. Supportive care has impacted positively these different treatment modalities. The advent of oral antifungals with activity against *Aspergillus* sp. (voriconazole, posaconazole) has allowed for continued outpatient therapy against these pathogens that represent the deadliest infectious culprit in SAA [194,205]. The use of granulocyte transfusion in selected cases has allowed for better control of fungal infections when neutropenia is severe and persistent and antifungals are not resolutive alone [206,207]. Growth factors can be effective in some patients with increments in neutrophil and/or hemoglobin levels lessening transfusion requirements and the risk associated with these cytopenias. Oral iron chelators have allowed for continued long-term transfusion programs in very refractory patients minimizing the risks associated with iron accumulation in the long term [208]. In the aggregate these advances have allowed for patients who failed IST and are not HSCT eligible to be supported for long periods of time with few complications associated with the persistent cytopenias. In general, the neutrophil count is an important determinant in how these patients who are kept on supportive care fare long term.

CONCLUSIONS

Nature of evidence, hematopoiesis, immune pathophysiology, genetic risk factors, and risk factors for clonal evolution for AA were summarized. Better understanding of hematopoiesis and immune mechanisms in AA may provide new therapeutic strategies, and novel mechanistic insights into normal hematopoiesis, relevant to leukemia and normal aging.

References

[1] Gerull S, Stern M, Apperley J, Beelen D, Brinch L, Bunjes D, et al. Syngeneic transplantation in aplastic anemia: pre-transplant conditioning and peripheral blood are associated with improved engraftment: an observational study on behalf of the Severe Aplastic Anemia and Pediatric Diseases Working Parties of the European Group for Blood and Marrow Transplantation. Haematologica 2013;98(11):1804–9.

[2] Young NS, Calado RT, Scheinberg P. Current concepts in the pathophysiology and treatment of aplastic anemia. Blood 2006;108(8):2509–19.

[3] Young NS, Bacigalupo A, Marsh JC. Aplastic anemia: pathophysiology and treatment. Biology of blood and marrow transplantation: journal of the American Society for Blood and Marrow Transplantation 2010; 16(1 Suppl.):S119–125.

[4] Killick SB, Bown N, Cavenagh J, Dokal I, Foukaneli T, Hill A, et al. Guidelines for the diagnosis and management of adult aplastic anaemia. Br J Haematol 2016;172(2):187–207.

[5] Scopes J, Bagnara M, Gordon-Smith EC, Ball SE, Gibson FM. Haemopoietic progenitor cells are reduced in aplastic anaemia. Br J Haematol 1994;86(2):427–30.

[6] Marsh JC, Chang J, Testa NG, Hows JM, Dexter TM. The hematopoietic defect in aplastic anemia assessed by long-term marrow culture. Blood 1990;76(9):1748–57.

[7] Maciejewski JP, Selleri C, Sato T, Anderson S, Young NS. A severe and consistent deficit in marrow and circulating primitive hematopoietic cells (long-term culture-initiating cells) in acquired aplastic anemia. Blood 1996;88(6):1983–91.

[8] Matsui WH, Brodsky RA, Smith BD, Borowitz MJ, Jones RJ. Quantitative analysis of bone marrow CD34 cells in aplastic anemia and hypoplastic myelodysplastic syndromes. Leukemia 2006;20(3):458–62.

[9] Rizzo S, Scopes J, Elebute MO, Papadaki HA, Gordon-Smith EC, Gibson FM. Stem cell defect in aplastic anemia: reduced long term culture-initiating cells (LTC-IC) in CD34+ cells isolated from aplastic anemia patient bone marrow. Hematol J 2002;3(5):230–6.

[10] Schrezenmeier H, Jenal M, Herrmann F, Heimpel H, Raghavachar A. Quantitative analysis of cobblestone area-forming cells in bone marrow of patients with aplastic anemia by limiting dilution assay. Blood 1996;88(12):4474–80.

[11] Sun J, Ramos A, Chapman B, Johnnidis JB, Le L, Ho YJ, et al. Clonal dynamics of native haematopoiesis. Nature 2014;514(7522):322–7.

[12] Busch K, Klapproth K, Barile M, Flossdorf M, Holland-Letz T, Schlenner SM, et al. Fundamental properties of unperturbed haematopoiesis from stem cells in vivo. Nature 2015;518(7540):542–6.

[13] Notta F, Zandi S, Takayama N, Dobson S, Gan OI, Wilson G, et al. Distinct routes of lineage development reshape the human blood hierarchy across ontogeny. Science 2016;351(6269):aab2116.

[14] Scheinberg P, Wu CO, Nunez O, Young NS. Predicting response to immunosuppressive therapy and survival in severe aplastic anaemia. Br J Haematol 2009;144(2):206–16.

[15] Takaku T, Malide D, Chen J, Calado RT, Kajigaya S, Young NS. Hematopoiesis in 3 dimensions: human and murine bone marrow architecture visualized by confocal microscopy. Blood 2010;116(15):e41–55.

[16] Naveiras O, Nardi V, Wenzel PL, Hauschka PV, Fahey F, Daley GQ. Bone-marrow adipocytes as negative regulators of the haematopoietic microenvironment. Nature 2009;460(7252):259–63.

[17] Sato K, Feng X, Chen J, Li J, Muranski P, Desierto MJ, et al. PPARγ antagonist attenuates mouse immune-mediated bone marrow failure by inhibition of T cell function. Haematologica 2016;101(1):57–67.

[18] Young NS, Maciejewski JP, Sloand E, Chen G, Zeng W, Risitano A, et al. The relationship of aplastic anemia and PNH. Int J Hematol 2002;76(Suppl. 2):168–72.

[19] Takeda J, Miyata T, Kawagoe K, Iida Y, Endo Y, Fujita T, et al. Deficiency of the GPI anchor caused by a somatic mutation of the PIG-A gene in paroxysmal nocturnal hemoglobinuria. Cell 1993;73(4):703–11.

[20] Dunn DE, Tanawattanacharoen P, Boccuni P, Nagakura S, Green SW, Kirby MR, et al. Paroxysmal nocturnal hemoglobinuria cells in patients with bone marrow failure syndromes. Ann Intern Med 1999;131(6):401–8.

[21] Scheinberg P, Marte M, Nunez O, Young NS. Paroxysmal nocturnal hemoglobinuria clones in severe aplastic anemia patients treated with horse antithymocyte globulin plus cyclosporine. Haematologica 2010;95(7):1075–80.

[22] Wang H, Chuhjo T, Yasue S, Omine M, Nakao S. Clinical significance of a minor population of paroxysmal nocturnal hemoglobinuria-type cells in bone marrow failure syndrome. Blood 2002;100(12):3897–902.

[23] Shen W, Clemente MJ, Hosono N, Yoshida K, Przychodzen B, Yoshizato T, et al. Deep sequencing reveals stepwise mutation acquisition in paroxysmal nocturnal hemoglobinuria. J Clin Invest 2014;124(10):4529–38.

[24] Katagiri T, Kawamoto H, Nakakuki T, Ishiyama K, Okada-Hatakeyama M, Ohtake S, et al. Individual hematopoietic stem cells in human bone marrow of patients with aplastic anemia or myelodysplastic syndrome stably give rise to limited cell lineages. Stem Cells 2013;31(3):536–46.

[25] Sloand EM, Fuhrer M, Keyvanfar K, Mainwaring L, Maciejewski J, Wang Y, et al. Cytogenetic abnormalities in paroxysmal nocturnal haemoglobinuria usually occur in haematopoietic cells that are glycosylphosphatidylinositol-anchored protein (GPI-AP) positive. Br J Haematol 2003;123(1):173–6.

[26] Yoshizato T, Dumitriu B, Hosokawa K, Makishima H, Yoshida K, Townsley D, et al. Somatic mutations and clonal hematopoiesis in aplastic anemia. N Engl J Med 2015;373(1):35–47.

[27] Babushok DV, Perdigones N, Perin JC, Olson TS, Ye W, Roth JJ, et al. Emergence of clonal hematopoiesis in the majority of patients with acquired aplastic anemia. Cancer Genet 2015;208(4):115–28.

[28] Lane AA, Odejide O, Kopp N, Kim S, Yoda A, Erlich R, et al. Low frequency clonal mutations recoverable by deep sequencing in patients with aplastic anemia. Leukemia 2013;27(4):968–71.

[29] Heuser M, Schlarmann C, Dobbernack V, Panagiota V, Wiehlmann L, Walter C, et al. Genetic characterization of acquired aplastic anemia by targeted sequencing. Haematologica 2014;99(9):e165–167.

[30] Alexandrov LB, Nik-Zainal S, Wedge DC, Aparicio SA, Behjati S, Biankin AV, et al. Signatures of mutational processes in human cancer. Nature 2013;500(7463):415–21.

[31] Welch JS, Ley TJ, Link DC, Miller CA, Larson DE, Koboldt DC, et al. The origin and evolution of mutations in acute myeloid leukemia. Cell 2012;150(2):264–78.

[32] Jaiswal S, Fontanillas P, Flannick J, Manning A, Grauman PV, Mar BG, et al. Age-related clonal hematopoiesis associated with adverse outcomes. N Engl J Med 2014;371(26):2488–98.

[33] Genovese G, Jaiswal S, Ebert BL, McCarroll SA. Clonal hematopoiesis and blood-cancer risk. N Engl J Med 2015;372(11):1071–2.

[34] Challen GA, Sun D, Jeong M, Luo M, Jelinek J, Berg JS, et al. Dnmt3a is essential for hematopoietic stem cell differentiation. Nat Genet 2012;44(1):23–31.

[35] Mayle A, Yang L, Rodriguez B, Zhou T, Chang E, Curry CV, et al. Dnmt3a loss predisposes murine hematopoietic stem cells to malignant transformation. Blood 2015;125(4):629–38.

[36] Abdel-Wahab O, Gao J, Adli M, Dey A, Trimarchi T, Chung YR, et al. Deletion of Asxl1 results in myelodysplasia and severe developmental defects in vivo. J Exp Med 2013;210(12):2641–59.

[37] Wang J, Li Z, He Y, Pan F, Chen S, Rhodes S, et al. Loss of Asxl1 leads to myelodysplastic syndrome-like disease in mice. Blood 2014;123(4):541–53.

[38] Gargiulo L, Papaioannou M, Sica M, Talini G, Chaidos A, Richichi B, et al. Glycosylphosphatidylinositol-specific, CD1d-restricted T cells in paroxysmal nocturnal hemoglobinuria. Blood 2013;121(14):2753–61.

[39] Murakami Y, Kosaka H, Maeda Y, Nishimura J, Inoue N, Ohishi K, et al. Inefficient response of T lymphocytes to glycosylphosphatidylinositol anchor-negative cells: implications for paroxysmal nocturnal hemoglobinuria. Blood 2002;100(12):4116–22.

[40] Hinterberger W, Rowlings PA, Hinterberger-Fischer M, Gibson J, Jacobsen N, Klein JP, et al. Results of transplanting bone marrow from genetically identical twins into patients with aplastic anemia. Ann Intern Med 1997;126(2):116–22.

[41] Young NS. Hematopoietic cell destruction by immune mechanisms in acquired aplastic anemia. Semin Hematol 2000;37(1):3–14.

[42] Nakao S, Takami A, Takamatsu H, Zeng W, Sugimori N, Yamazaki H, et al. Isolation of a T-cell clone showing HLA-DRB1*0405-restricted cytotoxicity for hematopoietic cells in a patient with aplastic anemia. Blood 1997;89(10):3691–9.

[43] Maciejewski J, Selleri C, Anderson S, Young NS. Fas antigen expression on CD34+ human marrow cells is induced by interferon gamma and tumor necrosis factor alpha and potentiates cytokine-mediated hematopoietic suppression in vitro. Blood 1995;85(11):3183–90.

[44] Zoumbos NC, Gascon P, Djeu JY, Young NS. Interferon is a mediator of hematopoietic suppression in aplastic anemia in vitro and possibly in vivo. Proc Natl Acad Sci USA 1985;82(1):188–92.

[45] Zoumbos NC, Gascon P, Djeu JY, Trost SR, Young NS. Circulating activated suppressor T lymphocytes in aplastic anemia. N Engl J Med 1985;312(5):257–65.

[46] Gascon P, Zoumbos NC, Scala G, Djeu JY, Moore JG, Young NS. Lymphokine abnormalities in aplastic anemia: implications for the mechanism of action of antithymocyte globulin. Blood 1985;65(2):407–13.

[47] Sloand E, Kim S, Maciejewski JP, Tisdale J, Follmann D, Young NS. Intracellular interferon-gamma in circulating and marrow T cells detected by flow cytometry and the response to immunosuppressive therapy in patients with aplastic anemia. Blood 2002;100(4):1185–91.

[48] Sato T, Selleri C, Young NS, Maciejewski JP. Inhibition of interferon regulatory factor-1 expression results in predominance of cell growth stimulatory effects of interferon-gamma due to phosphorylation of Stat1 and Stat3. Blood 1997;90(12):4749–58.

[49] Selleri C, Maciejewski JP, Sato T, Young NS. Interferon-gamma constitutively expressed in the stromal microenvironment of human marrow cultures mediates potent hematopoietic inhibition. Blood 1996;87(10):4149–57.

[50] Hosokawa K, Muranski P, Feng X, Townsley DM, Liu B, Knickelbein J, et al. Memory stem T cells in autoimmune disease: high frequency of circulating CD8+ memory stem cells in acquired aplastic anemia. J Immunol 2016;196(4):1568–78.

[51] Risitano AM, Maciejewski JP, Green S, Plasilova M, Zeng W, Young NS. In-vivo dominant immune responses in aplastic anaemia: molecular tracking of putatively pathogenetic T-cell clones by TCR beta-CDR3 sequencing. Lancet 2004;364(9431):355–64.

[52] Kook H, Risitano AM, Zeng W, Wlodarski M, Lottemann C, Nakamura R, et al. Changes in T-cell receptor VB repertoire in aplastic anemia: effects of different immunosuppressive regimens. Blood 2002;99(10):3668–75.

[53] Zeng W, Maciejewski JP, Chen G, Young NS. Limited heterogeneity of T cell receptor BV usage in aplastic anemia. J Clin Invest 2001;108(5):765–73.

[54] Solomou EE, Rezvani K, Mielke S, Malide D, Keyvanfar K, Visconte V, et al. Deficient CD4+ CD25+ FOXP3+ T regulatory cells in acquired aplastic anemia. Blood 2007;110(5):1603–6.

[55] de Latour RP, Visconte V, Takaku T, Wu C, Erie AJ, Sarcon AK, et al. Th17 immune responses contribute to the pathophysiology of aplastic anemia. Blood 2010;116(20):4175–84.

[56] Kordasti S, Marsh J, Al-Khan S, Jiang J, Smith A, Mohamedali A, et al. Functional characterization of CD4+ T cells in aplastic anemia. Blood 2012;119(9):2033–43.

[57] Solomou EE, Keyvanfar K, Young NS. T-bet, a Th1 transcription factor, is up-regulated in T cells from patients with aplastic anemia. Blood 2006;107(10):3983–91.

[58] Roderick JE, Gonzalez-Perez G, Kuksin CA, Dongre A, Roberts ER, Srinivasan J, et al. Therapeutic targeting of NOTCH signaling ameliorates immune-mediated bone marrow failure of aplastic anemia. J Exp Med 2013;210(7):1311–29.

[59] Tang Y, Desierto MJ, Chen J, Young NS. The role of the Th1 transcription factor T-bet in a mouse model of immune-mediated bone-marrow failure. Blood 2010;115(3):541–8.

[60] Omokaro SO, Desierto MJ, Eckhaus MA, Ellison FM, Chen J, Young NS. Lymphocytes with aberrant expression of Fas or Fas ligand attenuate immune bone marrow failure in a mouse model. J Immunol 2009;182(6):3414–22.

[61] Chen J, Ellison FM, Eckhaus MA, Smith AL, Keyvanfar K, Calado RT, et al. Minor antigen h60-mediated aplastic anemia is ameliorated by immunosuppression and the infusion of regulatory T cells. J Immunol 2007;178(7):4159–68.

[62] Chen J, Lipovsky K, Ellison FM, Calado RT, Young NS. Bystander destruction of hematopoietic progenitor and stem cells in a mouse model of infusion-induced bone marrow failure. Blood 2004;104(6):1671–8.

[63] Bloom ML, Wolk AG, Simon-Stoos KL, Bard JS, Chen J, Young NS. A mouse model of lymphocyte infusion-induced bone marrow failure. Exp Hematol 2004;32(12):1163–72.

[64] Maciejewski JP, Selleri C, Sato T, Anderson S, Young NS. Increased expression of Fas antigen on bone marrow CD34+ cells of patients with aplastic anaemia. Br J Haematol 1995;91(1):245–52.

[65] Hirayama Y, Sakamaki S, Matsunaga T, Kuga T, Kuroda H, Kusakabe T, et al. Concentrations of thrombopoietin in bone marrow in normal subjects and in patients with idiopathic thrombocytopenic purpura, aplastic anemia, and essential thrombocythemia correlate with its mRNA expression of bone marrow stromal cells. Blood 1998;92(1):46–52.

[66] Marsh JC, Gibson FM, Prue RL, Bowen A, Dunn VT, Hornkohl AC, et al. Serum thrombopoietin levels in patients with aplastic anaemia. Br J Haematol 1996;95(4):605–10.

[67] Kojima S, Matsuyama T, Kodera Y, Nishihira H, Ueda K, Shimbo T, et al. Measurement of endogenous plasma granulocyte colony-stimulating factor in patients with acquired aplastic anemia by a sensitive chemiluminescent immunoassay. Blood 1996;87(4):1303–8.

[68] Kojima S, Matsuyama T, Kodera Y, Tahara T, Kato T. Measurement of endogenous plasma thrombopoietin in patients with acquired aplastic anaemia by a sensitive enzyme-linked immunosorbent assay. Br J Haematol 1997;97(3):538–43.

[69] Feng X, Scheinberg P, Wu CO, Samsel L, Nunez O, Prince C, et al. Cytokine signature profiles in acquired aplastic anemia and myelodysplastic syndromes. Haematologica 2011;96(4):602–6.

[70] Du HZ, Wang Q, Ji J, Shen BM, Wei SC, Liu LJ, et al. Expression of IL-27, Th1 and Th17 in patients with aplastic anemia. J Clin Immunol 2013;33(2):436–45.

[71] Gu Y, Hu X, Liu C, Qv X, Xu C. Interleukin (IL)-17 promotes macrophages to produce IL-8, IL-6 and tumour necrosis factor-alpha in aplastic anaemia. Br J Haematol 2008;142(1):109–14.

[72] Elmahdi S, Hama A, Manabe A, Hasegawa D, Muramatsu H, Narita A, et al. A cytokine-based diagnostic program in pediatric aplastic anemia and hypocellular refractory cytopenia of childhood. Pediatr Blood Cancer 2016;63(4):652–8.

[73] Chapuis B, Von Fliedner VE, Jeannet M, Merica H, Vuagnat P, Gratwohl A, et al. Increased frequency of DR2 in patients with aplastic anaemia and increased DR sharing in their parents. Br J Haematol 1986;63(1):51–7.

[74] Maciejewski JP, Follmann D, Nakamura R, Saunthararajah Y, Rivera CE, Simonis T, et al. Increased frequency of HLA-DR2 in patients with paroxysmal nocturnal hemoglobinuria and the PNH/aplastic anemia syndrome. Blood 2001;98(13):3513–9.

[75] Sugimori C, Yamazaki H, Feng X, Mochizuki K, Kondo Y, Takami A, et al. Roles of DRB1 *1501 and DRB1 *1502 in the pathogenesis of aplastic anemia. Exp Hematol 2007;35(1):13–20.

[76] Nakao S, Takamatsu H, Chuhjo T, Ueda M, Shiobara S, Matsuda T, et al. Identification of a specific HLA class II haplotype strongly associated with susceptibility to cyclosporine-dependent aplastic anemia. Blood 1994;84(12):4257–61.

[77] Demeter J, Messer G, Schrezenmeier H. Clinical relevance of the TNF-alpha promoter/enhancer polymorphism in patients with aplastic anemia. Ann Hematol 2002;81(10):566–9.

[78] Dufour C, Capasso M, Svahn J, Marrone A, Haupt R, Bacigalupo A, et al. Homozygosis for (12) CA repeats in the first intron of the human IFN-gamma gene is significantly associated with the risk of aplastic anaemia in Caucasian population. Br J Haematol 2004;126(5):682–5.

[79] Svahn J, Capasso M, Lanciotti M, Marrone A, Haupt R, Bacigalupo A, et al. The polymorphisms -318C > T in the promoter and 49A > G in exon 1 of CTLA4 and the risk of aplastic anemia in a Caucasian population. Bone Marrow Transplant 2005;35(Suppl. 1):S89–92.

[80] Nakao S, Sugimori C, Yamazaki H. Clinical significance of a small population of paroxysmal nocturnal hemoglobinuria-type cells in the management of bone marrow failure. Int J Hematol 2006;84(2):118–22.

[81] Sugimori C, Chuhjo T, Feng X, Yamazaki H, Takami A, Teramura M, et al. Minor population of CD55-CD59-blood cells predicts response to immunosuppressive therapy and prognosis in patients with aplastic anemia. Blood 2006;107(4):1308–14.

[82] Kulagin A, Lisukov I, Ivanova M, Golubovskaya I, Kruchkova I, Bondarenko S, et al. Prognostic value of paroxysmal nocturnal haemoglobinuria clone presence in aplastic anaemia patients treated with combined immunosuppression: results of two-centre prospective study. Br J Haematol 2014;164(4):546–54.

[83] Hosokawa K, Sugimori N, Katagiri T, Sasaki Y, Saito C, Seiki Y, et al. Increased glycosylphosphatidylinositol-anchored protein-deficient granulocytes define a benign subset of bone marrow failures in patients with trisomy 8. Eur J Haematol 2015;95(3):230–8.

[84] Sugimori C, Mochizuki K, Qi Z, Sugimori N, Ishiyama K, Kondo Y, et al. Origin and fate of blood cells deficient in glycosylphosphatidylinositol-anchored protein among patients with bone marrow failure. Br J Haematol 2009;147(1):102–12.

[85] Chen G, Zeng W, Maciejewski JP, Kcyvanfar K, Billings EM, Young NS. Differential gene expression in hematopoietic progenitors from paroxysmal nocturnal hemoglobinuria patients reveals an apoptosis/immune response in 'normal' phenotype cells. Leukemia 2005;19(5):862–8.

[86] van Bijnen ST, Withaar M, Preijers F, van der Meer A, de Witte T, Muus P, et al. T cells expressing the activating NK-cell receptors KIR2DS4, NKG2C and NKG2D are elevated in paroxysmal nocturnal hemoglobinuria and cytotoxic toward hematopoietic progenitor cell lines. Exp Hematol 2011;39(7). 751-62 e1–751-62 e3.

[87] Katagiri T, Sato-Otsubo A, Kashiwase K, Morishima S, Sato Y, Mori Y, et al. Frequent loss of HLA alleles associated with copy number-neutral 6pLOH in acquired aplastic anemia. Blood 2011;118(25):6601–9.

[88] Babushok DV, Xie HM, Roth JJ, Perdigones N, Olson TS, Cockroft JD, et al. Single nucleotide polymorphism array analysis of bone marrow failure patients reveals characteristic patterns of genetic changes. Br J Haematol 2014;164(1):73–82.

[89] Afable MG II, Wlodarski M, Makishima H, Shaik M, Sekeres MA, Tiu RV, et al. SNP array-based karyotyping: differences and similarities between aplastic anemia and hypocellular myelodysplastic syndromes. Blood 2011;117(25):6876–84.

[90] Ueda Y, Nishimura J, Murakami Y, Kajigaya S, Kinoshita T, Kanakura Y, et al. Paroxysmal nocturnal hemoglobinuria with copy number-neutral 6pLOH in GPI (+) but not in GPI (−) granulocytes. Eur J Haematol 2014;92(5):450–3.

[91] Inaguma Y, Akatsuka Y, Hosokawa K, Maruyama H, Okamoto A, Katagiri T, et al. Induction of HLA-B*40:02-restricted T cells possessing cytotoxic and suppressive functions against haematopoietic progenitor cells from a patient with severe aplastic anaemia. Br J Haematol 2016;172(1):131–4.

[92] Loughran TP Jr. Clonal diseases of large granular lymphocytes. Blood 1993;82(1):1–14.

[93] Saunthararajah Y, Molldrem JL, Rivera M, Williams A, Stetler-Stevenson M, Sorbara L, et al. Coincident myelodysplastic syndrome and T-cell large granular lymphocytic disease: clinical and pathophysiological features. Br J Haematol 2001;112(1):195–200.

[94] Koskela HL, Eldfors S, Ellonen P, van Adrichem AJ, Kuusanmaki H, Andersson EI, et al. Somatic STAT3 mutations in large granular lymphocytic leukemia. N Engl J Med 2012;366(20):1905–13.

[95] Jerez A, Clemente MJ, Makishima H, Koskela H, Leblanc F, Peng Ng K, et al. STAT3 mutations unify the pathogenesis of chronic lymphoproliferative disorders of NK cells and T-cell large granular lymphocyte leukemia. Blood 2012;120(15):3048–57.

[96] Zeng W, Kajigaya S, Chen G, Risitano AM, Nunez O, Young NS. Transcript profile of CD4+ and CD8+ T cells from the bone marrow of acquired aplastic anemia patients. Exp Hematol 2004;32(9):806–14.

[97] Poggi A, Negrini S, Zocchi MR, Massaro AM, Garbarino L, Lastraioli S, et al. Patients with paroxysmal nocturnal hemoglobinuria have a high frequency of peripheral-blood T cells expressing activating

[98] isoforms of inhibiting superfamily receptors. Blood 2005;106(7):2399–408.

[98] Howe EC, Wlodarski M, Ball EJ, Rybicki L, Maciejewski JP. Killer immunoglobulin-like receptor genotype in immune-mediated bone marrow failure syndromes. Exp Hematol 2005;33(11):1357–62.

[99] Hosokawa K, Muranski P, Feng X, Keyvanfar K, Townsley DM, Dumitriu B, et al. Identification of novel microRNA signatures linked to acquired aplastic anemia. Haematologica 2015;100(12):1534–45.

[100] Takamatsu H, Feng X, Chuhjo T, Lu X, Sugimori C, Okawa K, et al. Specific antibodies to moesin, a membrane-cytoskeleton linker protein, are frequently detected in patients with acquired aplastic anemia. Blood 2007;109(6):2514–20.

[101] Feng X, Chuhjo T, Sugimori C, Kotani T, Lu X, Takami A, et al. Diazepam-binding inhibitor-related protein 1: a candidate autoantigen in acquired aplastic anemia patients harboring a minor population of paroxysmal nocturnal hemoglobinuria-type cells. Blood 2004;104(8):2425–31.

[102] Hirano N, Butler MO, Guinan EC, Nadler LM, Kojima S. Presence of antikinectin and anti-PMS1 antibodies in Japanese aplastic anaemia patients. Br J Haematol 2005;128(2):221–3.

[103] Qi Z, Takamatsu H, Espinoza JL, Lu X, Sugimori N, Yamazaki H, et al. Autoantibodies specific to hnRNP K: a new diagnostic marker for immune pathophysiology in aplastic anemia. Ann Hematol 2010;89(12):1255–63.

[104] Goto M, Kuribayashi K, Takahashi Y, Kondoh T, Tanaka M, Kobayashi D, et al. Identification of autoantibodies expressed in acquired aplastic anaemia. Br J Haematol 2013;160(3):359–62.

[105] de Bruin AM, Voermans C, Nolte MA. Impact of interferon-γ on hematopoiesis. Blood 2014;124(16):2479–86.

[106] de Bruin AM, Demirel O, Hooibrink B, Brandts CH, Nolte MA. Interferon-γ impairs proliferation of hematopoietic stem cells in mice. Blood 2013;121(18):3578–85.

[107] Chen J, Feng X, Desierto MJ, Keyvanfar K, Young NS. IFN-γ-mediated hematopoietic cell destruction in murine models of immune-mediated bone marrow failure. Blood 2015;126(24):2621–31.

[108] Lin FC, Karwan M, Saleh B, Hodge DL, Chan T, Boelte KC, et al. IFN-γ causes aplastic anemia by altering hematopoietic stem/progenitor cell composition and disrupting lineage differentiation. Blood 2014;124(25):3699–708.

[109] Scheinberg P, Chen J. Aplastic anemia: what have we learned from animal models and from the clinic. Semin Hematol 2013;50(2):156–64.

[110] Arieta Kuksin C, Gonzalez-Perez G, Minter LM. CXCR4 expression on pathogenic T cells facilitates

their bone marrow infiltration in a mouse model of aplastic anemia. Blood 2015;125(13):2087–94.

[111] Velasco Lezama R, Barrera Escorcia E, Munoz Torres A, Tapia Aguilar R, Gonzalez Ramirez C, Garcia Lorenzana M, et al. A model for the induction of aplastic anemia by subcutaneous administration of benzene in mice. Toxicology 2001;162(3):179–91.

[112] Morley A, Blake J. An animal model of chronic aplastic marrow failure. I. Late marrow failure after busulfan. Blood 1974;44(1):49–56.

[113] Gibson FM, Andrews CM, Diamanti P, Rizzo S, Macharia G, Gordon-Smith EC, et al. A new model of busulphan-induced chronic bone marrow aplasia in the female BALB/c mouse. Int J Exp Pathol 2003;84(1): 31–48.

[114] Szostak JW, Blackburn EH. Cloning yeast telomeres on linear plasmid vectors. Cell 1982;29(1):245–55.

[115] Greider CW, Blackburn EH. Identification of a specific telomere terminal transferase activity in *Tetrahymena* extracts. Cell 1985;43(2 Pt 1):405–13.

[116] Calado RT, Young NS. Telomere maintenance and human bone marrow failure. Blood 2008;111(9):4446–55.

[117] Calado RT, Young NS. Telomere diseases. N Engl J Med 2009;361(24):2353–65.

[118] Heiss NS, Knight SW, Vulliamy TJ, Klauck SM, Wiemann S, Mason PJ, et al. X-linked dyskeratosis congenita is caused by mutations in a highly conserved gene with putative nucleolar functions. Nat Genet 1998;19(1):32–8.

[119] Armanios M, Chen JL, Chang YP, Brodsky RA, Hawkins A, Griffin CA, et al. Haploinsufficiency of telomerase reverse transcriptase leads to anticipation in autosomal dominant dyskeratosis congenita. Proc Natl Acad Sci USA 2005;102(44):15960–4.

[120] Vulliamy T, Marrone A, Goldman F, Dearlove A, Bessler M, Mason PJ, et al. The RNA component of telomerase is mutated in autosomal dominant dyskeratosis congenita. Nature 2001;413(6854):432–5.

[121] Walne AJ, Vulliamy T, Marrone A, Beswick R, Kirwan M, Masunari Y, et al. Genetic heterogeneity in autosomal recessive dyskeratosis congenita with one subtype due to mutations in the telomerase-associated protein NOP10. Hum Mol Genet 2007;16(13):1619–29.

[122] Savage SA, Giri N, Baerlocher GM, Orr N, Lansdorp PM, Alter BP. TINF2, a component of the shelterin telomere protection complex, is mutated in dyskeratosis congenita. Am J Hum Genet 2008;82(2):501–9.

[123] Ball SE, Gibson FM, Rizzo S, Tooze JA, Marsh JC, Gordon-Smith EC. Progressive telomere shortening in aplastic anemia. Blood 1998;91(10):3582–92.

[124] Brummendorf TH, Maciejewski JP, Mak J, Young NS, Lansdorp PM. Telomere length in leukocyte subpopulations of patients with aplastic anemia. Blood 2001;97(4):895–900.

[125] Yamaguchi H, Calado RT, Ly H, Kajigaya S, Baerlocher GM, Chanock SJ, et al. Mutations in TERT, the gene for telomerase reverse transcriptase, in aplastic anemia. N Engl J Med 2005;352(14):1413–24.

[126] Yamaguchi H, Baerlocher GM, Lansdorp PM, Chanock SJ, Nunez O, Sloand E, et al. Mutations of the human telomerase RNA gene (TERC) in aplastic anemia and myelodysplastic syndrome. Blood 2003;102(3):916–8.

[127] Dror Y. Shwachman-Diamond syndrome. Pediatr Blood Cancer 2005;45(7):892–901.

[128] Calado RT, Graf SA, Wilkerson KL, Kajigaya S, Ancliff PJ, Dror Y, et al. Mutations in the SBDS gene in acquired aplastic anemia. Blood 2007;110(4):1141–6.

[129] Calado RT, Yewdell WT, Wilkerson KL, Regal JA, Kajigaya S, Stratakis CA, et al. Sex hormones, acting on the TERT gene, increase telomerase activity in human primary hematopoietic cells. Blood 2009;114(11): 2236–43.

[130] Ziegler P, Schrezenmeier H, Akkad J, Brassat U, Vankann L, Panse J, et al. Telomere elongation and clinical response to androgen treatment in a patient with aplastic anemia and a heterozygous hTERT gene mutation. Ann Hematol 2012;91(7):1115–20.

[131] Hahn CN, Chong CE, Carmichael CL, Wilkins EJ, Brautigan PJ, Li XC, et al. Heritable GATA2 mutations associated with familial myelodysplastic syndrome and acute myeloid leukemia. Nat Genet 2011;43(10):1012–7.

[132] Vinh DC, Patel SY, Uzel G, Anderson VL, Freeman AF, Olivier KN, et al. Autosomal dominant and sporadic monocytopenia with susceptibility to mycobacteria, fungi, papillomaviruses, and myelodysplasia. Blood 2010;115(8):1519–29.

[133] Hsu AP, Sampaio EP, Khan J, Calvo KR, Lemieux JE, Patel SY, et al. Mutations in GATA2 are associated with the autosomal dominant and sporadic monocytopenia and mycobacterial infection (MonoMAC) syndrome. Blood 2011;118(10):2653–5.

[134] Ostergaard P, Simpson MA, Connell FC, Steward CG, Brice G, Woollard WJ, et al. Mutations in GATA2 cause primary lymphedema associated with a predisposition to acute myeloid leukemia (Emberger syndrome). Nat Genet 2011;43(10):929–31.

[135] Dickinson RE, Griffin H, Bigley V, Reynard LN, Hussain R, Haniffa M, et al. Exome sequencing identifies GATA-2 mutation as the cause of dendritic cell, monocyte, B and NK lymphoid deficiency. Blood 2011;118(10):2656–8.

[136] Bigley V, Haniffa M, Doulatov S, Wang XN, Dickinson R, McGovern N, et al. The human syndrome of dendritic cell, monocyte, B and NK lymphoid deficiency. J Exp Med 2011;208(2):227–34.

[137] Pasquet M, Bellanne-Chantelot C, Tavitian S, Prade N, Beaupain B, Larochelle O, et al. High frequency

of GATA2 mutations in patients with mild chronic neutropenia evolving to MonoMac syndrome, myelodysplasia, and acute myeloid leukemia. Blood 2013;121(5):822–9.

[138] Townsley DM, Hsu A, Dumitriu B, Holland SM, Young NS. Regulatory mutations in GATA2 associated with aplastic anemia. Blood 2012;120(21).

[139] Ganapathi KA, Townsley DM, Hsu AP, Arthur DC, Zerbe CS, Cuellar-Rodriguez J, et al. GATA2 deficiency-associated bone marrow disorder differs from idiopathic aplastic anemia. Blood 2015;125(1):56–70.

[140] Grossman J, Cuellar-Rodriguez J, Gea-Banacloche J, Zerbe C, Calvo K, Hughes T, et al. Nonmyeloablative allogeneic hematopoietic stem cell transplantation for GATA2 deficiency. Biol Blood Marrow Transplant 2014;20(12):1940–8.

[141] Cuellar-Rodriguez J, Gea-Banacloche J, Freeman AF, Hsu AP, Zerbe CS, Calvo KR, et al. Successful allogeneic hematopoietic stem cell transplantation for GATA2 deficiency. Blood 2011;118(13):3715–20.

[142] Walne AJ, Dokal A, Plagnol V, Beswick R, Kirwan M, de la Fuente J, et al. Exome sequencing identifies MPL as a causative gene in familial aplastic anemia. Haematologica 2012;97(4):524–8.

[143] Ballmaier M, Germeshausen M, Schulze H, Cherkaoui K, Lang S, Gaudig A, et al. c-mpl mutations are the cause of congenital amegakaryocytic thrombocytopenia. Blood 2001;97(1):139–46.

[144] Khincha PP, Savage SA. Genomic characterization of the inherited bone marrow failure syndromes. Semin Hematol 2013;50(4):333–47.

[145] Socie G, Rosenfeld S, Frickhofen N, Gluckman E, Tichelli A. Late clonal diseases of treated aplastic anemia. Semin Hematol 2000;37(1):91–101.

[146] Maciejewski JP, Risitano A, Sloand EM, Nunez O, Young NS. Distinct clinical outcomes for cytogenetic abnormalities evolving from aplastic anemia. Blood 2002;99(9):3129–35.

[147] Ishiyama K, Karasawa M, Miyawaki S, Ueda Y, Noda M, Wakita A, et al. Aplastic anaemia with 13q-: a benign subset of bone marrow failure responsive to immunosuppressive therapy. Br J Haematol 2002;117(3):747–50.

[148] Hosokawa K, Katagiri T, Sugimori N, Ishiyama K, Sasaki Y, Seiki Y, et al. Favorable outcome of patients who have 13q deletion: a suggestion for revision of the WHO 'MDS-U' designation. Haematologica 2012;97(12):1845–9.

[149] Holbro A, Jotterand M, Passweg JR, Buser A, Tichelli A, Rovo A. Comment to "Favorable outcome of patients who have 13q deletion: a suggestion for revision of the WHO 'MDS-U' designation". Haematologica 2012;97(12):1845–9. Haematologica. 2013;98(4):e46–e47.

[150] Scheinberg P, Cooper JN, Sloand EM, Wu CO, Calado RT, Young NS. Association of telomere length of peripheral blood leukocytes with hematopoietic relapse, malignant transformation, and survival in severe aplastic anemia. JAMA 2010;304(12):1358–64.

[151] Calado RT, Cooper JN, Padilla-Nash HM, Sloand EM, Wu CO, Scheinberg P, et al. Short telomeres result in chromosomal instability in hematopoietic cells and precede malignant evolution in human aplastic anemia. Leukemia 2012;26(4):700–7.

[152] Dumitriu B, Feng X, Townsley DM, Ueda Y, Yoshizato T, Calado RT, et al. Telomere attrition and candidate gene mutations preceding monosomy 7 in aplastic anemia. Blood 2015;125(4):706–9.

[153] Marsh JC, Mufti GJ. Clinical significance of acquired somatic mutations in aplastic anaemia. Int J Hematol 2016;.

[154] Ogawa S. Clonal hematopoiesis in acquired aplastic anemia. Blood 2016;.

[155] Kulasekararaj AG, Jiang J, Smith AE, Mohamedali AM, Mian S, Gandhi S, et al. Somatic mutations identify a subgroup of aplastic anemia patients who progress to myelodysplastic syndrome. Blood 2014;124(17): 2698–704.

[156] Betensky M, Babushok D, Roth JJ, Mason PJ, Biegel JA, Busse TM, et al. Clonal evolution and clinical significance of copy number neutral loss of heterozygosity of chromosome arm 6p in acquired aplastic anemia. Cancer Genet 2016;209(1–2):1–10.

[157] Scheinberg P, Young NS. How I treat acquired aplastic anemia. Blood 2012;.

[158] Scheinberg P, Wu CO, Nunez O, Young NS. Long-term outcome of pediatric patients with severe aplastic anemia treated with antithymocyte globulin and cyclosporine. J Pediatr 2008;153(6):814–9.

[159] Scheinberg P, Young NS. How I treat acquired aplastic anemia. Blood 2012;120(6):1185–96.

[160] Gupta V, Eapen M, Brazauskas R, Carreras J, Aljurf M, Gale RP, et al. Impact of age on outcomes after bone marrow transplantation for acquired aplastic anemia using HLA-matched sibling donors. Haematologica 2010;95(12):2119–25.

[161] Socie G. Allogeneic BM transplantation for the treatment of aplastic anemia: current results and expanding donor possibilities. Hematol Am Soc Hematol Educ Prog 2013;2013:82–6.

[162] Bacigalupo A, Marsh JC. Unrelated donor search and unrelated donor transplantation in the adult aplastic anaemia patient aged 18–40 years without an HLA-identical sibling and failing immunosuppression. Bone Marrow Transplant 2013;48(2):198–200.

[163] Deeg HJ, Amylon ID, Harris RE, Collins R, Beatty PG, Feig S, et al. Marrow transplants from unrelated donors for patients with aplastic anemia: minimum

effective dose of total body irradiation. Biol Blood Marrow Transplant 2001;7(4):208–15.

[164] Kojima S, Matsuyama T, Kato S, Kigasawa H, Kobayashi R, Kikuta A, et al. Outcome of 154 patients with severe aplastic anemia who received transplants from unrelated donors: the Japan Marrow Donor Program. Blood 2002;100(3):799–803.

[165] Maury S, Balere-Appert ML, Chir Z, Boiron JM, Galambrun C, Yakouben K, et al. Unrelated stem cell transplantation for severe acquired aplastic anemia: improved outcome in the era of high-resolution HLA matching between donor and recipient. Haematologica 2007;92(5):589–96.

[166] Viollier R, Socie G, Tichelli A, Bacigalupo A, Korthof ET, Marsh J, et al. Recent improvement in outcome of unrelated donor transplantation for aplastic anemia. Bone Marrow Transplant 2008;41(1):45–50.

[167] Bacigalupo A, Socie G, Hamladji RM, Aljurf M, Maschan A, Kyrcz-Krzemien S, et al. Current outcome of HLA identical sibling versus unrelated donor transplants in severe aplastic anemia: an EBMT analysis. Haematologica 2015;100(5):696–702.

[168] Marsh JC, Kulasekararaj AG. Management of the refractory aplastic anemia patient: what are the options? Blood 2013;122(22):3561–7.

[169] Ciceri F, Lupo-Stanghellini MT, Korthof ET. Haploidentical transplantation in patients with acquired aplastic anemia. Bone Marrow Transplant 2013;48(2):183–5.

[170] Dezern AE, Luznik L, Fuchs EJ, Jones RJ, Brodsky RA. Post-transplantation cyclophosphamide for GVHD prophylaxis in severe aplastic anemia. Bone Marrow Transplant 2011;46(7):1012–3.

[171] DeZern AE, Dorr D, Luznik L, Bolanos-Meade J, Gamper C, Symons HJ, et al. Using haploidentical (haplo) donors and high-dose post-transplant cyclophosphamide (PTCy) for refractory severe aplastic anemia (SAA). Blood 2015;126(23).

[172] Esteves I, Bonfim C, Pasquini R, Funke V, Pereira NF, Rocha V, et al. Haploidentical BMT and post-transplant Cy for severe aplastic anemia: a multicenter retrospective study. Bone Marrow Transplant 2015;50(5):685–9.

[173] Peffault de Latour R, Purtill D, Ruggeri A, Sanz G, Michel G, Gandemer V, et al. Influence of nucleated cell dose on overall survival of unrelated cord blood transplantation for patients with severe acquired aplastic anemia: a study by eurocord and the aplastic anemia working party of the European group for blood and marrow transplantation. Biol Blood Marrow Transplant 2011;17(1):78–85.

[174] Purev E, Aue G, Kotecha R, Wilder J, Khuu HM, Stroncek DF, et al. Excellent engraftment and long-term survival in patients with severe aplastic anemia (SAA) Undergoing allogeneic hematopoietic stem cell transplantation (HSCT) with Haplo-identical CD34 + cells

combined with a single umbilical cord blood unit. Blood 2015;126(23).

[175] Barnes DW, Mole RH. Aplastic anaemia in sublethally irradiated mice given allogeneic lymph node cells. Br J Haematol 1967;13(4):482–91.

[176] Frickhofen N, Kaltwasser JP, Schrezenmeier H, Raghavachar A, Vogt HG, Herrmann F, et al. Treatment of aplastic anemia with antilymphocyte globulin and methylprednisolone with or without cyclosporine. The German Aplastic Anemia Study Group. N Engl J Med 1991;324(19):1297–304.

[177] Bacigalupo A, Bruno B, Saracco P, Di Bona E, Locasciulli A, Locatelli F, et al. Antilymphocyte globulin, cyclosporine, prednisolone, and granulocyte colony-stimulating factor for severe aplastic anemia: an update of the GITMO/EBMT study on 100 patients. European Group for Blood and Marrow Transplantation (EBMT) Working Party on Severe Aplastic Anemia and the Gruppo Italiano Trapianti di Midolio Osseo (GITMO). Blood 2000;95(6):1931–4.

[178] Kojima S, Hibi S, Kosaka Y, Yamamoto M, Tsuchida M, Mugishima H, et al. Immunosuppressive therapy using antithymocyte globulin, cyclosporine, and danazol with or without human granulocyte colony-stimulating factor in children with acquired aplastic anemia. Blood 2000;96(6):2049–54.

[179] Rosenfeld SJ, Kimball J, Vining D, Young NS. Intensive immunosuppression with antithymocyte globulin and cyclosporine as treatment for severe acquired aplastic anemia. Blood 1995;85(11):3058–65.

[180] Scheinberg P, Wu CO, Nunez O, Scheinberg P, Boss C, Sloand EM, et al. Treatment of severe aplastic anemia with a combination of horse antithymocyte globulin and cyclosporine, with or without sirolimus: a prospective randomized study. Haematologica 2009;94(3):348–54.

[181] Scheinberg P, Nunez O, Wu C, Young NS. Treatment of severe aplastic anaemia with combined immunosuppression: anti-thymocyte globulin, ciclosporin and mycophenolate mofetil. Br J Haematol 2006;133(6):606–11.

[182] Tisdale JF, Dunn DE, Maciejewski J. Cyclophosphamide and other new agents for the treatment of severe aplastic anemia. Semin Hematol 2000;37(1):102–9.

[183] Scheinberg P, Townsley D, Dumitriu B, Scheinberg P, Weinstein B, Daphtary M, et al. Moderate-dose cyclophosphamide for severe aplastic anemia has significant toxicity and does not prevent relapse and clonal evolution. Blood 2014;124(18):2820–3.

[184] Scheinberg P, Nunez O, Weinstein B, Scheinberg P, Wu CO, Young NS. Activity of alemtuzumab monotherapy in treatment-naive, relapsed, and refractory severe acquired aplastic anemia. Blood 2012;119(2):345–54.

[185] Scheinberg P, Nunez O, Weinstein B, Scheinberg P, Biancotto A, Wu CO, et al. Horse versus rabbit antithymocyte globulin in acquired aplastic anemia. N Engl J Med 2011;365(5):430–8.

[186] Gaber AO, First MR, Tesi RJ, Gaston RS, Mendez R, Mulloy LL, et al. Results of the double-blind, randomized, multicenter, phase III clinical trial of Thymoglobulin versus Atgam in the treatment of acute graft rejection episodes after renal transplantation. Transplantation 1998;66(1):29–37.

[187] Feng X, Kajigaya S, Solomou EE, Keyvanfar K, Xu X, Raghavachari N, et al. Rabbit ATG but not horse ATG promotes expansion of functional CD4 + CD-25highFOXP3+ regulatory T cells in vitro. Blood 2008;111(7):3675–83.

[188] Scheinberg P, Nunez O, Young NS. Retreatment with rabbit anti-thymocyte globulin and ciclosporin for patients with relapsed or refractory severe aplastic anaemia. Br J Haematol 2006;133(6):622–7.

[189] Scheinberg P, Fischer SH, Li L, Nunez O, Wu CO, Sloand EM, et al. Distinct EBV and CMV reactivation patterns following antibody-based immunosuppressive regimens in patients with severe aplastic anemia. Blood 2007;109(8):3219–24.

[190] Marsh JC, Bacigalupo A, Schrezenmeier H, Tichelli A, Risitano AM, Passweg JR, et al. Prospective study of rabbit antithymocyte globulin and ciclosporin for aplastic anemia from the EBMT Severe Aplastic Anemia Working Party. Blood 2012;.

[191] Bacigalupo A, Socie G, Hamladji RM, Aljurf M, Maschan A, Kyrcz-Krzemien S, et al. Current outcome of HLA identical sibling versus unrelated donor transplants in severe aplastic anemia: an EBMT analysis. Haematologica 2015;100(5):696–702.

[192] Marsh JC, Pearce RM, Koh MB, Lim Z, Pagliuca A, Mufti GJ, et al. Retrospective study of alemtuzumab vs ATG-based conditioning without irradiation for unrelated and matched sibling donor transplants in acquired severe aplastic anemia: a study from the British Society for Blood and Marrow Transplantation. Bone Marrow Transplant 2014;49(1):42–8.

[193] Valdez JM, Scheinberg P, Nunez O, Wu CO, Young NS, Walsh TJ. Decreased infection-related mortality and improved survival in severe aplastic anemia in the past two decades. Clin Infect Dis 2010;52(6):726–35.

[194] Valdez JM, Scheinberg P, Nunez O, Wu CO, Young NS, Walsh TJ. Decreased infection-related mortality and improved survival in severe aplastic anemia in the past two decades. Clin Infect Dis 2011;52(6):726–35.

[195] Scheinberg P, Rios O, Scheinberg P, Weinstein B, Wu CO, Young NS. Prolonged cyclosporine administration after antithymocyte globulin delays but does not prevent relapse in severe aplastic anemia. Am J Hematol 2014;89(6):571–4.

[196] Townsley DM, Desmond R, Dunbar CE, Young NS. Pathophysiology and management of thrombocytopenia in bone marrow failure: possible clinical applications of TPO receptor agonists in aplastic anemia and myelodysplastic syndromes. Int J Hematol 2013;98(1):48–55.

[197] Kaushansky K. The molecular mechanisms that control thrombopoiesis. J Clin Invest 2005;115(12):3339–47.

[198] Zeigler FC, de Sauvage F, Widmer HR, Keller GA, Donahue C, Schreiber RD, et al. In vitro megakaryocytopoietic and thrombopoietic activity of c-mpl ligand (TPO) on purified murine hematopoietic stem cells. Blood 1994;84(12):4045–52.

[199] Marsh JC, Ganser A, Stadler M. Hematopoietic growth factors in the treatment of acquired bone marrow failure states. Semin Hematol 2007;44(3):138–47.

[200] Kuter DJ. Biology and chemistry of thrombopoietic agents. Semin Hematol 2010;47(3):243–8.

[201] Feng X, Scheinberg P, Samsel L, Rios O, Chen J, McCoy JP Jr, et al. Decreased plasma cytokines associate with low platelet counts in aplastic anemia and immune thrombocytopenic purpura. J Thromb Haemost 2012;.

[202] Olnes MJ, Scheinberg P, Calvo KR, Desmond R, Tang Y, Dumitriu B, et al. Eltrombopag and improved hematopoiesis in refractory aplastic anemia. N Engl J Med 2012;367(1):11–9.

[203] Desmond R, Townsley DM, Dumitriu B, Olnes MJ, Scheinberg P, Bevans M, et al. Eltrombopag restores trilineage hematopoiesis in refractory severe aplastic anemia that can be sustained on discontinuation of drug. Blood 2014;123(12):1818–25.

[204] Townsley DM, Dumitriu B, Scheinberg P, Desmond R, Feng XM, Rios O, et al. Eltrombopag added to standard immunosuppression for aplastic anemia accelerates count recovery and increases response rates. Blood 2015;126(23).

[205] Valdez JM, Scheinberg P, Young NS, Walsh TJ. Infections in patients with aplastic anemia. Semin Hematol 2009;46(3):269–76.

[206] O'Donghaile D, Childs RW, Leitman SF. Blood consult: granulocyte transfusions to treat invasive aspergillosis in a patient with severe aplastic anemia awaiting mismatched hematopoietic progenitor cell transplantation. Blood 2012;119(6):1353–5.

[207] Quillen K, Wong E, Scheinberg P, Young NS, Walsh TJ, Wu CO, et al. Granulocyte transfusions in severe aplastic anemia: an eleven-year experience. Haematologica 2009;94(12):1661–8.

[208] Kohgo Y, Urabe A, Kilinc Y, Agaoglu L, Warzocha K, Miyamura K, et al. Deferasirox decreases liver iron concentration in iron-overloaded patients with myelodysplastic syndromes, aplastic anemia and other rare anemias. Acta Haematol 2015;134(4):233–42.

Diagnosis of Acquired Aplastic Anemia[a]

A. Rovó*, C. Dufour**, A. Tichelli[†]

*Hematology, University Hospital of Bern, Bern, Switzerland; **Hematology Unit,
G. Gaslini Children's Hospital; Unità di Ematologia Istituto Giannina Gaslini,
Genova, Italy; [†]Hematology, University Hospital of Basel, Basel, Switzerland

INTRODUCTION

Aplastic anemia (AA) is a rare, severe, non-malignant disease caused by autoimmune destruction of early hematopoietic cells. AA presents with geographic rate variability, with a global incidence rate range 0.7–7.4 cases per million inhabitants per year, with two- to three-fold higher rates in Asia than Europe and the United States [1]. The low incidence of its occurrence as well as the overlapping with other bone marrow failure syndromes makes its diagnosis a real challenge. The lack of experience approaching such a patient can lead to unnecessary delay for diagnosis and treatment initiation. During the last years, the continuous increase of new information has contributed to a better understanding of the current landscape of bone marrow failures; we now know that constitutional and acquired diseases are not so clearly delineated [2]. The extent of the knowl-edge stresses thus the need of expertise to establish the diagnosis accurately without missing relevant aspects which may eventually impact patient management.

AA is defined by the presence of pancytopenia with an empty bone marrow (Figs. 3.1 and 3.2), by exclusion of marrow aplasia due to direct effect of chemotherapy or radiotherapy [3]. Diagnosis difficulty is due to the number of marrow failure syndromes having a similar presentation. Furthermore, acquired AA has no specific disease markers; the bone marrow histology showing reduced cellularity with fat replacement provides the basis of the diagnosis, but the final diagnosis is mainly reached by exclusion of other entities. The better comprehension of AA will lead to significant revision in the near future: accelerated telomere attrition, acquired mutations of myeloid-related genes, somatic mutations, and cytokine profiles are some of the new players highlighting profound

---

[a]All the authors have worked on the behalf of the SAA-WP EBMT.

Congenital and Acquired Bone Marrow Failure
http://dx.doi.org/10.1016/B978-0-12-804152-9.00003-8

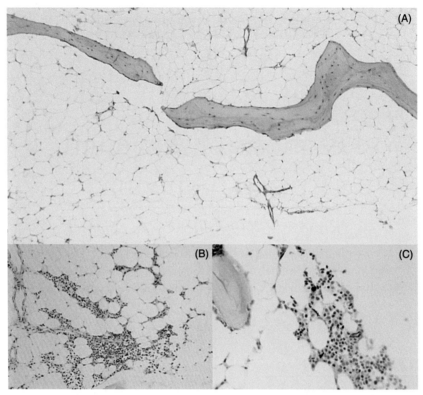

FIGURE 3.1   **Histology and immunohistology of bone marrow from a patient with SAA.** (A) PAS staining of a patients with SAA; (B) HE staining, hotspot of a patient with SAA; (C) immunostaining with CD34 of the hotspot, showing absence of CD34 positive cells. *Source: Picture courtesy of Professor Stephan Dirnhofer and Professor Alexandar Tzankov, Institute of Pathology, University Hospital of Basel, Basel, Switzerland.*

changes in the way to approach diagnosis of bone marrow failures. In this chapter, we will focus on the current approach to make diagnosis of AA in adults, future diagnostic challenges will be also discussed.

APPROACH TO DIAGNOSIS OF APLASTIC ANEMIA

In adult patients during the diagnostic work-up process, a number of diseases such as paroxysmal nocturnal hemoglobinuria (PNH), hypoplastic myelodysplastic syndromes (MDS), hypoplastic acute leukemia, large granular lymphocyte (LGL) leukemia, and autoimmune diseases should be systematically considered. A congenital marrow failure syndrome, such as Fanconi anemia (FA) and dyskeratosis congenita (DKC) as cause for the aplasia is more likely in younger adults (generally below 50 years) presenting with suggestive clinical findings or having a positive family history. Patient age, family history, exposure to toxic substances or medicaments, infections, as well as occupation, clinical presentation, and comorbidities are relevant information which helps to set priorities in the diagnostic process. During this process, first the diagnosis has to be confirmed, second the marrow failure syndrome needs to be characterized

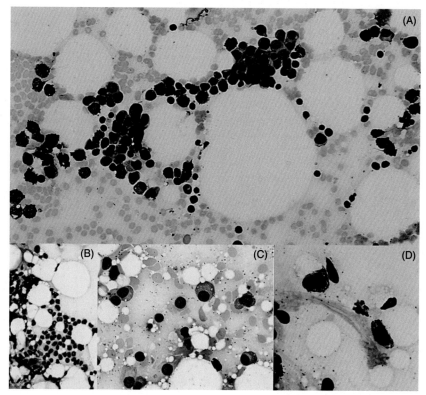

FIGURE 3.2 Cytomorphology of an aspiration of bone marrow from a patients with aplastic anemia. (A) Hotspot with a nest of erythroid cells; (B) reactive lymphoid infiltrate; (C) reactive plasma cells; (D) mast cells.

and finally the severity of the disease must be settled (Fig. 3.3).

DIAGNOSIS CONFIRMATION

Clinical Examination

Clinical examination is a part of the diagnostic procedure. Except for findings related to bleeding or infections, the examination presents mainly negative characteristics: absence of lymphadenopathy, no enlarged spleen or liver, and no infiltration of any other organ. Such findings, if present, would render the diagnosis of AA most unlikely.

Complete Blood Count

Pancytopenia is the main manifestation in the peripheral blood, but at least two cell lines should be decreased for the diagnosis. In early stages isolated cytopenia, particularly thrombocytopenia can be seen. Anemia is the most frequent cytopenia, this anemia is due to decreased red cell production presenting with reticulocytopenia. Macrocytosis is a common feature; red blood cells (RBCs) classically do not show relevant anisocytosis and/or poikilocytosis. The association with iron deficiency might generate, however, other changes in the peripheral blood. However, iron deficiency is rare in AA, and should suggest the presence of an active PNH

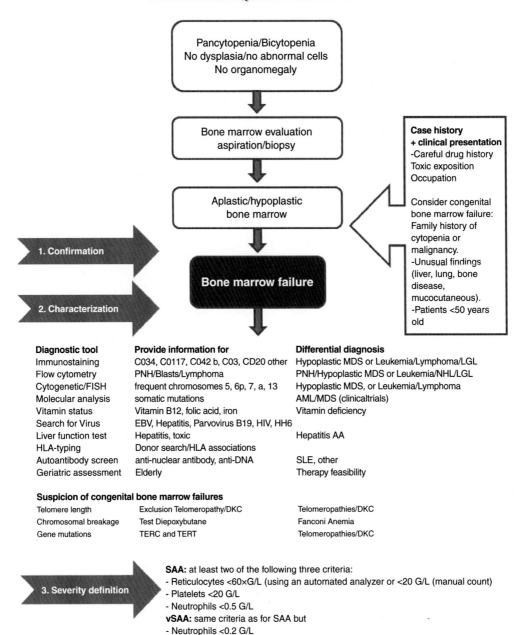

FIGURE 3.3   Stepwise diagnostic phases of acquired aplastic anemia.

clone. Leukopenia is variable, but the total leukocyte count can be normal. For the evaluation of leukocyte differentiation absolute numbers and not relative percentage should be analyzed. Neutropenia is frequent, it can occur at different degree of severity. Lymphocyte count is usually preserved. Monocytopenia can be present and imposes the differential diagnosis of hairy cell leukemia. A careful examination of blood film is needed to assess morphology abnormalities of RBCs, dysplastic changes of the neutrophils, the presence of erythro- and/or myelopoesis precursors and of blasts, abnormal platelets, or other abnormal cells, such as hairy cells. Any of these findings would be a strong argument against the diagnosis of AA. Fetal hemoglobin (Hb) can be increased in AA [4]; in children a pretransfusional increase of fetal Hb imposes the differential diagnosis of a myeloproliferative/MDS like juvenile myelomonocytic leukemia or other subtype of MDS [5,6].

Bone Marrow Examination

Bone marrow (BM) aspiration and biopsy are of paramount relevance to perform the diagnosis of AA. Both evaluations are complementary; aspiration allows a better discrimination of cellular morphology, particularly by the assessment of dysplasia. Trephine biopsy is crucial to assess overall cellularity, topography of hematopoietic cells, and abnormal infiltrates. A BM biopsy containing at least five to six intertrabecular spaces, providing a core of 20–30 mm length is considered as representative [7]. Dry tap is unusual and suggests diagnoses other than AA. Overall cellularity is more often reduced rather than completely absent. Aplastic and hypocellular BM is defined by <10% (empty marrow) and 30% of hematopoietic cells, respectively. This cut off has been established mainly for children and young adults. Diagnostic in elderly patients is discussed later in this chapter. Beside the variable amounts of residual hematopoietic cell, there are prominent fat spaces. Increase in

stromal cells, such as plasma cells, lymphocytes forming follicles, and mast cells are frequent but confounding findings. The increment of stromal cells can mimic the picture of a marrow with normal cellularity, and thus, stromal cells have to be excluded in the global estimation for cellularity. The presence of nests of erythropoiesis [4], conforming the so-called "hot spots" are frequent in AA marrow; they may show a certain degree of dyserythropoiesis, mainly with macro/megaloblastic changes (Fig. 3.2). The overinterpretation of this finding particularly in aspirates may guide falsely to the diagnosis of MDS. Megakaryocytes are usually strongly decreased or absent. Immunostaining allows the identification and assessment of the topographical distribution of blasts, megakaryocytes, abnormal cells, and infiltrates (Fig. 3.1). It might eventually identify the unusual association with lymphoma [8,9]. Fibrosis is not present in AA. Repeat BM biopsy, may be necessary and is recommended in any unclear case. Flow cytometry of BM aspiration may contribute identifying abnormal populations. Immunostained BM tissues can be successfully mapped by multicolor immunofluorescence using confocal reflection microscopy. BM architecture through three-dimensional images can be assessed qualitatively and quantitatively to appreciate the distribution of cell types and their interrelationship, with minimal perturbations. Confocal microscopy is currently only used for basic laboratory investigation and their potential contribution for clinical diagnosis in BM failure is promising [10].

Cytogenetic

Cytogenetic abnormalities can be present in up to 12–15% of otherwise typical AA patients; therefore, cytogenetic investigations should be systematically performed in all AA patients [11]. Due to hypocellular bone marrow, it frequently occurs that there are insufficient metaphases for an adequate analysis. In such cases, FISH targeting specific anomalies should be considered.

Most frequent anomalies in AA include trisomy 8, uniparental disomy of the 6p (6pUPD) [12], 5q-, anomalies of chromosome 7 and 13. Abnormal cytogenetic clones often are small at diagnosis, and may arise during the course of the disease or may be transient and disappear after immunosuppression (IST) [13,14]. While abnormalities of chromosome 7 and 5 even in absence of dysplasia turn the diagnosis more likely in MDS, other cytogenetic findings are less categorical. An abnormal cytogenetic clone does however not necessarily imply the diagnosis of MDS or AML. Del(13q) has been reported in patients with MDS and other hematologic malignancies cases [15] as well as in patients with bone marrow failure syndrome. Patients presented with del(13q) were reported mainly associated with good response to IST [16,17]. Transformation to MDS and AML may occurs usually many months to years after the diagnosis of AA, therefore cytogenetic monitoring during follow-up is recommended.

Molecular Analysis

Molecular analysis including next-generation sequencing (NGS) is increasingly used to understand disease pathophysiology. Kulasek-araraj et al. postulated that somatic mutations are present in a subset of AA and might predict malignant transformation. Clonal hematopoiesis was identified in a fifth of evaluated AA patients showing specific gene mutations associated with transformation to MDS [18]. A larger report evaluating 668 blood samples obtained from 439 patients with AA showed that somatic mutations in myeloid cancer candidate genes were present in one third of the evaluated AA patients. Furthermore clonal hematopoiesis was detected in 47% of the patients, most frequently as acquired mutations. The prevalence of the mutations increased with age, and mutations had an age-related signature. This study also showed that mutations in PIGA and BCOR and BCORL1 correlated with a better response

to IST, and had longer and a higher rate of overall and progression-free survival; in contrast, mutations in a subgroup of genes that included DNMT3A and ASXL1 were associated with worse outcomes. The pattern of somatic clones in individual patients was however variable and frequently unpredictable [19]. The emerging data showing clonal hematopoietic expansion in pediatric and young adult AA patients is strikingly different from healthy hematopoiesis. It also seems that between pediatric patients and older adults, there are potential differences in the mutational spectrum since adults are more likely to carry age-related somatic mutations associated with malignancy [20]. In clinical practice, the finding of somatic mutations tends to be interpreted as signature for malignancy; this belief might be true for certain mutations, but not necessary for all of them. Further data are needed to clarify the mutational profiles and disease outcomes in AA.

Telomere Length Measurement

Telomere attrition offers an interesting prognostic tool in acquired AA. Telomere shortening in AA patients was associated with both, numerical and structural chromosome abnormalities. Patients with shorter telomeres were at higher risk of malignant transformation. Shorter average telomere lengths inversely correlated with monosomy 7 at diagnosis [21]. Thirteen SAA patients were analyzed for acquired mutations in myeloid cells at the time of evolution to −7 and all had a dominant hematopoietic stem and progenitor cell clones bearing specific acquired mutations. Mutations in genes associated with MDS/AML were however present in only four cases. Patients who evolved to MDS and AML showed marked progressive telomere attrition before the emergence of −7 [22]. This result might have clinical implications because affected individual may benefit from therapies that eventually eliminate the shortest (and dysfunctional) telomeres. Telomere length measure, unless there is a suspicion of congenital bone marrow failure does not belong yet to standard screening in acquired AA. Due to constraints on health care budgets telomere length measurement will be not systematically covered by insurance. Inclusion of patients in clinical trials addressing this aspect is strongly recommended.

HLA-Typing

Human leukocyte antigen (HLA) typing of the patient and his family belongs to the diagnostics of patients with marrow failure syndrome. Today, HLA typing should no longer be restricted to children and younger adults. Indeed, early knowledge of the HLA-type of the patient and identification of a possible sibling donor allows to include early transplantation in the treatment decision process. Even patients without sibling donor and older fit patients should be nowadays HLA-tested. Matched unrelated donor transplantation looks consistent front line option in children [23]. Wider applicability of alternative-donor transplantation for AA is under investigation this will be discussed in several chapters of the book [24,25].

CHARACTERIZATION OF APLASTIC ANEMIA

Exclusion of Congenital Bone Marrow Failures

A positive family history including other members affected with cytopenia or malignancy, suggests an inherited bone marrow failure syndrome. The presence of unusual clinical features (liver, lung, bone disease) should alert the possibility of a congenital form of bone marrow failure. However, a normal clinical examination does not definitively rule out "cryptic" telomeropathies [26,27] or a nonclassical FA [28]. Bone marrow morphology in such a disease cannot be distinguished from acquired bone marrow

failure. To exclude FA, tests to demonstrate increased sensitivity to chromosomal breakage with mitomycin C or diepoxybutane should be performed. All patients with bone marrow failures who are candidates to hematopoietic stem cell transplantation should undergo this test, since conditioning regimen in FA patients has to be adapted because of the defective DNA repair mechanisms, and therefore the higher susceptibility to chemotherapy. However, even for IST-treated patients it is reasonable to perform the tests, first in order to be aware due to the higher risk of secondary cancers in these patients and second because many of them do not respond to standard IST with antithymocyte globulin (ATG) and cyclosporine (CSA), but sometimes show response to androgens. Screening should also include sibling donors of FA patients. Telomere shortening is a consistent and typical finding of telomeropathies [29]. Telomere length measurement of leukocytes from peripheral blood can be performed as screening test by suspicion of a telomeropathy. TERC and TERT gene mutations cause telomeropathies in both children and adults [30,31]. In adult patients, symptoms and clinical signs are often milder than in children, mucocutaneous findings and other physical anomalies are infrequent. The pattern of organ abnormalities is extremely variable among affected individuals. Indeed, a family history with blood count abnormalities or hematologic disease is often lacking. Given the higher frequency of congenital bone marrow failure syndromes in children, an age tailored diagnostic work up for AA in pediatric age is included in Chapter 11.

Differential Diagnosis of AA From the Hypocellular Variant of MDS

The distinction between AA and hypoplastic MDS is certainly the most difficult diagnostic task. Both diseases present with markedly hypocellular bone marrow, and increased fat cells [32,33]. Dysplasia of erythropoiesis may be present in both entities and therefore may not help for diagnosis differentiation. The absence of dysplasia in the megakaryopoiesis and myelopoiesis, and the absence of blast cells confirmed by immunostaining with CD34 and/or CD117, and/or flow cytometry are the most conspicuous arguments supporting the diagnosis of AA. Findings like bone marrow fibrosis or splenomegaly favor the diagnosis of MDS. PNH clones are present in 50% of AA patients [34,35] but are unusual in MDS. The significance of abnormal cytogenetic clones seen in around 12–15% of patients with AA is controversial and may not necessarily indicate a diagnosis of MDS [15]. The presence of some clones like +8 or del13q [16,17] are associated with a good response to IST and consequently linked more closely to AA. Cases presenting with isolated monosomy 7, even lacking dysplastic changes will be frequently classified as hypoplastic MDS; even if the cytogenetic abnormality is the only criterion to support this diagnosis, the poor prognosis of monosomy 7 [36] is the strongest argument to avoid classification as benign disease. Moreover, despite similar clinical presentations, distinct cytokine profiles were observed between AA and hypocellular MDS. Characteristic pattern of cytokines, such as thrombopoietin and chemokine (C–C motif) ligand 3 might in future be valuable tools to better discriminate both entities in clinical practice. [37]. Furthermore, mutational profiles might in the next future further contribute to define disease outcomes (see Section "Molecular Analysis") [18,19]. In vitro cultures of colony-forming cells can be useful in the differential diagnosis between AA and MDS [38,39]. Limitations of these assays in clinical practice are given by the scarce availability in most centers, and the lack of standardization of the analysis.

AA and PNH

There is a close correlation between AA and PNH [34]. PNH is a clonal hematopoietic stem cell disorder with features including hemolytic anemia, marrow failure, and thrombosis. This disease will be discussed in detail in Chapter 14. Patients with typical PNH can develop AA during the course of their disease [40] and patients with AA often present a PNH clone [41]. Even the presence of a very small PNH clone is a strong argument for a marrow failure syndrome. Flow cytometry is the gold standard method for screening and diagnosis of PNH. This is currently best achieved by analysis of glycophosphatidylinositol (GPI)-linked antigen using monoclonal antibodies and fluorescent aerolysin [42]. About 40–50% of patients with acquired AA have a detectable PNH clone [43]. Most clones are small and patients do not have symptoms related to PNH. In some patients, the PNH clone can increase after immunosuppressive treatment. In such cases, the patient may be followed for eventual apparition of typical symptoms and complications of the disease. PNH clone size measurements should be performed at presentation and follow-up on serial monitoring every 6–12 months even when initial testing was negative.

AA and Viral Infections

Viruses can affect various cell lineages in the bone marrow causing uncommonly aplasia. Epstein–Barr virus, dengue, parvovirus B19, human herpes virus 6, HIV, and disseminated adenovirus infections have been reported to cause marrow suppression mainly in patients with chronic hemolytic anemia, immunocompromised patients, and transplant recipients [44–49]. Hemophagocytic lymphohistiocytosis (HLH) is a rare life-threatening hyperinflammatory syndrome, occurring as either a familial disorder or a sporadic condition often triggered by viral infections. The disease is characterized by activation of macrophages and hemophagocytosis; pancytopenia is noted in the majority of these patients who present with febrile illness. Extremely high ferritin and elevated lactate dehydrogenase level might help to differentiate this entity from AA [50]. Association

AA and Hepatitis

The association between seronegative hepatitis is described in 5–10% of patients with acquired AA [52–54]. This association affects mainly young, healthy males with severe but self-limited liver inflammation. Liver function tests and viral hepatitis studies (serological and DNA/RNA) looking for HCV, HDV, HEV, HGV, parvovirus B19, EBV, and CMV seeking for association with viral infection should be performed. Patients with posthepatitis AA were reported as having similar response to IST compared to patients with idiopathic acquired AA [53]. Some recent data showed that patients with such association had in a Chinese cohort, a higher early infection rate and more infection-related mortality in the first 2 years after diagnosis compared to AA patients without hepatitis. The 2-year overall survival rate was also lower, suggesting that patients with this association have a more severe T-cell imbalance and a poorer prognosis [55].

Drug-Induced AA

Since years a number of drugs, particularly some nonsteroidal antiinflammatory medicaments, chloramphenicol, gold, antiepileptic drugs, nifedipine, and sulfonamides has been associated with AA. Most patients exposed to these drugs will not develop a marrow failure syndrome; the reason for the appearance of idiosyncratic reactions remains unclear. Regularly drug-induced AA involving new drugs is reported; recently isolated cases of AA were described associated with levamisole contaminated cocaine [56], methimazole [57], and leflunomide used as immunosuppressive therapy to treat rheumatoid arthritis [58]. There are also several reports about severe hematologic complications including AA using mesalazine [59].

Some anecdotal reports about drug-induced AA without recovery using 1st- and 2nd-generation tyrosine kinase inhibitors to treat chronic myeloid leukemia were also communicated; the underlying responsible mechanism has not yet been clarified [60,61]. Most cases of AA are idiopathic, nevertheless a careful drug history must be taken and any putative causative drug should be discontinued.

AA Associated With Chemicals

There is some evidence that AA risk is increased by certain industrial chemicals [62,63]. Results from occupational epidemiology studies have shown increased risk of AA associated with pesticides and benzene exposures [64,65]. The higher incidence among working-age adults suggests that environmental or occupational factors may play a role in the development of AA [65–67]. A large-scale case-control study reported a higher proportion of AA patients exposed to pesticides, solvents, glues, paints, and fuel. However, due to the diversity of compounds, a clear relationship was not conclusive [67]. Risk of AA from occupational exposures to pesticides and industrial chemicals was also reported in a hospital-based case control study in Thailand, where 541 cases of AA and 2261 controls were evaluated. This study suggested an increased risk of AA among those exposed to organophosphate, carbamate, organochlorine, and paraquat pesticides. However, due to confounding factors it was not possible to define whether the risk was associated with one or the concomitance of more than one of these agents [68].

AA and Association With Autoimmune Diseases

There is an increasing evidence to support an autoimmune basis for AA. Therefore the association between AA and other autoimmune diseases is not surprising. AA is a very rare complication of systemic lupus erythematosus (SLE) and can also appear after IST that is used to treat the SLE. The underlying mechanism for the occurrence of the aplasia is postulated as immune mediated through autoantibodies against bone marrow precursors [69,70]. Autoimmune myelofibrosis is an extremely uncommon cause of cytopenia in patients with SLE. It can be diagnosed with a bone marrow biopsy, responds well to IST, and has a favorable prognosis compared to primary myelofibrosis, which is a myeloproliferative disorder [71,72]. Associations of AA with other autoimmune disease have been shown in single-case reports [73–75]. In a single-center report, 5.3% of the patients had an autoimmune disease before the diagnosis of AA. This study also showed that 4.5% of AA developed an autoimmune disease after diagnosis and treatment of AA [76]. The occurrence of autoimmune disease was reported at any time before or after the AA. Autoimmune diseases affected more frequently older AA patients (>50 years). In a large multicenter study of the Severe Aplastic Anemia Working Party of the European Group for Blood and Marrow Transplantation, 50 of 1251 AA patients had an autoimmune disease [77]. Whether or not the IST applied to treat the AA has an influence on the outcome of the autoimmune disease still remains a controversial topic. Autoantibody screen panel according to clinical presentation, including antinuclear antibody and anti-DNA antibody should be investigated when SLE is suspected as underlying disease.

AA and HLA-DR Typing

The HLA system has been reported to be involved in the development of AA. HLA-DR typing might be useful for predicting a response to IST in AA patients. AA patients having HLA-DR15, and especially the frequency of DRB1(*)1502 was markedly higher in patients 40 years of age and older (52.4%) than that in those below 40 years [78] Ethnicity seems to play a role in the relationship between HLA-related

factors and AA. In Japanese patients, DRB1*1501 seems to be associated with the presence of a small population of PNH-type cells and a good response to the IST [79]. In a study on 37 Korean patients with severe AA, responders to IST had a significantly higher HLA-DR15 and a lower DR4 frequency compared with nonresponders [80]. A study in which HLA-A, B, C, DRB1, and DQB1 alleles were compared between 96 Chinese severe AA patients and 600 healthy people, showed a different gene profile in both groups. Comparison among AA patients with various severity exhibited also significant differences of specific alleles, identifying thus several risk and protective HLA alleles among Chinese AA patients [81]. An analysis on gene frequencies of HLA-DRB1 alleles in Mexican mestizo patients with AA showed in coincidence with previous reported data, a positive association of the DRB1(*)15 allele and AA [82].

AA and LGL

LGL is a clonal lymphoid disorder characterized by cytopenia and clonal expansion of either CD3-positive cytotoxic T lymphocytes or CD3-negative natural killer (NK) cells. Although LGL can be associated with other entities, such as AA or MDS, it is a distinct clinical entity with a specific diagnostic pathway [83,84]. Abnormal T-cell populations expressing NK markers and dowregulation of normal T-cell markers (i.e., CD5, CD7) suggest this diagnosis [85]. STAT3 mutations are frequent in LGL leukemia suggesting a similar molecular dysregulation in malignant chronic expansions of NK cells and cytotoxic T lymphocytes origin. STAT3 mutations may distinguish truly malignant lymphoproliferations involving T and NK cells from reactive expansions [86]. Since LGL clones can accompany a number of diseases, additional investigations are necessary to confirm the LGL as etiology of the bone marrow failure. First diagnostic step would be search for an LGL population by flow cytometry. The second step would include tests, allowing distinction between a neoplastic and a reactive LGL proliferation.

Diagnostic of AA in the Elderly

The diagnostic procedure of AA does not differ essentially in elderly patient. There are however aspects that need to be considered. Cellularity of the bone marrow, which is of critical importance for the diagnosis of AA, changes with the age of an individual. Bone marrow cellularity is highest at birth, and declines progressively with age. There is approximately a 10% decrease of cellularity by decade, although with huge variations. Overall, it is 40–70% in adult and declines below 30% in healthy elderly [87]. The quality of trephine biopsy is more critical in the elderly. A short piece of subcortical bone marrow is totally inadequate for assessment because even in a healthy elderly person it is likely to be of low cellularity [88]. A second relevant point in the elderly is the distinction between AA and hypoplastic MDS. The frequency of MDS increases with advanced age, and its distinction from AA may be difficult. Recent data showed that the prevalence of the mutations increased with age and mutations had an age-related signature [19]. AA adults patients are more likely to carry age-related somatic mutations associated with malignancy [20]. As for younger adults, AA and MDS requires comprehensive morphological examination of blood and bone marrow. Information obtained by additional studies, such as karyotype, flow-cytometry, and molecular genetics provide further relevant information. Incomplete assessment due economic reasons might have direct consequences in the accuracy of diagnosis and lead to incorrect treatment decision. Overtreatment because of incomplete assessment could lead to harmful consequences, particularly in an elderly patient. Finally, in elderly patients diagnosis is not limited to disease specific items, but includes also a comprehensive geriatric assessment (Chapter 12).

Defining Severity of AA [47,48]

Once the diagnosis of AA is confirmed, and the AA is characterized, the severity has to be defined. Severity definition of the disease is *crucial* for therapy decisions. The severity is based exclusively on values of the peripheral blood, hence three groups of AA are defined: severe AA (SAA) [89], when at least two of the following three criteria are fulfilled: (1) reticulocytes $<60 \times 10^9 \, L^{-1}$ (using an automated analyzer) or $<20 \times 10^9 \, L^{-1}$ (using manual count), (2) platelets $<20 \times 10^9 \, L^{-1}$, (3) neutrophils $<0.5 \times 10^9 \, L^{-1}$; very severe SAA (vSAA), when SAA criteria are fulfilled but neutrophil count is $<0.2 \times 10^9 \, L^{-1}$; nonsevere AA, when the criteria for SAA and vSAA are not fulfilling.

Defining Response After Treatment

The definition of the response criteria to treatment belongs to the diagnostic tasks. The definition of response is particularly critical after IST. The normalization of blood values is considered as complete response. For the definition of complete remission (CR) several criteria are used worldwide [89,90], however with minor impact on the clinical outcome. Partial remission (PR) covers a wide range of blood values; it includes all patients who do not meet criteria for SAA any longer and are independent of transfusions, but are not in CR.

FUTURE CHALLENGES IN THE DIAGNOSTICS OF AA

There is a current better understanding of the underlying mechanisms involved in bone marrow failures, however diagnostic delimitation between the different entities still needs improvement. New advances in diagnostic technology provide permanently additional fascinating information, which might solve unmet diagnostic problems. Modern medicine is economically regulated in part by health insurance decisions, which will only cover diagnostic methods once they are validated. Therefore many diagnostic innovations should primarily be reserved for research or clinical trials until results' confirmation is available.

Another diagnostic challenge is to identify AA patients at risk of failing IST, and therefore be candidates to early HSCT, even when no matched sibling donor is available. During the past years, a number of studies focused on parameters to predict response after IST, but none of these parameters was able to recognize upfront refractoriness. Baseline absolute reticulocyte and lymphocyte counts were defined as simple predictors of response following IST [91]. In children, lower white blood cell count ($<2.0 \times 10^9 \text{ L}^{-1}$) was the most significant predictive marker of better response [92]. In patients receiving G-CSF, the lack of a neutrophils response ($<0.5 \times 10^9 \text{ L}^{-1}$) by day 30 was associated with significantly lower response rate and survival [89]. However, none of these factors predicts patients who will definitively not respond to IST. Patients refractory to two consecutive courses of ATG have a very low chance to respond to a third course of ATG.

Future studies in respect of diagnostics should focus on identification of true refractory patients, since these patients may be suitable candidates for early novel therapeutic options [93,94]. These factors may be patient-specific (i.e., HLA system, polymorphisms, age, gender, telomere length, others), disease-specific (i.e., blood values, mutational profile), and treatment-specific factors (factors who make that IST is not responsive, i.e., hereditary vs. acquired AA).

Acknowledgment

Conflict of interest: the authors declare no conflict of interest for this manuscript.

References

[1] Young NS, Kaufman DW. The epidemiology of acquired aplastic anemia. Haematologica 2008;93:489–92.

[2] Townsley DM, Dumitriu B, Young NS. Bone marrow failure and the telomeropathies. Blood 2014;124:2775–83.

[3] Young NS. Acquired aplastic anemia. Ann Intern Med 2002;136:534–46.

[4] Tichelli A, Gratwohl A, Nissen C, Signer E, Stebler GC, Speck B. Morphology in patients with severe aplastic anemia treated with antilymphocyte globulin. Blood 1992;80:337–45.

[5] Hasle H, Baumann I, Bergstrasser E, Fenu S, Fischer A, Kardos G, Kerndrup G, Locatelli F, Rogge T, Schultz KR, Stary J, Trebo M, van den Heuvel-Eibrink MM, Harbott J, Nollke P, Niemeyer CM. The International Prognostic Scoring System (IPSS) for childhood myelodysplastic syndrome (MDS) and juvenile myelomonocytic leukemia (JMML). Leukemia 2004;18:2008–14.

[6] Hasle H, Niemeyer CM. Advances in the prognostication and management of advanced MDS in children. Br J Haematol 2011;154:185–95.

[7] Bain BJ, Clark DM, Wilkins BS. Bone Marrow Pathology. 4th ed. United States:Wiley-Blackwell; 2009.

[8] Medinger M, Buser A, Stern M, Heim D, Halter J, Rovo A, Tzankov A, Tichelli A, Passweg J. Aplastic anemia in association with a lymphoproliferative neoplasm: coincidence or causality? Leuk Res 2012;36:250–1.

[9] Zonder JA, Keating M, Schiffer CA. Chronic lymphocytic leukemia presenting in association with aplastic anemia. Am J Hematol 2002;71:323–7.

[10] Takaku T, Malide D, Chen J, Calado RT, Kajigaya S, Young NS. Hematopoiesis in 3 dimensions: human and murine bone marrow architecture visualized by confocal microscopy. Blood 2010;116:e41–55.

[11] Gupta V, Brooker C, Tooze JA, Yi QL, Sage D, Turner D, Kangasabapathy P, Marsh JC. Clinical relevance of cytogenetic abnormalities at diagnosis of acquired aplastic anaemia in adults. Br J Haematol 2006;134: 95–9.

[12] Katagiri T, Sato-Otsubo A, Kashiwase K, Morishima S, Sato Y, Mori Y, Kato M, Sanada M, Morishima Y, Hosokawa K, Sasaki Y, Ohtake S, Ogawa S, Nakao S. Frequent loss of HLA alleles associated with copy number-neutral 6pLOH in acquired aplastic anemia. Blood 2011;118:6601–9.

[13] Tichelli A, Gratwohl A, Nissen C, Speck B. Late clonal complications in severe aplastic anemia. Leuk Lymphoma 1994;12:167–75.

[14] Socie G, Rosenfeld S, Frickhofen N, Gluckman E, Tichelli A. Late clonal diseases of treated aplastic anemia. Semin Hematol 2000;37:91–101.

[15] Steensma DP, Dewald GW, Hodnefield JM, Tefferi A, Hanson CA. Clonal cytogenetic abnormalities in bone marrow specimens without clear morphologic evidence of dysplasia: a form fruste of myelodysplasia? Leuk Res 2003;27:235–42.

[16] Hosokawa K, Katagiri T, Sugimori N, Ishiyama K, Sasaki Y, Seiki Y, Sato-Otsubo A, Sanada M, Ogawa S, Nakao S. Favorable outcome of patients who have 13q

deletion: a suggestion for revision of the WHO 'MDS-U' designation. Haematologica 2012;97:1845–9.

[17] Holbro A, Jotterand M, Passweg JR, Buser A, Tichelli A, Rovo A. Comment to "Favorable outcome of patients who have 13q deletion: a suggestion for revision of the WHO 'MDS-U' designation. Haematologica 2012;97(12):1845–9." Haematologica 2013;98:e46–e47.

[18] Kulasekararaj AG, Jiang J, Smith AE, Mohamedali AM, Mian S, Gandhi S, Gaken J, Czepulkowski B, Marsh JC, Mufti GJ. Somatic mutations identify a subgroup of aplastic anemia patients who progress to myelodysplastic syndrome. Blood 2014;124:2698–704.

[19] Yoshizato T, Dumitriu B, Hosokawa K, Makishima H, Yoshida K, Townsley D, Sato-Otsubo A, Sato Y, Liu D, Suzuki H, Wu CO, Shiraishi Y, Clemente MJ, Kataoka K, Shiozawa Y, Okuno Y, Chiba K, Tanaka H, Nagata Y, Katagiri T, Kon A, Sanada M, Scheinberg P, Miyano S, Maciejewski JP, Nakao S, Young NS, Ogawa S. Somatic mutations and clonal hematopoiesis in aplastic anemia. N Engl J Med 2015;373:35–47.

[20] Babushok DV, Perdigones N, Perin JC, Olson TS, Ye W, Roth JJ, Lind C, Cattier C, Li Y, Hartung H, Paessler ME, Frank DM, Xie HM, Cross S, Cockroft JD, Podsakoff GM, Monos D, Biegel JA, Mason PJ, Bessler M. Emergence of clonal hematopoiesis in the majority of patients with acquired aplastic anemia. Cancer Genet 2015;208:115–28.

[21] Calado RT, Cooper JN, Padilla-Nash HM, Sloand EM, Wu CO, Scheinberg P, Ried T, Young NS. Short telomeres result in chromosomal instability in hematopoietic cells and precede malignant evolution in human aplastic anemia. Leukemia 2011;26(4):700–7.

[22] Dumitriu B, Feng X, Townsley DM, Ueda Y, Yoshizato T, Calado RT, Yang Y, Wakabayashi Y, Kajigaya S, Ogawa S, Zhu J, Young NS. Telomere attrition and candidate gene mutations preceding monosomy 7 in aplastic anemia. Blood 2015;125:706–9.

[23] Dufour C, Veys P, Carraro E, Bhatnagar N, Pillon M, Wynn R, Gibson B, Vora AJ, Steward CG, Ewins AM, Hough RE, de la Fuente J, Velangi M, Amrolia PJ, Skinner R, Bacigalupo A, Risitano AM, Socie G, Peffault de LR, Passweg J, Rovo A, Tichelli A, Schrezenmeier H, Hochsmann B, Bader P, van BA, Aljurf MD, Kulasekararaj A, Marsh JC, Samarasinghe S. Similar outcome of upfront-unrelated and matched sibling stem cell transplantation in idiopathic paediatric aplastic anaemia. A study on behalf of the UK Paediatric BMT Working Party, Paediatric Diseases Working Party and Severe Aplastic Anaemia Working Party of EBMT. Br J Haematol 2015;171:585–94.

[24] Maury S, Balere-Appert ML, Chir Z, Boiron JM, Galambrun C, Yakouben K, Bordigoni P, Marie-Cardine A, Milpied N, Kanold J, Maillard N, Socie G. Unrelated stem cell transplantation for severe acquired aplastic anemia: improved outcome in the era of high-resolution HLA matching between donor and recipient. Haematologica 2007;92:589–96.

[25] Peffault de LR, Purtill D, Ruggeri A, Sanz G, Michel G, Gandemer V, Maury S, Kurtzberg J, Bonfim C, Aljurf M, Gluckman E, Socie G, Passweg J, Rocha V. Influence of nucleated cell dose on overall survival of unrelated cord blood transplantation for patients with severe acquired aplastic anemia: a study by eurocord and the aplastic anemia working party of the European group for blood and marrow transplantation. Biol Blood Marrow Transplant 2011;17:78–85.

[26] Walne AJ, Dokal I. Dyskeratosis congenita: a historical perspective. Mech Ageing Dev 2008;129:48–59.

[27] Walne AJ, Dokal I. Advances in the understanding of dyskeratosis congenita. Br J Haematol 2009;145:164–72.

[28] Auerbach AD. Fanconi anemia and its diagnosis. Mutat Res 2009;668:4–10.

[29] Shimamura A. Clinical approach to marrow failure. Hematol Am Soc Hematol Educ Program 2009;329–37.

[30] Yamaguchi H, Baerlocher GM, Lansdorp PM, Chanock SJ, Nunez O, Sloand E, Young NS. Mutations of the human telomerase RNA gene (TERC) in aplastic anemia and myelodysplastic syndrome. Blood 2003;102:916–8.

[31] Yamaguchi H, Calado RT, Ly H, Kajigaya S, Baerlocher GM, Chanock SJ, Lansdorp PM, Young NS. Mutations in TERT, the gene for telomerase reverse transcriptase, in aplastic anemia. N Engl J Med 2005;352:1413–24.

[32] Barrett J, Saunthararajah Y, Molldrem J. Myelodysplastic syndrome and aplastic anemia: distinct entities or diseases linked by a common pathophysiology? Semin Hematol 2000;37:15–29.

[33] Bennett JM, Orazi A. Diagnostic criteria to distinguish hypocellular acute myeloid leukemia from hypocellular myelodysplastic syndromes and aplastic anemia: recommendations for a standardized approach. Haematologica 2009;94:264–8.

[34] Young NS, Maciejewski JP, Sloand E, Chen G, Zeng W, Risitano A, Miyazato A. The relationship of aplastic anemia and PNH. Int J Hematol 2002;76(Suppl 2):168–72.

[35] Pu JJ, Mukhina G, Wang H, Savage WJ, Brodsky RA. Natural history of paroxysmal nocturnal hemoglobinuria clones in patients presenting as aplastic anemia. Eur J Haematol 2011;87:37–45.

[36] Greenberg PL, Tuechler H, Schanz J, Sanz G, Garcia-Manero G, Sole F, Bennett JM, Bowen D, Fenaux P, Dreyfus F, Kantarjian H, Kuendgen A, Levis A, Malcovati L, Cazzola M, Cermak J, Fonatsch C, Le Beau MM, Slovak ML, Krieger O, Luebbert M, Maciejewski J, Magalhaes SM, Miyazaki Y, Pfeilstocker M, Sekeres M, Sperr WR, Stauder R, Tauro S, Valent P, Vallespi T, van de Loosdrecht AA, Germing U, Haase D. Revised international prognostic scoring system for myelodysplastic syndromes. Blood 2012;120:2454–65.

[37] Feng X, Scheinberg P, Wu CO, Samsel L, Nunez O, Prince C, Ganetzky RD, McCoy JP Jr, Maciejewski JP, Young NS. Cytokine signature profiles in acquired aplastic anemia and myelodysplastic syndromes. Haematologica 2011;96:602–6.

[38] Zhang TJ, Feng M, Zheng YZ, Li XX, Xu ZF, Qin TJ, Zhang Y, Yang YB, Liu JX, Ren YS, Xiao ZJ. Analysis of in vitro characteristics of colony-forming cells in myelodysplastic syndrome and comparison with that in non-severe aplastic anemia. Zhonghua Xue Ye Xue Za Zhi 2012;33:516–21.

[39] Nissen C, Wodnar-Filipowicz A, Slanicka Krieger MS, Slanicka GA, Tichelli A, Speck B. Persistent growth impairment of bone marrow stroma after antilymphocyte globulin treatment for severe aplastic anaemia and its association with relapse. Eur J Haematol 1995;55:255–61.

[40] Pu JJ, Mukhina G, Wang H, Savage WJ, Brodsky RA. Natural history of paroxysmal nocturnal hemoglobinuria clones in patients presenting as aplastic anemia. Eur J Haematol 2011;87:37–45.

[41] Pu JJ, Brodsky RA. Paroxysmal nocturnal hemoglobinuria from bench to bedside. Clin Transl Sci 2011;4:219–24.

[42] Brodsky RA, Mukhina GL, Li S, Nelson KL, Chiurazzi PL, Buckley JT, Borowitz MJ. Improved detection and characterization of paroxysmal nocturnal hemoglobinuria using fluorescent aerolysin. Am J Clin Pathol 2000;114:459–66.

[43] Sachdeva MU, Varma N, Chandra D, Bose P, Malhotra P, Varma S. Multiparameter FLAER-based flow cytometry for screening of paroxysmal nocturnal hemoglobinuria enhances detection rates in patients with aplastic anemia. Ann Hematol 2015;94:721–8.

[44] Parra D, Mekki Y, Durieu I, Broussolle C, Seve P. Clinical and biological manifestations in primary parvovirus B19 infection in immunocompetent adult: a retrospective study of 26 cases. Rev Med Interne 2014;35:289–96.

[45] Kaptan K, Beyan C, Ural AU, Ustun C, Cetin T, Avcu F, Kubar A, Alis M, Yalcin A. Successful treatment of severe aplastic anemia associated with human parvovirus B19 and Epstein–Barr virus in a healthy subject with allo-BMT. Am J Hematol 2001;67:252–5.

[46] Agrawal M, Paul RT, Pamu P, Avmr N. Parvovirus B19 induced transient aplastic crisis in an immunocompetent child. Turk Patoloji Derg 2015;31:158–60.

[47] Kurtzman G, Young N. Viruses and bone marrow failure. Baillieres Clin. Haematol. 1989;2:51–67.

[48] Kurtzman GJ, Ozawa K, Cohen B, Hanson G, Oseas R, Young NS. Chronic bone marrow failure due to persistent B19 parvovirus infection. N Engl J Med 1987;317:287–94.

[49] Ramzan M, PrakashYadav S, Sachdeva A. Post-dengue fever severe aplastic anemia: a rare association. Hematol Oncol Stem Cell Ther 2012;5:122–4.

[50] Jordan MB, Allen CE, Weitzman S, Filipovich AH, McClain KL. How I treat hemophagocytic lymphohistiocytosis. Blood 2011;118:4041–52.

[51] Min KW, Jung HY, Han HS, Hwang TS, Kim SY, Kim WS, Lim SD, Kim WY. Ileal mass-like lesion induced by Epstein–Barr virus-associated hemophagocytic lymphohistiocytosis in a patient with aplastic anemia. APMIS 2015;123:81–6.

[52] Young NS. Flaviviruses and bone marrow failure. JAMA 1990;263:3065–8.

[53] Locasciulli A, Bacigalupo A, Bruno B, Montante B, Marsh J, Tichelli A, Socie G, Passweg J. Hepatitis-associated aplastic anaemia: epidemiology and treatment results obtained in Europe. A report of the EBMT Aplastic Anaemia Working Party. Br J Haematol 2010;149:890–5.

[54] Brown KE, Tisdale J, Barrett AJ, Dunbar CE, Young NS. Hepatitis-associated aplastic anemia. N Engl J Med 1997;336:1059–64.

[55] Wang H, Tu M, Fu R, Wu Y, Liu H, Xing L, Shao Z. The clinical and immune characteristics of patients with hepatitis-associated aplastic anemia in China. PLoS One 2014;9:e98142.

[56] Karch SB, Mari F, Bartolini V, Bertol E. Aminorex poisoning in cocaine abusers. Int J Cardiol 2012;158:344–6.

[57] Josol CV, Buenaluz-Sedurante M, Sandoval MA, Castillo G. Successful treatment of methimazole-induced severe aplastic anaemia in a diabetic patient with other co-morbidities. BMJ Case Rep 2010;2010.

[58] Wusthof M, Smirnova A, Bacher U, Kroger N, Zander AR, Schuch G, Bokemeyer C. Severe aplastic anaemia following leflunomide therapy. Rheumatology (Oxford) 2010;49:1016–7.

[59] Wiesen A, Wiesen J, Limaye S, Kaushik N. Mesalazine-induced aplastic anemia. Am J Gastroenterol 2009;104:1063.

[60] Chng WJ, Tan LH. Late-onset marrow aplasia due to imatinib in newly diagnosed chronic phase chronic myeloid leukaemia. Leuk Res 2005;29:719–20.

[61] Song MK, Choi YJ, Seol YM, Shin HJ, Chung JS, Cho GJ, Lee EY. Nilotinib-induced bone marrow aplasia. Eur J Haematol 2009;83:161–2.

[62] Issaragrisil S, Chansung K, Kaufman DW, Sirijirachai J, Thamprasit T, Young NS. Aplastic anemia in rural Thailand: its association with grain farming and agricultural pesticide exposure. Aplastic Anemia Study Group. Am J Public Health 1997;87:1551–4.

[63] Maluf EM, Pasquini R, Eluf JN, Kelly J, Kaufman DW. Aplastic anemia in Brazil: incidence and risk factors. Am J Hematol 2002;71:268–74.

[64] Muir KR, Chilvers CE, Harriss C, Coulson L, Grainge M, Darbyshire P, Geary C, Hows J, Marsh J, Rutherford T, Taylor M, Gordon-Smith EC. The role of occupational and environmental exposures in the aetiology of

acquired severe aplastic anaemia: a case control investigation. Br J Haematol 2003;123:906–14.

[65] Issaragrisil S, Kaufman DW, Anderson T, Chansung K, Leaverton PE, Shapiro S, Young NS. The epidemiology of aplastic anemia in Thailand. Blood 2006;107:1299–307.

[66] Fleming LE, Timmeny W. Aplastic anemia and pesticides. An etiologic association? J Occup Med 1993;35:1106–16.

[67] Guiguet M, Baumelou E, Mary JY. A case-control study of aplastic anaemia: occupational exposures. The French Cooperative Group for Epidemiological Study of Aplastic Anaemia. Int J Epidemiol 1995;24:993–9.

[68] Prihartono N, Kriebel D, Woskie S, Thetkhathuek A, Sripaung N, Padungtod C, Kaufman D. Risk of aplastic anemia and pesticide and other chemical exposures. Asia Pac J Public Health 2011;23:369–77.

[69] Newman K, Owlia MB, El-Hemaidi I, Akhtari M. Management of immune cytopenias in patients with systemic lupus erythematosus—Old and new. Autoimmun Rev 2013;12:784–91.

[70] Chalayer E, Ffrench M, Cathebras P. Aplastic anemia as a feature of systemic lupus erythematosus: a case report and literature review. Rheumatol Int 2014;35(6):1073–82.

[71] Paquette RL, Meshkinpour A, Rosen PJ. Autoimmune myelofibrosis. A steroid-responsive cause of bone marrow fibrosis associated with systemic lupus erythematosus. Medicine 1994;73:145–52.

[72] Ungprasert P, Chowdhary VR, Davis MD, Makol A. Autoimmune myelofibrosis with pancytopenia as a presenting manifestation of systemic lupus erythematosus responsive to mycophenolate mofetil. Lupus 2016;25:427–30.

[73] Antic M, Lautenschlager S, Itin PH. Eosinophilic fasciitis 30 years after—what do we really know? Report of 11 patients and review of the literature. Dermatology 2006;213:93–101.

[74] Grey-Davies E, Hows JM, Marsh JC. Aplastic anaemia in association with coeliac disease: a series of three cases. Br J Haematol 2008;143:258–60.

[75] Hinterberger-Fischer M, Kier P, Forstinger I, Lechner K, Kornek G, Breyer S, Ogris H, Pont J, Hinterberger W. Coincidence of severe aplastic anaemia with multiple sclerosis or thyroid disorders. Report of 5 cases. Acta Haematol 1994;92:136–9.

[76] Stalder MP, Rovo A, Halter J, Heim D, Silzle T, Passweg J, Rischewski J, Stern M, Arber C, Buser A, Meyer-Monard S, Tichelli A, Gratwohl A. Aplastic anemia and concomitant autoimmune diseases. Ann Hematol 2009;88:659–65.

[77] Cesaro S, Marsh J, Tridello G, Rovo A, Maury S, Montante B, Masszi T, Van Lint MT, Afanasyev A, Iriondo AA, Bierings M, Carbone C, Doubek M, Lanino E, Sarhan M, Risitano A, Steinerova K, Wahlin A, Pegoraro A, Passweg J. Retrospective survey on the prevalence and outcome of prior autoimmune diseases in patients with aplastic anemia reported to the registry of the European Group for Blood and Marrow Transplantation. Acta Haematol 2010;124:19–22.

[78] Sugimori C, Yamazaki H, Feng X, Mochizuki K, Kondo Y, Takami A, Chuhjo T, Kimura A, Teramura M, Mizoguchi H, Omine M, Nakao S. Roles of DRB1 *1501 and DRB1 *1502 in the pathogenesis of aplastic anemia. Exp Hematol 2007;35:13–20.

[79] Yoshida N, Yagasaki H, Takahashi Y, Yamamoto T, Liang J, Wang Y, Tanaka M, Hama A, Nishio N, Kobayashi R, Hotta N, Asami K, Kikuta A, Fukushima T, Hirano N, Kojima S. Clinical impact of HLA-DR15, a minor population of paroxysmal nocturnal haemoglobinuria-type cells, and an aplastic anaemia-associated autoantibody in children with acquired aplastic anaemia. Br J Haematol 2008;142:427–35.

[80] Song EY, Kang HJ, Shin HY, Ahn HS, Kim I, Yoon SS, Park S, Kim BK, Park MH. Association of human leukocyte antigen class II alleles with response to immunosuppressive therapy in Korean aplastic anemia patients. Hum Immunol 2010;71:88–92.

[81] Wang M, Nie N, Feng S, Shi J, Ge M, Li X, Shao Y, Huang J, Zheng Y. The polymorphisms of human leukocyte antigen loci may contribute to the susceptibility and severity of severe aplastic anemia in Chinese patients. Hum Immunol 2014;75:867–72.

[82] Fernandez-Torres J, Flores-Jimenez D, Arroyo-Perez A, Granados J, Lopez-Reyes A. The ancestry of the HLA-DRB1*15 allele predisposes the Mexican mestizo to the development of aplastic anemia. Hum Immunol 2012;73:840–3.

[83] Poullot E, Zambello R, Leblanc F, Bareau B, De ME, Roussel M, Boulland ML, Houot R, Renault A, Fest T, Semenzato G, Loughran T, Lamy T. Chronic natural killer lymphoproliferative disorders: characteristics of an international cohort of 70 patients. Ann Oncol 2014;25:2030–5.

[84] Lamy T, Loughran TP Jr. How I treat LGL leukemia. Blood 2011;117:2764–74.

[85] Yang W, Qi J, Li Z, Liu W, Yi S, Xu Y, Zhao Y, Qiu L. Analysis of clinical characteristics for large granular lymphocytic leukemia. Zhonghua Yi Xue Za Zhi 2014;94:276–9.

[86] Jerez A, Clemente MJ, Makishima H, Koskela H, Leblanc F, Peng NK, Olson T, Przychodzen B, Afable M, Gomez-Segui I, Guinta K, Durkin L, Hsi ED, McGraw K, Zhang D, Wlodarski MW, Porkka K, Sekeres MA, List A, Mustjoki S, Loughran TP, Maciejewski JP. STAT3 mutations unify the pathogenesis of chronic lymphoproliferative disorders of NK cells and T-cell large granular lymphocyte leukemia. Blood 2012;120:3048–57.

[87] Hartsock RJ, Smith EB, Petty CS. Normal variations with aging of the amount of hematopoietic tissue in bone marrow from the anterior iliac crest. a study made from 177 cases of sudden death examined by necropsy. Am J Clin Pathol 1965;43:326–331.

[88] Bain BJ. Bone marrow trephine biopsy. J Clin Pathol 2001;54:737–42.

[89] Tichelli A, Schrezenmeier H, Socie G, Marsh J, Bacigalupo A, Duhrsen U, Franzke A, Hallek M, Thiel E, Wilhelm M, Hochsmann B, Barrois A, Champion K, Passweg JR. A randomized controlled study in patients with newly diagnosed severe aplastic anemia receiving antithymocyte globulin (ATG), cyclosporine, with or without G-CSF: a study of the SAA Working Party of the European Group for Blood and Marrow Transplantation. Blood 2011;117:4434–41.

[90] Scheinberg P, Nunez O, Young NS. Retreatment with rabbit anti-thymocyte globulin and ciclosporin for patients with relapsed or refractory severe aplastic anaemia. Br J Haematol 2006;133:622–7.

[91] Scheinberg P, Wu CO, Nunez O, Young NS. Predicting response to immunosuppressive therapy and survival in severe aplastic anaemia. Br J Haematol 2009;144:206–16.

[92] Yoshida N, Yagasaki H, Hama A, Takahashi Y, Kosaka Y, Kobayashi R, Yabe H, Kaneko T, Tsuchida M, Ohara A, Nakahata T, Kojima S. Predicting response to immunosuppressive therapy in childhood aplastic anemia. Haematologica 2011;96:771–4.

[93] Desmond R, Townsley DM, Dumitriu B, Olnes MJ, Scheinberg P, Bevans M, Parikh AR, Broder K, Calvo KR, Wu CO, Young NS, Dunbar CE. Eltrombopag restores trilineage hematopoiesis in refractory severe aplastic anemia that can be sustained on discontinuation of drug. Blood 2014;123:1818–25.

[94] Olnes MJ, Scheinberg P, Calvo KR, Desmond R, Tang Y, Dumitriu B, Parikh AR, Soto S, Biancotto A, Feng X, Lozier J, Wu CO, Young NS, Dunbar CE. Eltrombopag and improved hematopoiesis in refractory aplastic anemia. N Engl J Med 2012;367:11–9.

Acquired Overlap Bone Marrow Failure Disorders

J.R. Passweg

Division of Hematology, University Hospital Basel, Basel, Switzerland

INTRODUCTION

Diagnosis of aplastic anemia (AA) is a diagnosis of exclusion. This includes the notion that bone marrow failure with hypocellular marrow will be diagnosed as AA, if no other disease entity can be diagnosed [1–6]. Whereas some of the competing diagnosis, that is hereditary marrow failure, may be diagnosed based on tests that give a clear indication, such as positive chromosomal breakage stress test, others are less conclusive, for example, in the case of marrow failure with short telomeres but without confirmed genetic diagnosis [e.g., 25–30% of dyskeratosis congenita (DKC) cases will not have a mutation identified in the telomerase complex] and without other clinical diagnostic features. AA may result in short telomeres based on replicative stress on the remaining hematopoietic stem cells [7].

Hypoplastic myelodysplastic syndrome (MDS) is equally a diagnosis of exclusion; a set of criteria is used to separate hypoplastic MDS from AA but there is a large gray zone of overlapping entities [8–10]. These have been termed idiopathic cytopenia of undetermined signifi-

cance (ICUS), that is, a pre-MDS without sufficient alterations to warrant the diagnosis of MDS and clonal hematopoiesis of indeterminate potential (CHIP), as it has been increasingly recognized that clonality may be found in patients with AA but also in healthy persons without signs of marrow failure [11–14].

All these reflections have to be included in the diagnostic workup of the cytopenic patient. Whereas diagnosis may be straightforward in the young patient with very severe AA and a totally acellular marrow the older patient with pancytopenia, a mildly hypocellular marrow, some dysplasia in erythropoesis, no increase of blasts, no cytogenetic abnormality, and no other salient feature may pose a real challenge and repetitive marrow examination may not carry the diagnosis forward as disease progress may be slow.

HYPOPLASTIC MDS

In particular in older persons the most important differential diagnosis is AA versus MDS. MDS is 80–100 times more common than AA and often diagnosis is clear. According to WHO classification, MDS is currently categorized according to the number of dysplastic lineages, the presence of ring sideroblasts, and the percentage of blasts in the bone marrow and the peripheral blood. The thresholds to define the percentage of dysplastic cells and of ringed sideroblasts as significant for diagnosis are somewhat arbitrary. A proportion of MDS patients will present with hypocellular marrow. When differentiating hypoplastic MDS from AA, features diagnostic of MDS are usually considered to be: marrow fibrosis, increased myeloid blasts, increased dysplastic megakaryopoesis of myelopoesis, and splenomegaly. Morphologic hot spots of erythropoesis with some degree of dyserythropoesis, as well as, the presence of a paroxysmal nocturnal hemoglobinuria (PNH) clone can be seen in both diseases [6].

AA is known to have late clonal complications of which MDS is the most frequently diagnosed disease [8]. Whether this represents a true transformation event or is an unmasking of preexisting MDS by immunosuppressive therapy and, probably more importantly, by the lapse of time is not known.

The WHO provisional entities of ICUS and CHIP are not very helpful in my view as these include cases that are probably bona fide autoimmune cytopenias, as well as, clonal hematologic disease. The provisional category, ICUS, has been introduced. To be classified in this, patients must have cytopenia in one or more lineage (hemoglobin <110 g/L, neutrophil count <1.5 × 10^9 L^{-1}, and platelet count <100 × 10^9 L^{-1}) that is persistent for at least 6 months that cannot be explained by other disease and does not meet diagnostic criteria of MDS. Recent attempts of clarification by using the technology of next generation sequencing have shown in AA patients that some mutations cluster around autoimmunity whereas others around clonal myeloid disorders and that the latter do have a poorer prognosis. Recurrent somatic mutations have been identified in several genes, including RNA splicing machinery (SF3B1, SRSF2, U2AF1, and ZRSR2); DNA methylation (TET2, DNMT3A, and IDH1/2); histone modification (ASXL1 and EZH2); transcription regulation (RUNX1); DNA repair (TP53); signal transduction (CBL, NRAS, and KRAS), and others. However, currently there is not enough evidence and no general agreement to reclassify all AA cases with clonal hematopoesis as hypoplastic MDS [12,15]. In the past patients with otherwise typical AA but clonal cytogenetic markers, such as trisomy 8 or monosomy 7, have been diagnosed as AA and treated as AA even though it was known that they were at an increased risk of MDS and acute myeloid leukemia (AML) transformation.

In fact, over 40% of patients with acquired AA have expanded clones of PNH cells, as a result of somatic PIG-A gene mutations arising in hematopoietic stem cells. In addition, cytogenetic abnormalities, sometimes transient, have been reported in AA without MDS. In a study with over 100 AA patients with no morphologic evidence of MDS, somatic mutations were detected in over 20% of cases. Most frequently mutated genes included ASXL1, DNMT3A, and BCOR, with a median mutant allele burden of 20%. Patients with somatic mutations had shorter telomere lengths and a higher risk of transformation to MDS compared with patients without mutations [15]. In a study [12] already mentioned in the introduction, approximately one-third of patients had mutations in genes commonly affected in myeloid neoplasms. Most frequently mutated genes included PIGA, BCOR/BCORL1, DNMT3A, and ASXL1. However, a substantial diversity was observed in mutation frequencies compared with myeloid neoplasms, mutations in PIGA and BCOR/BCORL1 being more common in AA [12]. The number of driver mutations per person and the mutant allele burden are increasing from CHIP and AA with clonal hematopoiesis to MDS, implicating that these small hematopoietic clones may have the potential to progress or represent small malignant clones at a preclinical stage. However, the mechanism by which a specific mutated gene is driving the transformation and the role of external factors needs more research.

Finally in the aging hematopoiesis evidence of clonal hematopoiesis resulting from an expansion of cells harboring an initiating driver mutation may be found, and this suggests that these hematopoietic clones may represent a premalignant state. The spectrum of mutations is, however similar.

Patients with MDS may respond to immunosuppressive therapy in a similar way as patients with AA, possibly more frequently if the marrow is hypoplastic, HLA DR 15 is present, and the transfusion requirements are low [16–20]. This highlights the implication of the immune system in the pathophysiology of MDS and, hence, some similarities between AA and MDS.

SINGLE LINEAGE CYTOPENIAS (PURE RED CELL APLASIA OR IMMUNE THROMBOCYTOPENIA)

AA may present initially as a single lineage disorder with predominant hyporegenerative anemia or thrombocytopenia. Differentiating AA from pure red cell aplasia (PRCA) and immune thrombocytopenia (ITP) is important as treatment strategies may differ. ITP is defined as megakaryocytic thrombocytopenia with no other lineage abnormalities. Due to bleeding there may be, however, anemia at initial presentation. Marrow aspirate and biopsy should be diagnostic if it is of importance, however, one needs to be aware that in the elderly, subcortical marrow may be severely hypocellullar and that only a biopsy of sufficient depth may clarify the situation.

PCRA is defined as aregenerative anemia with missing erythropoesis in the marrow. Parvovirus infection has to be excluded. Causes of PRCA vary and may include parainfectious state, as well as, an equivalent of "single lineage"

AA with preservation of the other lineages; in addition, the congenital form Diamond–Blackfan anemia has to be excluded in young patients.

Agranulocytosis has a vast differential diagnosis covering toxic, autoimmune, and congenital causes. Discussion of all these entities is beyond the scope of this chapter.

T CELL LARGE GRANULAR LYMPHOCYTES

Large granular lymphocyte (LGL) leukemia is a rare lymphoproliferative neoplasia defined by the clonal expansion of CD3 cytotoxic T lymphocytes. It is subgroup of mature peripheral T-cell neoplasms. Clinical features include neutropenia, anemia, and rheumatoid arthritis. Approximately one-third of patients are asymptomatic at diagnosis. Eighty-five percent of patients with LGL leukemia experience neutropenia and 45% develop agranulocytosis. Recurrent bacterial infections are a hallmark of the disease. Transfusion-dependent hyporegenerative anemia is found more rarely. Thrombocytopenia may be found in some patients. Rheumatoid arthritis may occur and often precedes the diagnosis of T-LGL. Diagnosis of LGL leukemia is established by documentation of an increased number of clonal LGLs. Phenotypic analyses show terminal effector memory T cells, most patients with T-LGL leukemia show a CD3+ CD8+ CD57+ CD56− CD28−, TCRα/β phenotype. Clonality is confirmed by TCR-gene rearrangement. LGL leukemia is thought to arise from chronic antigenic stimulation, with the long-term survival of LGL being promoted by constitutive activation of multiple survival signaling pathways. These lead to deregulation of apoptosis and resistance to normal pathways of cell death. Cytopenias are thought to be more often due to immune dysregulation and autoimmunity rather than infiltration of the marrow with consecutive displacement of hematopoiesis [21]. Overlap with other marrow failure disorders exists, in one study [22,23] clonal effector T-cell expansion was demonstrated in patients with PNH based on Vβ repertoires. Whether these clonal T-cell expansion reflect bone fide T-LGL or a reactive process is difficult to determine. Most common treatment options include methotrexate and cyclosporine. Detailed discussion of T-LGL leukemia is beyond the scope of this paragraph. Diagnosis of T-LGL requires attention to the differential diagnosis and the search for the underlying lymphoid neoplasia.

PAROXYSMAL NOCTURNAL HEMOGLOBINURIA

Obviously there is a large overlap between AA and PNH, with AA patients often having a PNH clone detectable and patients with classical PNH evolving into a PNH–AA syndrome. For further discussion on PNH, please refer to Chapter 14. Patients with AA with PNH clones have been shown in some studies to have better responses to immunosuppressive treatment [24], possibly due to the fact that PNH clones rescue hematopoiesis in the face of an autoimmune attack.

CONGENITAL MARROW FAILURE UNDIAGNOSED

Whereas congenital forms of marrow failure are discussed in detail in Part II of this book, there are some considerations worth mentioning here. It is important to consider that many different types of inherited disorders may have a common end-organ failure, which is hematopoiesis. Mutations are so diverse that they impact the machineries of DNA repair, telomere elongation, ribosome composition, and others and may lead to marrow failure. Distinction of inherited forms of marrow failure from acquired forms is often difficult. Inherited forms of marrow failure have a great variability in expression of clinical

manifestations and reliance on additional signs, for example, of Fanconi anemia such as *café au lait* spots, microcephaly, skeletal abnormalities, and growth retardation is not justified as these may be completely absent. Furthermore, marrow failure may occur at an "advanced" age and it is not unheard of to diagnose, for example Fanconi anemia, in patients between 35 and 40 years of age. It is nevertheless reasonable to limit the diagnostic search for congenital forms to a certain age, for example 35 years, and perform such tests in older individuals only if a specific suspicion exists. It is not known how many patients with congenital forms of marrow failure have been treated by immunosuppressive therapy assuming acquired AA, but many hematologists interested in marrow failure disorders have encountered such cases. Therefore maintaining a high degree of suspicion is of importance. In addition, there may be cases in whom a congenital marrow disorder is suspected based on the combination of marrow failure with other features. Once the common congenital marrow failures have been ruled out, diagnosis is difficult. NIH in the United States is maintaining specialty clinics for rare and undiagnosed disorders and such clinics may contribute to widen the scope of inherited and acquired disorders of the hematopoietic system.

CONCLUSIONS

As shown in Fig. 4.1 there is considerable overlap in the disorders discussed in this chapter. Some of the differential diagnoses have major impact on treatment and congenital marrow failure should not be treated by immunosuppressive therapies. Whereas it is of importance to know that hypocellular MDS, and for that matter also MDS, without increased blast counts may respond to immunosuppressive approaches similar to AA [16–18]. Ultimately ineffective hematopoesis in MDS may, in part, be due to autoimmune T-cell clones driving hematopoesis

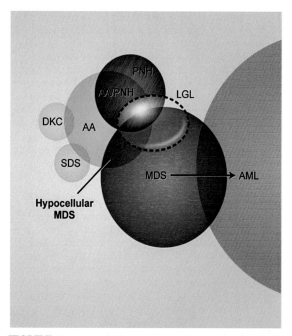

FIGURE 4.1 **The overlap of aplastic anemia (AA) with other clonal and congenital disorders.** *AML,* Acute myeloid leukemia; *DKC,* dyskeratosis congenita; *LGL,* large granular lymphocyte leukemia; *MDS,* myelodysplastic syndrome; *PNH,* paroxysmal nocturnal hemoglobinuria; *SDS,* Shwachman–Diamond syndrome. *Source: With permission from Young NS, Calado RT, Scheinberg P. Current concepts in the pathophysiology and treatment of aplastic anemia. Blood 2006;108(8):2509–19 [5].*

toward apoptosis. Marrow failure syndromes remain difficult to diagnose, often more than one biopsy is required, and careful attention to many details have to be given to narrow diagnostic possibilities as much as possible. Finally physicians need to be aware of the overlap in clinical presentation among many of these disease entities.

References

[1] Young NS. The problem of clonality in aplastic anemia: Dr. Dameshek's riddle, restated. Blood 1992;79(6): 1385–92.

[2] Young NS, Barrett AJ. The treatment of severe acquired aplastic anemia. Blood 1995;85(12):3367–77.

[3] Young NS, Maciejewski J. The pathophysiology of acquired aplastic anemia. N Engl J Med 1997;336(19): 1365–72.

[4] Young NS. Acquired aplastic anemia. JAMA 1999;282(3):271–8.

[5] Young NS, Calado RT, Scheinberg P. Current concepts in the pathophysiology and treatment of aplastic anemia. Blood 2006;108(8):2509–19.

[6] Rovó A, Tichelli A, Dufour C. SAA-WP EBMT. Diagnosis of acquired aplastic anemia. Bone Marrow Transplant 2013;48(2):162–7.

[7] Scheinberg P, Cooper JN, Sloand EM, Wu CO, Calado RT, Young NS. Association of telomere length of peripheral blood leukocytes with hematopoietic relapse, malignant transformation, and survival in severe aplastic anemia. JAMA 2010;304(12):1358–64.

[8] Socie G, Mary JY, Schrezenmeier H, Marsh J, Bacigalupo A, Locasciulli A, Fuehrer M, Bekassy A, Tichelli A, Passweg J. Granulocyte-stimulating factor and severe aplastic anemia: a survey by the European Group for Blood and Marrow Transplantation (EBMT). Blood 2007;109(7):2794–6.

[9] Barrett J, Saunthararajah Y, Molldrem J. Myelodysplastic syndrome and aplastic anemia: distinct entities or diseases linked by a common pathophysiology? Semin Hematol 2000;37(1):15–29.

[10] Malcovati L, Cazzola M. The shadowlands of MDS: idiopathic cytopenias of undetermined significance (ICUS) and clonal hematopoiesis of indeterminate potential (CHIP). Hematology Am Soc Hematol Educ Program 2015;2015(1):299–307.

[11] Maciejewski JP, Risitano A, Sloand EM, Nunez O, Young NS. Distinct clinical outcomes for cytogenetic abnormalities evolving from aplastic anemia. Blood 2002;99(9):3129–35.

[12] Yoshizato T, Dumitriu B, Hosokawa K, Makishima H, Yoshida K, Townsley D, Sato-Otsubo A, Sato Y, Liu D, Suzuki H, Wu CO, Shiraishi Y, Clemente MJ, Kataoka K, Shiozawa Y, Okuno Y, Chiba K, Tanaka H, Nagata Y, Katagiri T, Kon A, Sanada M, Scheinberg P, Miyano S, Maciejewski JP, Nakao S, Young NS, Ogawa S. Somatic mutations and clonal hematopoiesis in aplastic anemia. N Engl J Med 2015;373(1):35–47.

[13] Genovese G, Kähler AK, Handsaker RE, Lindberg J, Rose SA, Bakhoum SF, Chambert K, Mick E, Neale BM, Fromer M, Purcell SM, Svantesson O, Landén M, Höglund M, Lehmann S, Gabriel SB, Moran JL, Lander ES, Sullivan PF, Sklar P, Grönberg H, Hultman CM, McCarroll SA. Clonal hematopoiesis and blood-cancer risk inferred from blood DNA sequence. N Engl J Med 2014;371(26):2477–87.

[14] Jaiswal S, Fontanillas P, Flannick J, Manning A, Grauman PV, Mar BG, Lindsley RC, Mermel CH, Burtt N, Chavez A, Higgins JM, Moltchanov V, Kuo FC, Kluk

MJ, Henderson B, Kinnunen L, Koistinen HA, Ladenvall C, Getz G, Correa A, Banahan BF, Gabriel S, Kathiresan S, Stringham HM, McCarthy MI, Boehnke M, Tuomilehto J, Haiman C, Groop L, Atzmon G, Wilson JG, Neuberg D, Altshuler D, Ebert BL. Age-related clonal hematopoiesis associated with adverse outcomes. N Engl J Med 2014;371(26):2488–98.

[15] Kulasekararaj AG, Jiang J, Smith AE, et al. Somatic mutations identify a subgroup of aplastic anemia patients who progress to myelodysplastic syndrome. Blood 2014;124(17):2698–704.

[16] Sloand EM, Wu CO, Greenberg P, Young N, Barrett J. Factors affecting response and survival in patients with myelodysplasia treated with immunosuppressive therapy. J Clin Oncol 2008;26(15):2505–11.

[17] Passweg JR, Giagounidis AA, Simcock M, Aul C, Dobbelstein C, Stadler M, Ossenkoppele G, Hofmann WK, Schilling K, Tichelli A, Ganser A. Immunosuppressive therapy for patients with myelodysplastic syndrome: a prospective randomized multicenter phase III trial comparing antithymocyte globulin plus cyclosporine with best supportive care—SAKK 33/99. J Clin Oncol 2011;29(3):303–9.

[18] Sloand EM, Olnes MJ, Shenoy A, Weinstein B, Boss C, Loeliger K, Wu CO, More K, Barrett AJ, Scheinberg P, Young NS. Alemtuzumab treatment of intermediate-1 myelodysplasia patients is associated with sustained improvement in blood counts and cytogenetic remissions. J Clin Oncol 2010;28(35):5166–73.

[19] Saunthararajah Y, Nakamura R, Nam JM, Robyn J, Loberiza F, Maciejewski JP, Simonis T, Molldrem J, Young

NS, Barrett AJ. HLA-DR15 (DR2) is overrepresented in myelodysplastic syndrome and aplastic anemia and predicts a response to immunosuppression in myelodysplastic syndrome. Blood 2002;100(5):1570–4.

[20] Sloand EM, Kim S, Fuhrer M, Risitano AM, Nakamura R, Maciejewski JP, Barrett AJ, Young NS. Fas-mediated apoptosis is important in regulating cell replication and death in trisomy 8 hematopoietic cells but not in cells with other cytogenetic abnormalities. Blood 2002;100(13):4427–32.

[21] Zhang D, Loughran TP Jr. Large granular lymphocytic leukemia: molecular pathogenesis, clinical manifestations, and treatment. Hematology Am Soc Hematol Educ Program 2012;2012:652–9.

[22] Saunthararajah Y, Molldrem JL, Rivera M, Williams A, Stetler-Stevenson M, Sorbara L, Young NS, Barrett JA. Coincident myelodysplastic syndrome and T-cell large granular lymphocytic disease: clinical and pathophysiological features. Br J Haematol 2001;112(1):195–200.

[23] Risitano AM, Maciejewski JP, Muranski P, Wlodarski M, O'Keefe C, Sloand EM, Young NS. Large granular lymphocyte (LGL)-like clonal expansions in paroxysmal nocturnal hemoglobinuria (PNH) patients. Leukemia 2005;19(2):217–22.

[24] Sugimori C, Chuhjo T, Feng X, Yamazaki H, Takami A, Teramura M, Mizoguchi H, Omine M, Nakao S. Minor population of CD55– CD59– blood cells predicts response to immunosuppressive therapy and prognosis in patients with aplastic anemia. Blood 2006;107(4):1308–14.

Supportive Care in Aplastic Anemia

B. Höchsmann, H. Schrezenmeier

Institute of Transfusion Medicine, University of Ulm, Ulm, Germany;
Institute of Clinical Transfusion Medicine and Immunogenetics, German Red
Cross Blood Transfusion Service Baden–Württemberg–Hessia, and University
Hospital Ulm, Ulm, Germany

INTRODUCTION

Survival of patients with aplastic anemia (AA) after both immunosuppression and allogeneic bone marrow transplantation has improved substantially over the last 30 years [1]. For patients who received immunosuppressive treatment, this improvement was not restricted to patients who achieve partial or complete remission after specific treatment [2], but also the prognosis of nonresponders to initial immunosuppression substantially improved over time [2]. This most likely reflects the effect of improved supportive care on the overall survival of patients. The individual risk of a patient for life-threatening infectious complications is mainly determined by the neutrophil- and monocyte counts [2–8]. During the past two decades, infection-related mortality and the incidence of invasive fungal infections in patients with AA have decreased [2]. However, infections are still the major threat for patients with severe AA (SAA) or very severe AA (vSAA). In the mot recently published large clinical trials in AA the most prevalent complication and the major cause of death were still bacterial and fungal infections [3–5].

In this chapter we summarize recommendations on supportive care. Some aspects have been discussed in details in a previous review of the EBMT Working Party Aplastic Anemia [9] and therefore here will only be summarized in brief. In this chapter we focus on the role of granulocyte transfusions, transfusion strategy, treatment of iron overload, gender specific aspects, role of exercise and psychological support for AA, and related bone marrow failures.

PREVENTION OF INFECTIONS BY GENERAL MEDICAL MANAGEMENT OF APLASTIC ANEMIA PATIENTS

Infections are the major threat for patients with AA; bacterial and fungal infections are still the most prevalent complications and the main cause of death [3–6,10].

Otherwise infection-related mortality and invasive fungal infections decreased substantially since the availability of new antifungal and antibiotic drugs during the last 20–30 years [2].

This results in a relevant improvement of survival after both transplant and nontransplant treatment strategies, as well as, in treatment refractory or relapsed patients [1,2]. Therefore, prevention and treatment of infections might still be the key to further improvement of survival in AA. The individual risk for infections is mainly determined by the neutrophil counts of a patient [2–8]. Thus, especially neutropenic patients with neutrophil counts <0.5 g/L are in focus for strategies to prevent infections.

Besides antibiotic prophylaxis and individual hygienic rules, a protective environment, isolation, and special low bacterial diet were typical strategies but the evidence for most of them is weak.

Protective Environment

Prolonged neutropenia is a major risk factor for acquiring invasive mold infections. There are no evidence-based data supporting avoidance of situations with an increased risk for exposure to molds. Exceptions are well-documented outbreaks of *Aspergillus* infections in patients exposed to sites of construction. Therefore, exposure to construction areas should be avoided. Additionally, it seems reasonable to avoid contact with garbage, compost, or potted plants.

Protective Isolation

During hospitalization of neutropenic cancer patients no significant effect on mortality and the occurrence of infections has been shown by control of air quality and barrier isolation alone. Nevertheless, the combination of prophylactic antibiotics with protective isolation, including air quality control and barrier isolation, resulted in a significant reduction of all-cause mortality [RR 0.60 (95% CI 0.5–0.7)] in these patients [11].

Additionally, barrier precautions and single patient rooms might be helpful help to reduce spreading of (multi-)resistant bacteria [12]. In conclusion, barrier isolation and HEPA filtration, as well as, one or two beds rooms with en-suite facilities are not imperative but should be used if available.

Special Low Bacterial Diet

Although surveys showed that 43–78% of the hospitals use a special neutropenic diet, available data do not show a benefit by this approach [13–17]. Low bacterial diet did not significantly affect the incidence of neutropenic infections and fever, days of antibiotic use, survival, and stool colonization by Gram-negative bacilli, as well as, yeasts in hospitalized patients [13–15] and febrile admission in an outpatient setting [16].

Nevertheless, we recommend in particular SAA and vSAA patients to consider basic principles like: (1) washing hands before preparing food; (2) careful handling and consuming of raw meat, fish, seafood, and eggs; (3) consuming pasteurized juices and dairy products; and (4) avoiding uncooked and unwashed fruits and vegetables [16,18].

Individual Hygiene Rules

In spite of the mentioned considerations, adhering to the hygiene rules was proven effective [10,19]. Thus, teaching the staff, patients, and relatives is an important issue in preventing infections in neutropenic AA patients.

The recommendations should follow the local guidelines for clothing and hygienic routines, as well as, for the correct handling of central and peripheral venous catheters.

Hand washing and rubbing with alcohol-based disinfection solutions must be practiced before and after handling the patient by the staff and by visitors. Especially, optimal hand hygiene has been shown to be highly effective in reducing neutropenic infections. Furthermore, a good oral hygiene is necessary, as neutropenic infections are often caused by the transmission of the patient's microbial flora from the mucosa into the blood [10,18,19].

As mucositis and oral lesions are known portals of entry, soft toothbrushes should be used to avoid bleedings and lesions and the additional usage of antiseptic medical mouth rinse may be reasonable. In this context it is important to focus on gingival hyperplasia, which is a frequent cyclosporine-induced side effect.

Although plaque control and azithromycin, especially azithromycin-containing toothpaste, have been shown to be of benefit, gingival surgery is occasionally necessary [20,21].

PREVENTION OF INFECTIONS BY ANTIBIOTIC/ANTIMYCOTIC/ ANTIVIRAL PROPHYLAXIS

Antimycotic Prophylaxis

As SAA and vSAA patients with prolonged neutropenia have a high mortality by mold, especially *Aspergillus* infections, antimycotic prophylaxis is common in these patients [2,6,7,22]. A metaanalysis of 64 randomized controlled clinical trials in neutropenic patients after chemotherapy or stem cell transplantation (SCT) confirmed a significant reduction in fungal-related mortality and documented invasive fungal infection by antimycotic prophylaxis

[22]. Antimycotic drugs, such as voriconazole or posaconazole, with activity against *Aspergillus* and other mold species appear to be more effective than fluconazole.

Prophylactic antifungals with activity against *Aspergillus* or mold should be used in AA patients:

- with an absolute neutrophil count (ANC) below 0.2 G/L [22];
- during the first months after an intensified immunosuppressive therapy and after SCT at least until an ANC above 0.5 G/L; and
- after an invasive fungal infection, the secondary antimycotic prophylaxis should be continued until recovery of ANC.

A prophylaxis against *Pneumocystis jirovecii* pneumonia is routinely given only after SCT. Otherwise no common recommendation exists [23].

The usage of this prophylaxis after intensified immunosuppressive therapy (e.g., ATG or alemtuzumab) depends on the individual center but might be used in the first months at least until T-cell recovery. Usually, nebulized pentamidine or cotrimoxazole is used.

Antibiotic Prophylaxis

Prophylactic antibiotics may prevent Gram-negative sepsis in AA patients. In a metaanalysis of 95 randomized controlled clinical trials in afebrile neutropenic cancer patients, prophylactic antibiotics, especially quinolone antibiotics or nonabsorbable antibiotics, reduced mortality [24]. However, prophylactic use of fluoroquinolones increased the risk for the development of antibacterial resistance. Thus, centers should develop individual guidelines for antibiotic prophylaxis and treatment based on the infection rate and the predominant antibacterial resistance.

Antibiotic prophylaxis should be used in AA-patients:

- with an ANC below 0.2 G/L and
- with an ANC below 0.5 G/L after SCT or intensified immunosuppression.

As the risk for severe infections decreases rapidly with increasing neutrophil counts and in patients with ANC > 0.5 G/L, antibacterial prophylaxis is not recommended.

Antiviral Prophylaxis

Antiviral prophylaxis should be used in AA patients:

- after intensified immunosuppression for 3–6 months posttreatment until T-cell recovery,
- after allogeneic SCT according to the regular procedures, including CMV monitoring and preemptive therapy with acyclovir or valacyclovir for 3–6 months posttreatment until T-cell recovery [25], and
- after contact with influenza of patients with vSAA or with reduced T-cell number (e.g., early after ATG therapy); a postexposure prophylaxis with neuraminidase inhibitors should be considered.

In CMV-seronegative patients who are or might become candidates for allogeneic SCT, CMV-negative blood products should be transfused.

Vaccination

Vaccination strategies for AA patients are not well defined except for patients undergoing transplantation where recommendations are well established [26]. There have been concerns about immune activation. The support for these concerns is only based on case reports [27–29] while larger controlled studies in rheumatology patients have not shown an increased risk [30,31].

On the other hand, the response to vaccination in patients with vSAA is likely to be poor and is not recommended during profound leukopenia (neutrophils < 0.5 G/L, lymphocytes < 0.7 G/L) or intensified immunosuppressive therapy.

In less severely affected patients, vaccinations against pneumococci and influenza can be indicated, as the risk/benefit ratio of vaccination is likely to be positive. Children with AA should receive age-appropriate vaccines with the exception of live vaccines (varicella, MMR, and live influenza vaccine) while receiving immunosuppression.

Strategies are not well defined except for patients undergoing transplantation

HEMATOPOIETIC GROWTH FACTORS AS PROPHYLAXIS OF INFECTIONS OR IN COMBINATION WITH IMMUNOSUPPRESSION TO IMPROVE QUALITY OF RESPONSE

There is no indication for treatment of AA with growth factors alone [32–35].

The addition of G-CSF to triple immunosuppression with ATG, cyclosporine, and cortcosteroids accelerates the recovery of neutrophils and reduces the rate of early infections and the length of hospital stay [4,36]. Addition of G-CSF does not improve trilineage response, event-free survival, relapse rate, and overall survival [4,36]. One study from Japan comparing immunosuppession with or without G-CSF reported a lower relapse rate in the G-CSF group [5]. This was not confirmed in a large European multicenter trial [4]. One must also take into account the potential risk that G-CSF increases the incidence of secondary clonal disorders [37,38].

There are currently no data to support the routine use of growth factors in AA either alone or as adjunct to immunosuppression [32,34,37,39].

TREATMENT OF INFECTIONS

Neutropenic fever requires a fast and intense diagnostic and treatment and should be according the hospital individual guidelines.

Varicella zoster infections can become severe in severely immunocompromised patients and antiviral therapy with acyclovir or valacyclovir of either primary or secondary infections (chicken pox and herpes zoster) is recommended.

Influenza is also a potentially severe infection in immunocompromised patients and therapy with neuramidase inhibitors (oseltamivir and zanamivir) should be considered in patients with proven influenza.

However, CMV and EBV infection and reactivation have been described in AA patients undergoing intensive immunosuppression [40], especially after alemtuzumab and rabbit ATG, respectively. NAT testing for EBV and CMV DNA might be performed in patients with unexplained fever and antiviral therapy might be considered.

Granulocyte Transfusions

In life-threatening infections during neutropenia, the use of irradiated granulocyte transfusions should be discussed with the awareness of limited data to support this procedure and possible side effects. Some of the studies summarized further were conducted in patients with AA [41,42]. However, similar to antimicrobial treatment and prophylaxis, we include information from studies on other neutropenic states. Granulocyte transfusions may have an adjunctive role to bridge the gap between specific treatment and neutrophil recovery [41–44]. Two recently published single center reports on treatment of 32 and 56 patients with AA and infections presented promising results. The study from NIH used granulocyte concentrates from stimulated donors (most received G-CSF and dexamethasone) with a mean transfused dose of 6.8×10^{10} [41] while the center in Tianjin collected granulocytes from unstimulated donors (mean granulocyte dose 9.2×10^9) and administered G-CSF (5–10 μg/kg s.c.) to recipients. This approach was based on the hypothesis that this stimulated migration and phagocytosis of the transfused granulocytes [42].

These studies demonstrated a strong correlation between response and survival with hematopoietic recovery [41,42].

Adverse events of granulocyte infusions are fever [45], chills, pulmonary reactions, and alloimmunization [42,43,46]. Of patients not alloimmunized at baseline, 17% were reported to develop HLA antibodies after the start of granulocyte transfusions [41]. Fever and pulmonary complications occur more frequently in patients with HLA alloantibodies [47]. There has been the concern that HLA alloimmunization will inhibit therapeutic activity of granulocyte concentrates. Patients with or without HLA alloantibodies did not differ in mean posttransfusion neutrophil count [41]. Alloimmunization, however, remains a problem because of its negative impact on increments after platelet transfusion and potential increase of graft failure after SCT. Donor-specific HLA antibodies might be implicated in early graft failure [48]. Thus, if granulocyte transfusions are used prior to a planned unrelated donor transplantation, recipients should be monitored for the development of HLA antibodies and the search algorithm for the unrelated donor should take donor-specific antibodies into account [43]. While there are these risks of granulocyte transfusion in the context of allogeneic transplantation, there are also benefits. Granulocyte transfusions can help to control active fungal infections in a very high-risk population of patients who otherwise are denied by the transplant program [44]. One retrospective study of granulocyte transfusions in 26 pediatric patients undergoing allogeneic SCT (21% AA) even reported a low incidence of aGvHD and speculated that granulocyte transfusion might maintain mucosal integrity and thus reduce bacterial translocation and triggers for GvHD [49].

Recently, the results of the Resolving Infections in Neutropenia with Granulocytes (RING) study have been published [50]. This trial randomized patients with neutropenia (<0.5 G/L) and probable/presumed/proven infection to receive standard antimicrobial therapy or standard microbial therapy plus daily granulocyte transfusions from donors who were stimulated with G-CSF and dexamethasone. Ninety-seven patients were randomized (mainly with hematologic malignancies), only two patients had AA [50]. Patients in the granulocyte transfusion group received a median of five transfusions with a mean of 54×10^9 granulocytes. Primary endpoint was a composite of survival and microbial response at 42 days after randomization. Overall success rate for the control and granulocyte transfusion group were 41% and 49%, respectively (not significant) for the patients who had received their assigned treatment [50]. Also survival to day 42 differed significantly between the control and granulocyte group. The intention of this trial was to study the effect of high-dose granulocyte transfusions. However, more than 25% of patients received a dose below the target dose per protocol ($\geq 40 \times 10^9$ granulocytes per transfusion); some patients received rather low doses. Therefore the outcome of patients who received a mean dose of $\geq 0.6 \times 10^9$ granulocytes/kg per transfusion (high dose) and patients who received a mean dose of $<0.6 \times 10^9$ granulocytes/kg per transfusion (low dose) were compared in a posthoc analysis. Both primary success rate and survival up to day 42 was statistically better for patients in the high dose compared to the low-dose group. The primary success rate for high dose, low dose, or control group was 59, 15, and 37%, respectively [50].

Thus in this prospective randomized trial granulocyte, transfusions had no effect on the primary outcome [50]. However, patients who received a high dose ($\geq 0.6 \times 10^9$ granulocytes/kg per transfusion) fared better than patients who received lower doses. There are several limitations that prevent the study from being definitive: enrolment was slow and trial was terminated early and therefore power to detect a true difference was low [50]. Actual dose of granulocytes varied widely and substantial proportion of patients received (too) low dose [50].

As stated in the editorial to the RING study, questions remain [51]. If any conclusion can be deduced from this trial for AA patients, it is about the dose. If a decision to start granulocyte transfusions is made, the collections center should ensure a high dose concentrate is provided by appropriate donor selection, precollection stimulation, and apheresis techniques [51]. The optimal number of transfusions is unclear. Randomized trials [50], retrospective studies [42], and case reports [52] demonstrate that, if needed, large numbers of granulocyte transfusions can be given to a recipient.

Further studies, either prospective, randomized clinical trials, or at least a prospective registry, are needed to answer the remaining open questions on indication, optimal dose, and treatment schedule for granulocyte transfusions.

TRANSFUSION THERAPY

Many patients require regular transfusions of red blood cell (RBC) concentrates. While these maintain quality of life and physical activity, they are also associated with adverse events (e.g., alloimmunization and iron overload). Several studies in the 1980s and 1990s have demonstrated a negative impact of the number of pretransplant transfusions on outcome after transplantation [53–55]. However, the quality of blood products changed substantially over time. All blood products for AA patients must be leukodepleted [34,56]. Whether the association between number of transfusion and posttransplant outcome is still true for efficiently leukocyte-depleted blood products is not clear [57,58]. Universal leukoreduction of blood products has reduced patient alloimmunization [57,59–61]. Allosensitization is mostly due to residual leukocytes in transfused blood products, although a study in mice indicates that minor histocompatibility antigens on donor erythrocytes can contribute to allosensitization [58]. This eythrocyte-dependent immunization is not eliminated by leukocyte depletion [58]. Further studies on the impact of transfusions on outcome after SCT are needed.

A restrictive transfusion policy should be applied, in particular, in patients who are candidates for allogenic SCT [53–55,62]. However, a planned transplantation is no reason to withheld an otherwise indicated transfusion. The decision for transfusion should be based on clinical symptoms (signs of hypoxic anemia and signs of hemorrhage) [56].

The decision to transfuse RBCs depends on clinical symptoms, hemoglobin concentration, comorbidities, and quality of life. Based on this, the RBC transfusion trigger in most AA patients is around 80 g/L. Patients with cardiac, pulmonary, or vascular comorbidities may require a higher transfusion trigger [56]. To avoid hypoxic cell damage, hemoglobin levels ≤ 6.0 g/dL should be avoided.

AA patients should receive prophylactic platelet transfusions in case of platelets <10 G/L without fever, bleeding signs (including petechial bleeding), or history of major bleeding events [34]. Based on data from a single center study [63], some guidelines even suggest a trigger of 5 G/L for prophylactic transfusion in stable patients without bleeding signs (including petechial bleeding) [56]. This very low-transfusion trigger requires regular measurement of platelet count, daily physical examination, and immediate availability of platelet concentrates in case of major bleeding. Transfusion threshold for prophylactic transfusions should be increased to 20 G/L in case of fever, bleeding signs, or history of relevant bleeding (WHO grade 3 or 4 bleeding; e.g., cerebral bleeding). Also the bleeding history of a patient must be taken into account [64]. Immediate therapeutic transfusion is necessary in case of WHO grade 3 or 4 bleeding. Two randomized prospective clinical trials compared a prophylactic versus a therapeutic transfusion strategy in patients with hematologic malignancies undergoing chemotherapy or allogeneic transplantation [65,66]. Based on these trials,

the therapeutic strategy could become a new standard-of-care after autologous SCT, which is associated with a rather low-bleeding risk anyhow [65,67,68]. However, in other indications these studies support the need for the continued use of prophylaxis with platelet transfusion and show the benefit of such prophylaxis for reducing bleeding [65,66].

There are no data that the use of non-HLA-matched apheresis platelet concentrates are superior to pooled platelet concentrates in nonallosensitized patients [69].

In order to avoid a transfusion-associated GvHD, irradiation with 30 Gy is recommended for the following situations: (1) during ATG treatment and thereafter until lymphocyte count recovers to at least 1 G/L [56,70]; (2) during and after other intensive immunosuppressive treatment, for example, fludarabine or alemtuzumab [56]; (3) patients receiving allogeneic SCT [56], transfusion regimen has to be switched to irradiated blood products at the latest at start of the conditioning [56]; (4) HLA-matched apheresis platelet concentrates [56]; and (5) granulocyte concentrates [34,56].

It is also possible to use irradiated blood products for all AA patients, irrespective of treatment, to avoid alloimmunization [70,71]. An European survey demonstrated substantial heterogeneity in the current practice regarding the use of irradiated blood products for AA [70].

In general, there is no need for CMV-negative blood products if universal leukodepletion is applied [72–75]. However, some centers give only CMV-negative blood products for patients undergoing HSCT where both the patient and donor are CMV negative [76].

IRON CHELATION THERAPY

AA patients are at a risk of developing iron overload. Iron overload can significantly impact morbidity and mortality of AA [77–80]. A cohort study on 550 AA patients demonstrated that 13% of patients present with iron overload at diagnosis [81]. Multivariate analysis showed that number of transfusions, male gender, and age (adults vs. children) are risk factors for iron overload in AA. The number of RBC units (RBCU) required for increasing serum ferritin levels to greater than 1000 ng/mL in 50% of male and female patients were 23.5 and 30 RBCU, respectively [81].

The cumulative incidence of iron overload was significantly higher in nonresponders as compared to responders to immunosuppression [81]. Iron overload increased sharply in the first 2 years of treatment and was in particular high in patients who received more than 20 RBCU. Thus, in the first few months after diagnosis, serum ferritin concentrations and liver iron content do not increase to an extent, which required immediate start of chelation therapy. Decision on start of iron chelation therapy very much depends on the response to definitive treatment of AA. After initiation of immunosuppressive treatment, one should wait for at least 4–6 months. If a patient responds and achieves remission, it is likely that the serum ferritin will decline. However, this will take time. As shown by Jin et al. the serum ferritin remained almost unchanged in patients in the first 6 months after transfusion independence was achieved, then declined in the period 1–3 years after transfusion independence, and reached normal levels in most patients after 3 years [81]. It is also possible to treat iron overload by venisection.

In nonresponders to immunosuppression with ongoing RBC transfusion dependence, start of chelation therapy is recommended when serum ferritin concentration rise above 1000 ng/mL [1,34,82]. This applies, in particular, to candidates for allogeneic SCT, as iron overload is associated with higher transplant-related mortality and worse survival after transplantation [83–86]. However, the recommendation with a threshold of 1000 ng/mL has been adopted from other disorders, for example, myelodysplastic syndromes [87,88]. There are no studies that established a

specific treatment threshold for AA. An individual assessment should be performed for each AA patient, taking into account the individual risk of complications of iron overload, the potential adverse events of iron chelators, and the expected period with ongoing RBC transfusion dependence. Iron chelation can be performed with desferrioxamin or deferasirox [89]. A relatively high incidence of agranulocytosis has been reported with deferiprone [89]. Therefore, this is not recommended for AA patients. Only few studies are available on the use of desferrioxamine in AA [80,90,91]. A substudy of the EPIC trial [92] analyzed 116 AA patients with iron overload who were treated with deferasirox. In this subgroup the serum ferritin concentration fell significantly during 1 year of treatment and no drug-induced cytopenias were observed [93]. Decrease occurred in chelation naïve and previously chelated patients [93]. This is in line with other reports demonstrating stable and predictable pharmacokinetic and pharmacodynamic profile of deferasirox irrespective of underlying disease [94–97]. In AA patients, dose might be limited by nephrotoxicity, in particular, when used along with cyclosporine A (CSA) [93].

Reports on improved hematopoiesis in patients with myelodysplastic syndrome receiving iron chelation therapy, prompted interest whether similar improvements can occur in AA.

Another posthoc analysis of the EPIC trial [92,93] was conducted in 72 patients with evaluable hematologic response parameters [98]. Twenty-four of these patients received deferasirox without concomitant immunosuppression. Partial hematologic response was observed in 46% of these patients [98]. Mean serum ferritin levels at end of the deferasirox study were significantly reduced compared to baseline in partial responders. The reduction in serum ferritin was less pronounced in nonresponders [98].

Other studies and case reports also suggest an association between iron chelation therapy and hematopoietic response [91,99–103] or probability and timing of relapse [100].

Prospective randomized trials are needed to confirm the effect of iron chelation. However, already existing evidence supports that in non-responding patients with ongoing transfusion dependence, iron chelation must be considered not only to prevent severe end organ damage but, in addition, iron chelation might improve hematopoiesis.

PHYSICAL EXERCISE

Physical exercises with anemia training should be adapted according to the heart rate.

Early mobilization of the patient should avoid bedsores and assist pulmonary function. Physical exercises are recommended if the patient is well enough. Especially in patients with anemia, training should be adapted according to the heart rate. In thrombocytopenia, accident-prone sports should be avoided. Additionally, passive mobilization and breathing exercises by a physical therapist may be helpful for hospitalized patients who are not able to perform such activities.

GENDER-SPECIFIC ISSUES/SEX LIFE

Menstrual Cycle

In female patients with childbearing potential, thrombocytopenia-related menorrhagia has to be considered and the duration and extent of menstrual bleeding should be routinely asked for [104,105]. Women of reproductive age, at risk for thrombocytopenia, benefit from menses suppression. A number of effective medical regimens are available. The type of regimen selected depends on the patient's need for contraception and the ability to tolerate estrogen-containing medications. In patients who fail medical therapy, endometrial ablation appears to be effective in women with thrombocytopenia.

It is a safe and effective technique of treating acute menorrhagia in patients with AA. It can reduce vaginal bleeding and decrease transfusion dependence.

Women with AA might be at an increased risk for corpus luteum rupture due to thrombocytopenia and infection. Thrombocytopenia hemoperitoneum resulting from a ruptured corpus luteum could be a life-threatening condition in patients with AA. Prompt and appropriate evaluation of corpus luteum rupture and emergent therapy are needed [106,107]. Additionally, treatment-related effects have to be taken into account.

Fertility

First, cytopenia, fatigue, and disorder-related mental strain will affect sexual activity and can lead to a disturbance of the libido and erectile function. A possible differential diagnosis of an erectile dysfunction in this setting might be a hemolytic PNH. Furthermore, possible negative effects by an anemia-related reduced perfusion of reproductive organs or toxic effects by an iron overload are discussed.

Subsequently, some data of treatment-related negative influence on fertility:

- In male mice and rats, a dose-related negative effect of CSA on the weight of the reproductive organs, the testicular function, and sperm maturation up to sterility in high CSA doses has been observed [108–112]. In CSA-treated humans, a reduced mobility and viability of the spermatozoas in male, as well as, an increased risk for spontaneous abortion during the early phase of pregnancy was observed [113,114]. After HSCT, infertility is a major late effect. Although the fertility recovery rate is higher than after transplantation for other diseases, options for cryoconservation should be considered. Univariate analysis of risk for fertility recovery showed age group ($p = 0.03$) and GvHD ($p = 0.05$) are important factors. Neither gender of patients nor type of preparative regimens used for HSCT (cyclo/ATG vs. cyclo/flu) was a risk factor. In multivariate analysis, age group was the only confirmed independent risk factor for fertility recovery ($p = 0.02$) (HR= 2.02, CI = 1.012–3.64) [115].

Pregnancy

Pregnancies in AA patients have the risk of multiple maternal and fetal complications. Low-hemoglobin concentration (8.0 g/dL), platelet counts (<20 G/L), and neutrophils (<0.2 G/L) may be the primary risk factors for obstetric complications in pregnancies associated with AA.

In a retrospective analysis of 60 pregnancies with AA, 34 of these patients had obstetric complications. The major maternal complications in this study were premature labor, gestational diabetes, preeclampsia, acute heart failure, postpartum hemorrhage, and severe postpartum infection. Patients without complications had higher mean hemoglobin concentration (75.38 ± 16.19 g/L) and platelet counts (23.92 ± 14.82 G/L) than did women with complications (mean hemoglobin concentration, 61.47 ± 15.15 g/L, $p = 0.001$; mean platelet counts, 12.11 ± 7.87 G/L, $p < 0.001$) [116]. Additionally, Tichelli et al. reported 7 relapses in a series of 34 pregnancies in AA patients. Three of these seven relapses improved spontaneously after delivery. Further three patients improved after treatment and one patient died in relapse [117].

In summary:

- Anemia < 8.0 g/dL and thrombocytopenia < 20 G/L should be avoided during pregnancy.
- CSA should not be used during the first 3 months due to a higher risk of abortion.
- Blood counts should be carefully monitored due to a higher risk for relapse or progress of the cytopenia.

Pregnancy Prevention

Special risks of contraceptives according to the AA:

- Hormonal contraception could lead to higher risk of thromboembolism, which is especially in patients with an additional PNH clone of relevance.
- Diaphragm, coil, and cervical cap could have a higher risk for inflammation and infections.
- Sterilization and vasectomy could be difficult case of thrombocytopenia and neutropenia.
- Condom and femidom, if used as double barrier method, are not only useful to prevent a pregnancy but also to prevent sexually transmitted infections.

PSYCHOLOGICAL SUPPORT

Mental constitution is an important point for quality of life. AA patients have a special burden by the life-threatening, rare, and sometimes chronic character of the disease.

Thus, careful explanation about the nature of the disease, treatment, and the prognostic and social impact for the patients and their families is essential.

Improvement of quality of life should be defined as treatment aim. Unnecessary cuts of social contacts should be avoided. A life as normal as possible is an important support for psychological health.

As the diagnosis of AA is a life-changing experience, some patients will need professional psychological support. For some patients it is helpful to be in contact with other AA patients.

References

[1] Passweg JR, Marsh JC. Aplastic anemia: first-line treatment by immunosuppression and sibling marrow transplantation. Hematology Am Soc Hematol Educ Program 2010;2010:36–42.

[2] Valdez JM, Scheinberg P, Nunez O, Wu CO, Young NS, Walsh TJ. Decreased infection-related mortality and improved survival in severe aplastic anemia in the past two decades. Clin Infect Dis 2011;52:726–35.

[3] Scheinberg P, Nunez O, Weinstein B, Scheinberg P, Biancotto A, Wu CO, et al. Horse versus rabbit antithymocyte globulin in acquired aplastic anemia. New Engl J Med 2011;365:430–8.

[4] Tichelli A, Schrezenmeier H, Socie G, Marsh J, Bacigalupo A, Duhrsen U, et al. A randomized controlled study in patients with newly diagnosed severe aplastic anemia receiving antithymocyte globulin (ATG), cyclosporine, with or without G-CSF: a study of the SAA Working Party of the European Group for Blood and Marrow Transplantation. Blood 2011;117:4434–41.

[5] Teramura M, Kimura A, Iwase S, Yonemura Y, Nakao S, Urabe A, et al. Treatment of severe aplastic anemia with antithymocyte globulin and cyclosporin A with or without G-CSF in adults: a multicenter randomized study in Japan. Blood 2007;110:1756–61.

[6] Valdez JM, Scheinberg P, Young NS, Walsh TJ. Infections in patients with aplastic anemia. Semin Hematol 2009;46:269–76.

[7] Weinberger M, Elattar I, Marshall D, Steinberg SM, Redner RL, Young NS, et al. Patterns of infection in patients with aplastic anemia and the emergence of *Aspergillus* as a major cause of death. Medicine 1992;71:24–43.

[8] Bacigalupo A, Brand R, Oneto R, Bruno B, Socie G, Passweg J, et al. Treatment of acquired severe aplastic anemia: bone marrow transplantation compared with immunosuppressive therapy—The European Group for Blood and Marrow Transplantation experience. Semin Hematol 2000;37:69–80.

[9] Hochsmann B, Moicean A, Risitano A, Ljungman P, Schrezenmeier H. Supportive care in severe and very severe aplastic anemia. Bone Marrow Transplant 2013;48:168–73.

[10] Gould D. Nurses' hand decontamination practice: results of a local study. J Hosp Infect 1994;28:15–30.

[11] Schlesinger A, Paul M, Gafter-Gvili A, Rubinovitch B, Leibovici L. Infection-control interventions for cancer patients after chemotherapy: a systematic review and meta-analysis. Lancet Infect Dis 2009;9:97–107.

[12] Mattner F, Bange FC, Meyer E, Seifert H, Wichelhaus TA, Chaberny IF. Preventing the spread of multidrug-resistant gram-negative pathogens: recommendations of an expert panel of the german society for hygiene and microbiology. Dtsch Arztebl Int 2012;109:39–45.

[13] van TF, Harbers MM, Terporten PH, van Boxtel RT, Kessels AG, Voss GB, et al. Normal hospital and low-bacterial diet in patients with cytopenia after intensive chemotherapy for hematological malignancy: a study of safety. Ann Oncol 2007;18:1080–4.

[14] Gardner A, Mattiuzzi G, Faderl S, Borthakur G, Garcia-Manero G, Pierce S, et al. Randomized comparison of cooked and noncooked diets in patients undergoing remission induction therapy for acute myeloid leukemia. J Clin Oncol 2008;26:5684–8.

[15] Moody K, Finlay J, Mancuso C, Charlson M. Feasibility and safety of a pilot randomized trial of infection rate: neutropenic diet versus standard food safety guidelines. J Pediatr Hematol Oncol 2006;28:126–33.

[16] DeMille D, Deming P, Lupinacci P, Jacobs LA. The effect of the neutropenic diet in the outpatient setting: a pilot study. Oncol Nurs Forum 2006;33:337–43.

[17] Jubelirer SJ. The benefit of the neutropenic diet: fact or fiction? Oncologist 2011;16:704–7.

[18] Hayes-Lattin B, Leis JF, Maziarz RT. Isolation in the allogeneic transplant environment: how protective is it? Bone Marrow Transplant 2005;36:373–81.

[19] Khan SA, Wingard JR. Infection and mucosal injury in cancer treatment. J Natl Cancer Inst Monogr 2001;29:31–6.

[20] Gomez E, Sanchez-Nunez M, Sanchez JE, Corte C, Aguado S, Portal C, et al. Treatment of cyclosporin-induced gingival hyperplasia with azithromycin. Nephrol Dial Transplant 1997;12:2694–7.

[21] Argani H, Pourabbas R, Hassanzadeh D, Masri M, Rahravi H. Treatment of cyclosporine-induced gingival overgrowth with azithromycin-containing toothpaste. Exp Clin Transplant 2006;4:420–4.

[22] Robenshtok E, Gafter-Gvili A, Goldberg E, Weinberger M, Yeshurun M, Leibovici L, et al. Antifungal prophylaxis in cancer patients after chemotherapy or hematopoietic stem-cell transplantation: systematic review and meta-analysis. J Clin Oncol 2007;25:5471–89.

[23] Gafter-Gvili A, Ram R, Raanani P, Shpilberg O. Management of aplastic anemia: the role of systematic reviews and meta-analyses. Acta Haematol 2011;125:47–54.

[24] Gafter-Gvili A, Fraser A, Paul M, Leibovici L. Meta-analysis: antibiotic prophylaxis reduces mortality in neutropenic patients. Ann Intern Med 2005;142:979–95.

[25] Tomblyn M, Chiller T, Einsele H, Gress R, Sepkowitz K, Storek J, et al. Guidelines for preventing infectious complications among hematopoietic cell transplantation recipients: a global perspective. Biol Blood Marrow Transplant 2009;15:1143–238.

[26] Ljungman P, Small TN. Vaccination of SCT recipients. Bone Marrow Transplant 2011;46:621.

[27] Jeong SH, Lee HS, Hepatitis A. Clinical manifestations and management. Intervirology 2010;53:15–9.

[28] Viallard JF, Boiron JM, Parrens M, Moreau JF, Ranchin V, Reiffers J, et al. Severe pancytopenia triggered by recombinant hepatitis B vaccine. Br J Haematol 2000;110:230–3.

[29] Hendry CL, Sivakumaran M, Marsh JC, Gordon-Smith EC. Relapse of severe aplastic anaemia after influenza immunization. Br J Haematol 2002;119:283–4.

[30] Bengtsson C, Kapetanovic MC, Kallberg H, Sverdrup B, Nordmark B, Klareskog L, et al. Common vaccinations among adults do not increase the risk of developing rheumatoid arthritis: results from the Swedish EIRA study. Ann Rheum Dis 2010;69:1831–3.

[31] Salemi S, D'Amelio R. Are anti-infectious vaccinations safe and effective in patients with autoimmunity? Int Rev Immunol 2010;29:270–314.

[32] Marsh JC, Socie G, Schrezenmeier H, Tichelli A, Gluckman E, Ljungman P, et al. Haemopoietic growth factors in aplastic anaemia: a cautionary note. European Bone Marrow Transplant Working Party for Severe Aplastic Anaemia. Lancet 1994;344:172–3.

[33] Marsh JC, Ganser A, Stadler M. Hematopoietic growth factors in the treatment of acquired bone marrow failure states. Semin Hematol 2007;44:138–47.

[34] Marsh JC, Ball SE, Cavenagh J, Darbyshire P, Dokal I, Gordon-Smith EC, et al. Guidelines for the diagnosis and management of aplastic anaemia. Br J Haematol 2009;147:43–70.

[35] Gurion R, Gafter-Gvili A, Paul M, Vidal L, Ben-Bassat I, Yeshurun M, et al. Hematopoietic growth factors in aplastic anemia patients treated with immunosuppressive therapy-systematic review and meta-analysis. Haematologica 2009;94:712–9.

[36] Gluckman E, Rokicka-Milewska R, Hann I, Nikiforakis E, Tavakoli F, Cohen-Scali S, et al. Results and follow-up of a phase III randomized study of recombinant human-granulocyte stimulating factor as support for immunosuppressive therapy in patients with severe aplastic anaemia. Br J Haematol 2002;119:1075–82.

[37] Socie G, Mary JY, Schrezenmeier H, Marsh J, Bacigalupo A, Locasciulli A, et al. Granulocyte-stimulating factor and severe aplastic anemia: a survey by the European Group for Blood and Marrow Transplantation (EBMT). Blood 2007;109:2794–6.

[38] Ohara A, Kojima S, Hamajima N, Tsuchida M, Imashuku S, Ohta S, et al. Myelodysplastic syndrome and acute myelogenous leukemia as a late clonal complication in children with acquired aplastic anemia. Blood 1997;90:1009–13.

[39] Passweg JR, Marsh JC. Aplastic anemia: first-line treatment by immunosuppression and sibling marrow transplantation. Hematology Am Soc Hematol Educ Program 2010;2010:36–42.

[40] Scheinberg P, Nunez O, Weinstein B, Scheinberg P, Wu CO, Young NS. Activity of alemtuzumab monotherapy in treatment-naive, relapsed, and refractory severe acquired aplastic anemia. Blood 2012;119:345–54.

[41] Quillen K, Wong E, Scheinberg P, Young NS, Walsh TJ, Wu CO, et al. Granulocyte transfusions in severe aplastic anemia: an eleven-year experience. Haematologica 2009;94:1661–8.

[42] Wang H, Wu Y, Fu R, Qu W, Ruan E, Wang G, et al. Granulocyte transfusion combined with granulocyte colony stimulating factor in severe infection patients with severe aplastic anemia: a single center experience from China. PLoS ONE 2014;9:e88148.

[43] O'Donghaile D, Childs RW, Leitman SF. Blood consult: granulocyte transfusions to treat invasive aspergillosis in a patient with severe aplastic anemia awaiting mismatched hematopoietic progenitor cell transplantation. Blood 2012;119:1353–5.

[44] Yenicesu I, Sucak G, Dilsiz G, Aki SZ, Yegin ZA. Hematopoietic stem cell transplantation in a very high risk group of patients with the support of granulocyte transfusion. Indian J Hematol Blood Transfus 2011;27:146–51.

[45] Hubel K, Carter RA, Liles WC, Dale DC, Price TH, Bowden RA, et al. Granulocyte transfusion therapy for infections in candidates and recipients of HPC transplantation: a comparative analysis of feasibility and outcome for community donors versus related donors. Transfusion 2002;42:1414–21.

[46] Bow EJ, Schroeder ML, Louie TJ. Pulmonary complications in patients receiving granulocyte transfusions and amphotericin B. Can Med Assoc J 1984;130:593–7.

[47] Stroncek DF, Leonard K, Eiber G, Malech HL, Gallin JI, Leitman SF. Alloimmunization after granulocyte transfusions. Transfusion 1996;36:1009–15.

[48] Spellman S, Bray R, Rosen-Bronson S, Haagenson M, Klein J, Flesch S, et al. The detection of donor-directed, HLA-specific alloantibodies in recipients of unrelated hematopoietic cell transplantation is predictive of graft failure. Blood 2010;115:2704–8.

[49] Nikolajeva O, Mijovic A, Hess D, Tatam E, Amrolia P, Chiesa R, et al. Single-donor granulocyte transfusions for improving the outcome of high-risk pediatric patients with known bacterial and fungal infections undergoing stem cell transplantation: a 10-year single-center experience. Bone Marrow Transplant 2015;50:846–9.

[50] Price TH, Boeckh M, Harrison RW, McCullough J, Ness PM, Strauss RG, et al. Efficacy of transfusion with granulocytes from G-CSF/dexamethasone-treated donors in neutropenic patients with infection. Blood 2015;126:2153–61.

[51] Cancelas JA. Granulocyte transfusion: questions remain. Blood 2015;126:2082–3.

[52] Bozkaya IO, Kara A, Yarali N, Cagli A, Turgut S, Tunc B. Numerous granulocyte transfusions to a patient with severe aplastic anemia without severe complication. Transfus Apher Sci 2013;48:371–3.

[53] Champlin RE, Horowitz MM, van Bekkum DW, Camitta BM, Elfenbein GE, Gale RP, et al. Graft failure following bone marrow transplantation for severe aplastic anemia: risk factors and treatment results. Blood 1989;73:606–13.

[54] Piccin A, O'Marcaigh A, Smith O, O'Riordan J, Crowley M, Vandenberg E, et al. Outcome of bone marrow transplantation in acquired and inherited aplastic anaemia in the Republic of Ireland. Ir J Med Sci 2005;174:13–9.

[55] Hernandez-Boluda JC, Marin P, Carreras E, Aguilar JL, Granena A, Rozman C, et al. Bone marrow transplantation for severe aplastic anemia: the Barcelona Hospital Clinic experience. Haematologica 1999;84: 26–31.

[56] German Association of Physicians. Cross-sectional guidelines for therapy with blood components and plasma derivatives. Transf Med Hemother 2009;36:347–481.

[57] Killick SB, Win N, Marsh JC, Kaye T, Yandle A, Humphries C, et al. Pilot study of HLA alloimmunization after transfusion with pre-storage leucodepleted blood products in aplastic anaemia. Br J Haematol 1997;97:677–84.

[58] Desmarets M, Cadwell CM, Peterson KR, Neades R, Zimring JC. Minor histocompatibility antigens on transfused leukoreduced units of red blood cells induce bone marrow transplant rejection in a mouse model. Blood 2009;114:2315–22.

[59] Leukocyte reduction and ultraviolet B irradiation of platelets to prevent alloimmunization and refractoriness to platelet transfusions. The Trial to Reduce Alloimmunization to Platelets Study Group. New Engl J Med 1997;337:1861–9.

[60] Dzik WH. Leukoreduction of blood components. Curr Opin Hematol 2002;9:521–6.

[61] Cohen H, Mold D, Jones H, et al, on behalf of the Serious Hazards of Transfusion (SHOT) Steering Group. In: Taylor C, editor. The 2009 Annual SHOT Report. 2010. Available from: http://www.shotuk.org/wp-content/uploads/2010/07/SHOT2009.pdf

[62] Kaminski ER, Hows JM, Goldman JM, Batchelor JR. Pretransfused patients with severe aplastic anaemia exhibit high numbers of cytotoxic T lymphocyte precursors probably directed at non-HLA antigens. Br J Haematol 1990;76:401–5.

[63] Sagmeister M, Oec L, Gmur J. A restrictive platelet transfusion policy allowing long-term support of outpatients with severe aplastic anemia. Blood 1999;93:3124–6.

[64] Höchsmann B, Seidel H, Wiesneth M, Schrezenmeier H. Is the low platelet transfusion trigger in the new Cross-Sectional German Guidelines of risk in outpatient setting? Prophylactic platelet transfusions in aplastic anemia. Transf Med Hemother 2010;36 (Suppl.1):38.

[65] Wandt H, Schaefer-Eckart K, Wendelin K, Pilz B, Wilhelm M, Thalheimer M, et al. Therapeutic platelet transfusion versus routine prophylactic transfusion in patients with haematological malignancies: an open-label, multicentre, randomised study. Lancet 2012;380:1309–16.

[66] Stanworth SJ, Estcourt LJ, Powter G, Kahan BC, Dyer C, Choo L, et al. A no-prophylaxis platelet-transfusion strategy for hematologic cancers. New Engl J Med 2013;368:1771–80.

[67] Wandt H, Schaefer-Eckart K, Frank M, Birkmann J, Wilhelm M. A therapeutic platelet transfusion strategy is safe and feasible in patients after autologous peripheral blood stem cell transplantation. Bone Marrow Transplant 2006;37:387–92.

[68] Stanworth SJ, Estcourt LJ, Llewelyn CA, Murphy MF, Wood EM. Impact of prophylactic platelet transfusions on bleeding events in patients with hematologic malignancies: a subgroup analysis of a randomized trial. Transfusion 2014;54(10):2385–893.

[69] Schrezenmeier H, Seifried E. Buffy-coat-derived pooled platelet concentrates and apheresis platelet concentrates: which product type should be preferred? Vox Sang 2010;99:1–15.

[70] Marsh J, Socie G, Tichelli A, Schrezenmeier H, Hochsmann B, Risitano AM, et al. Should irradiated blood products be given routinely to all patients with aplastic anaemia undergoing immunosuppressive therapy with antithymocyte globulin (ATG)? A survey from the European Group for Blood and Marrow Transplantation Severe Aplastic Anaemia Working Party. Br J Haematol 2010;150:377–9.

[71] Bean MA, Graham T, Appelbaum FR, Deeg HJ, Schuening F, Sale GE, et al. Gamma-irradiation of pretransplant blood transfusions from unrelated donors prevents sensitization to minor histocompatibility antigens on dog leukocyte antigen-identical canine marrow grafts. Transplantation 1994;57:423–6.

[72] Vamvakas EC. Is white blood cell reduction equivalent to antibody screening in preventing transmission of cytomegalovirus by transfusion? A review of the literature and meta-analysis. Transfus Med Rev 2005;19:181–99.

[73] Bowden RA, Slichter SJ, Sayers MH, Mori M, Cays MJ, Meyers JD. Use of leukocyte-depleted platelets and cytomegalovirus- seronegative red blood cells for prevention of primary cytomegalovirus infection after marrow transplant. Blood 1991;78:246–50.

[74] Bowden RA, Slichter SJ, Sayers M, Weisdorf D, Cays M, Schoch G, et al. A comparison of filtered leukocyte-reduced and cytomegalovirus (CMV) seronegative blood products for the prevention of

transfusion-associated CMV infection after marrow transplant. Blood 1995;86:3598–603.

[75] Nichols WG, Price TH, Gooley T, Corey L, Boeckh M. Transfusion-transmitted cytomegalovirus infection after receipt of leukoreduced blood products. Blood 2003;101:4195–200.

[76] Pamphilon DH, Rider JR, Barbara JA, Williamson LM. Prevention of transfusion-transmitted cytomegalovirus infection. Transfus Med 1999;9:115–23.

[77] Lee JW. Iron chelation therapy in the myelodysplastic syndromes and aplastic anemia: a review of experience in South Korea. Int J Hematol 2008;88:16–23.

[78] Kim KH, Kim JW, Rhee JY, Kim MK, Kim BS, Kim I, et al. Cost analysis of iron-related complications in a single institute. Korean J Intern Med 2009;24:33–6.

[79] Kushner JP, Porter JP, Olivieri NF. Secondary iron overload. Hematology Am Soc Hematol Educ Program 2001;47–61.

[80] Takatoku M, Uchiyama T, Okamoto S, Kanakura Y, Sawada K, Tomonaga M, et al. Retrospective nationwide survey of Japanese patients with transfusion-dependent MDS and aplastic anemia highlights the negative impact of iron overload on morbidity/mortality. Eur J Haematol 2007;78:487–94.

[81] Jin P, Wang J, Li X, Wang M, Ge M, Zhang J, et al. Evolution of iron burden in acquired aplastic anemia: a cohort study of more than 3-year follow-up. Int J Hematol 2015;101:13–22.

[82] Marsh JC, Kulasekararaj AG. Management of the refractory aplastic anemia patient: what are the options? Blood 2013;122:3561–7.

[83] Armand P, Kim HT, Cutler CS, Ho VT, Koreth J, Alyea EP, et al. Prognostic impact of elevated pretransplantation serum ferritin in patients undergoing myeloablative stem cell transplantation. Blood 2007;109:4586–8.

[84] Pullarkat V, Blanchard S, Tegtmeier B, Dagis A, Patane K, Ito J, et al. Iron overload adversely affects outcome of allogeneic hematopoietic cell transplantation. Bone Marrow Transplant 2008;42:799–805.

[85] Koreth J, Antin JH. Iron overload in hematologic malignancies and outcome of allogeneic hematopoietic stem cell transplantation. Haematologica 2010;95: 364–6.

[86] Deeg HJ, Spaulding E, Shulman HM. Iron overload, hematopoietic cell transplantation, and graft-versus-host disease. Leuk Lymphoma 2009;50:1566–72.

[87] Gattermann N. Guidelines on iron chelation therapy in patients with myelodysplastic syndromes and transfusional iron overload. Leuk Res 2007;31(Suppl. 3):S10–5.

[88] Gattermann N. Overview of guidelines on iron chelation therapy in patients with myelodysplastic syndromes and transfusional iron overload. Int J Hematol 2008;88:24–9.

[89] Neufeld EJ. Oral chelators deferasirox and deferiprone for transfusional iron overload in thalassemia major: new data, new questions. Blood 2006;107:3436–41.

[90] Mwanda OW, Otieno CF, Abdalla FK. Transfusion haemosiderosis inspite of regular use of desferrioxamine: case report. East Afr Med J 2004;81:326–8.

[91] Park SJ, Han CW. Complete hematopoietic recovery after continuous iron chelation therapy in a patient with severe aplastic anemia with secondary hemochromatosis. J Korean Med Sci 2008;23:320–3.

[92] Cappellini MD, Porter J, El-Beshlawy A, Li CK, Seymour JF, Elalfy M, et al. Tailoring iron chelation by iron intake and serum ferritin: the prospective EPIC study of deferasirox in 1744 patients with transfusion-dependent anemias. Haematologica 2010;95:557–66.

[93] Lee JW, Yoon SS, Shen ZX, Ganser A, Hsu HC, Habr D, et al. Iron chelation therapy with deferasirox in patients with aplastic anemia: a subgroup analysis of 116 patients from the EPIC trial. Blood 2010;116:2448–54.

[94] Miyazawa K, Ohyashiki K, Urabe A, Hata T, Nakao S, Ozawa K, et al. A safety, pharmacokinetic and pharmacodynamic investigation of deferasirox (Exjade, ICL670) in patients with transfusion-dependent anemias and iron-overload: a Phase I study in Japan. Int J Hematol 2008;88:73–81.

[95] Porter J, Galanello R, Saglio G, Neufeld EJ, Vichinsky E, Cappellini MD, et al. Relative response of patients with myelodysplastic syndromes and other transfusion-dependent anaemias to deferasirox (ICL670): a 1-yr prospective study. Eur J Haematol 2008;80: 168–76.

[96] Kohgo Y, Urabe A, Kilinc Y, Agaoglu L, Warzocha K, Miyamura K, et al. Deferasirox decreases liver iron concentration in iron-overloaded patients with myelodysplastic syndromes, aplastic anemia and other rare anemias. Acta Haematol 2015;134:233–42.

[97] Kim IH, Moon JH, Lim SN, Sohn SK, Kim HG, Lee GW, et al. Efficacy and safety of deferasirox estimated by serum ferritin and labile plasma iron levels in patients with aplastic anemia, myelodysplastic syndrome, or acute myeloid leukemia with transfusional iron overload. Transfusion 2015;55:1613–20.

[98] Lee JW, Yoon SS, Shen ZX, Ganser A, Hsu HC, El-Ali A, et al. Hematologic responses in patients with aplastic anemia treated with deferasirox: a post hoc analysis from the EPIC study. Haematologica 2013;98:1045–8.

[99] Cheong JW, Kim HJ, Lee KH, Yoon SS, Lee JH, Park HS, et al. Deferasirox improves hematologic and hepatic function with effective reduction of serum ferritin and liver iron concentration in transfusional iron overload patients with myelodysplastic syndrome or aplastic anemia. Transfusion 2014;54:1542–51.

[100] Pawelec K, Salamonowicz M, Panasiuk A, Leszczynska E, Krawczuk-Rybak M, Demkow U, et al. Influence

of iron overload on immunosuppressive therapy in children with severe aplastic anemia. Adv Exp Med Biol 2015;866:83–9.

[101] Lee SE, Yahng SA, Cho BS, Eom KS, Kim YJ, Lee S, et al. Improvement in hematopoiesis after iron chelation therapy with deferasirox in patients with aplastic anemia. Acta Haematol 2013;129:72–7.

[102] Oliva EN, Ronco F, Marino A, Alati C, Pratico G, Nobile F. Iron chelation therapy associated with improvement of hematopoiesis in transfusion-dependent patients. Transfusion 2010;50:1568–70.

[103] Koh KN, Park M, Kim BE, Im HJ, Seo JJ. Restoration of hematopoiesis after iron chelation therapy with deferasirox in 2 children with severe aplastic anemia. J Pediatr Hematol Oncol 2010;32:611–4.

[104] Martin-Johnston MK, Okoji OY, Armstrong A. Therapeutic amenorrhea in patients at risk for thrombocytopenia. Obstet Gynecol Surv 2008;63: 395–402.

[105] Jacob S, Abdullah A, Hurwitz J, Stedman JK, Samuelson R, Shahabi S. Endometrial ablation for aplastic anemia-associated menorrhagia. Conn Med 2015;79:289–90.

[106] Wang H, Guo L, Shao Z. Hemoperitoneum from corpus luteum rupture in patients with aplastic anemia. Clin Lab 2015;61:427–30.

[107] Payne JH, Maclean RM, Hampton KK, Baxter AJ, Makris M. Haemoperitoneum associated with ovulation in women with bleeding disorders: the case for conservative management and the role of the contraceptive pill. Haemophilia 2007;13:93–7.

[108] Handelsman DJ, McDowell IF, Caterson ID, Tiller DJ, Hall BM, Turtle JR. Testicular function after renal transplantation: comparison of cyclosporin A with azathioprine and prednisone combination regimes. Clin Nephrol 1984;22:144–8.

[109] Seethalakshmi L, Menon M, Malhotra RK, Diamond DA. Effect of cyclosporine A on male reproduction in rats. J Urol 1987;138:991–5.

[110] Seethalakshmi L, Diamond DA, Malhotra RK, Mazanitis SG, Kumar S, Menon M. Cyclosporine-induced testicular dysfunction: a separation of the nephrotoxic component and an assessment of a 60-day recovery period. Transplant Proc 1988;20:1005–10.

[111] Seethalakshmi L, Menon M, Pallias JD, Khauli RB, Diamond DA. Cyclosporine: its harmful effects on testicular function and male fertility. Transplant Proc 1989;21:928–30.

[112] Seethalakshmi L, Flores C, Carboni AA, Bala R, Diamond DA, Menon M. Cyclosporine: its effects on testicular function and fertility in the prepubertal rat. J Androl 1990;11:17–24.

[113] Bantle JP, Nath KA, Sutherland DE, Najarian JS, Ferris TF. Effects of cyclosporine on the renin–angiotensin–aldosterone system and potassium excretion in renal transplant recipients. Arch Intern Med 1985;145:505–8.

[114] Misro MM, Chaki SP, Srinivas M, Chaube SK. Effect of cyclosporine on human sperm motility in vitro. Arch Androl 1999;43:215–20.

[115] Alfraih F, Ahmed S, Kim DDH, Rashee W, Aldawasari G. Fertility recovery following allogeneic bone marrow transplantation in aplastic anemia; a study of 157 patients. Blood 2015;126(23):4783.

[116] Bo L, Mei-Ying L, Yang Z, Shan-Mi W, Xiao-Hong Z. Aplastic anemia associated with pregnancy: maternal and fetal complications. J Matern Fetal Neonatal Med 2016;29:1120–4.

[117] Tichelli A, Socie G, Marsh J, Barge R, Frickhofen N, McCann S, et al. Outcome of pregnancy and disease course among women with aplastic anemia treated with immunosuppression. Ann Intern Med 2002;137:164–72.

6

Immunosuppressive Therapy for Aplastic Anemia

J.C.W. Marsh, S. Gandhi, G.J. Mufti

Department of Haematological Medicine, King's College Hospital/King's College London, London, United Kingdom

THE IMMUNE DEFECT IN APLASTIC ANEMIA AND THE RATIONALE FOR IMMUNOSUPPRESSIVE THERAPY

In acquired aplastic anemia (AA), although there is oligoclonal expansion of cytotoxic CD8 T cells, the more important component comprises the CD4 T cells, with expansion of Th1 (clonal) and Th17 cells, and reduced and dysfunctional Tregs (defined by CD4+ CD25high CD27+ FOXP3+) [1–5]. The reduction in Treg numbers correlates with disease severity, and the defect is most prominent in severe AA (SAA) and very severe AA (VSAA). The numbers of resting and activated Tregs are significantly reduced, whereas cytokine-secreting non-Tregs are increased. Tregs in AA are also dysfunctional as they show reduced ability to suppress autologous T-effector proliferation [2]. This results in oligoclonal expansion of CD8+ cytotoxic T cells (CTLs) and apoptosis of hematopoietic stem- and progenitor cells (HSC and HPC, respectively).

Understanding of human Tregs and Th17 cells and their plasticity has improved greatly following the discovery of distinct Treg subsets. These can be defined as resting Tregs (CD45RA+ FOXP3low), activated Tregs (CD45RA− FOXP3high), or cytokine-secreting CD4+ T cells (CD45RA− FOXP3low). Activated and resting Tregs were reduced in AA whereas cytokine-secreting non-Tregs were increased [2,6]. More recently, multidimensional mass cytometry (CyTOF), has been used to define a specific Treg signature that defines AA and that predicts response to immunosuppressive therapy (IST) [7,8].

TREATMENT OPTIONS AND INDICATIONS FOR IST

The choice of first line treatment of AA comprises essentially allogeneic hematopoietic stem cell transplantation (HSCT) or IST, and this will depend on disease severity, patient age, availability of a matched sibling donor (MSD),

Congenital and Acquired Bone Marrow Failure
http://dx.doi.org/10.1016/B978-0-12-804152-9.00006-3

individual patient performance status, and patient choice. Fig. 6.1 summarizes the current algorithm from the 2016 British Committee for Standards in Haematology (BCSH) AA guidelines [9]. Antithymocyte globulin (ATG) + cyclosporine (CSA) is indicated as first line therapy for (1) nonsevere AA (NSAA) patients who are transfusion dependent, bleeding, encountering infections, or for lifestyle activities, for example, a patient may have low blood counts but not quite low enough to warrant transfusions but who may be more at risk of bleeding due to their particular lifestyle activity; (2) SAA or VSAA in the absence of an HLA matched sibling; and (3) SAA/VSAA patients aged >35–50 years. This age range takes into account differences in comorbidities that occur between patients of the same age, thus enabling physician to decide if a particular patient is fit enough or not for upfront HSCT [9,10]. There is no upper age limit for ATG, but there is increased morbidity and mortality in patients aged >60 years treated with ATG, so elderly patients should be individually assessed

with respect to comorbidities and their specific wishes respected, as quality of life is paramount [11,12]. Indications for a second course are discussed in Section "Use of Rabbit ATG Instead of Horse ATG in First Line IST."

ATG: POSSIBLE MECHANISMS OF ACTION AND ADMINISTRATION

ATG is a polyclonal IgG-antibody preparation produced by immunizing horses or rabbits with human thymocytes. Possible mechanisms of action include T-cell depletion by complement-mediated lysis, destruction of activated CTLs by Fas-mediated apoptosis and antibody-dependent cellular cytotoxicity, reduced apoptosis and Fas expression on AA CD34+ bone marrow cells, direct stimulation of normal and AA CD34+ bone marrow cells, and a mitogenic effect in the absence of complement resulting in release of hematopoietic growth factors [13]. There are differences between horse- and rabbit ATG that

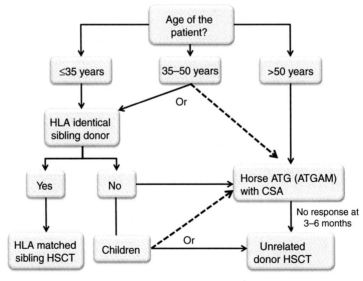

FIGURE 6.1 **Indications for first line immunosuppressive therapy (IST) for acquired severe aplastic anemia (SAA).** This treatment strategy follows EBMT [10] and BCSH 2016 guidelines [9]. *Source: Modified with permission from Killick SB, Bown N, Cavenagh J, Dokal I, Foukaneli T, Hill A, Hillmen P, Ireland R, Kulasekararaj A, Mufti G, Snowden JA, Marsh JCW. Guidelines for the diagnosis and management of adult aplastic anaemia. Br J Haematol 2016;172(2):187–207 [9].*

may explain their different response rates and effects. Rabbit ATG remains in the blood longer than horse ATG and reduces neutrophils and CD4+ T cells to a greater degree than horse ATG. Although the frequency of Tregs (as defined by CD4+ FOXP3+ CD127−) in the blood is higher after rabbit ATG compared to horse ATG, the absolute number is lower due to the lower number of total CD4+ T cells. Horse- and rabbit ATG both induce a similar cytokine storm except that CCL4 levels are higher after rabbit ATG compared to horse ATG [14,15]. Some of these findings may contribute to a higher risk of infections and a lower rate of early response with rabbit ATG, as discussed later in this chapter.

ATG must be given as an in-patient, and should only be used in centers that are familiar with using the drug and its side effects. As ATG induces a fall in the platelet count during administration and platelet consumption may occur during the period of serum sickness, prior to starting ATG platelet count, increments should be performed to exclude platelet refractoriness and patients screened for HLA antibodies. For older patients, especially those aged >60 years, careful assessment must be made of comorbidities and medical fitness for ATG, as there is increased mortality from infection and bleeding after ATG in this age group (see Chapter 12). The dose of horse ATG (ATGAM) is 40 mg/kg/day for 4 days, given as an intravenous infusion over 12–18 h, following a "test dose" due to the risk of anaphylaxis (the test dose usually comprises the first 100 mL of the first dose of ATG infused very slowly over 1 h). Each dose of ATG is preceded with intravenous methyl prednisolone 1 mg/kg, chlorpheniramine to reduce immediate allergic reactions, and platelet transfusions aiming to keep the platelet count >20–30 × 10^9 L^{-1} if possible. Fluid retention occurs commonly during ATG treatment, especially in older patients; careful attention to fluid balance is important. Prednisolone 1 mg/kg/day is commenced at the start of ATG for 2 weeks to reduce the risk of serum sickness, followed by rapid taper over the following 2 weeks.

HISTORICAL DEVELOPMENT OF THE CURRENT STANDARD ATG PROTOCOL (HORSE ATG COMBINED WITH CYCLOSPORINE)

A summary of the developmental steps in the IST protocol for treatment of AA is shown in Fig. 6.2. ATG was initially used as sole IST to treat SAA with response rates of around 40%. High dose methyl prednisolone (up to 10 mg/kg) was given concurrently in the mistaken belief that this had efficacy in the treatment of AA [16,17]. Subsequently, the dose of prednisolone was reduced to 1 mg/kg with the sole aim of preventing the immediate allergic reactions and late (serum sickness) side effects of ATG, and given for the shortest period of time, usually 2 weeks, to reduce serious side effects, particularly, bacterial and fungal infection and avascular necrosis of bone [18].

A randomized EBMT trial of ATG with or without androgens showed significantly improved response in the androgen arm at 4 months, especially in females with neutrophil count < 0.5 × 10^9 L^{-1}. However, the addition of androgens (oxymetholone) conferred no survival advantage and was associated with hepatotoxicity and masculinizing side effects in female patients [22]. There is renewed interest in use of the semisynthetic androgen, danazol, in constitutional AA, as it is less toxic than oxymetholone, but for acquired idiopathic AA it is now only occasionally used in refractory SAA patients who are not HSCT candidates [9]. As a cautionary note for patients with presumed acquired AA who may later require HSCT, the use of androgens in constitutional AA is an adverse risk factor for survival following subsequent HSCT [23].

A major advance in the treatment of AA was the use of CSA in combination with ATG, which significantly increased the response rate compared to ATG alone (70% vs. 41%, respectively, and for SAA 65% vs. 31%, respectively), in a German prospective randomized multicenter study of AA [24,25]. Complete response (CR), with

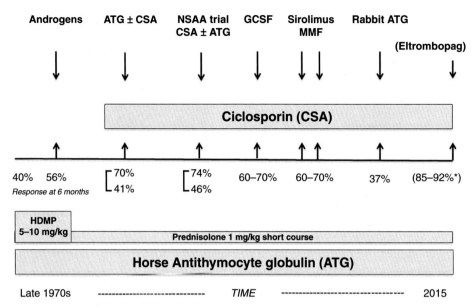

FIGURE 6.2 **Prospective randomized trials of antithymocyte globulin (ATG)-based first line treatment of aplastic anemia (AA).** The nonsevere aplastic anemia (NSAA) EBMT trial reported response rates of 74% and 46% for the combination of ATG with cyclosporine (CSA) and CSA alone, respectively [19]. *Responses of 85–92% with addition of eltrombopag to ATG and CSA have been reported from a preliminary analysis of a single center Phase II study from NIH [20]. The randomized prospective "RACE" EBMT study of horse ATG and CSA with or without eltrombopag is currently in progress [21]. *HDMP*, High dose methyl prednisolone; *MMF*, mycophenolate mofetil.

normalization of blood counts, was observed in 10% of patients, confirming that in most responding patients, hematologic response is incomplete. Patients receiving CSA with ATG responded more rapidly but it was noted that 26% of patients who responded were dependent on CSA and complete withdrawal of CSA was not possible.

ALTERNATIVE STRATEGIES USED IN AN ATTEMPT TO IMPROVE RESPONSE TO STANDARD IST WITH ATG + CSA (FIG. 6.2)

ATG and CSA Combined With GCSF

The addition of daily GCSF to the combination of ATG and CSA was explored with the rationale of (1) reducing early infections during the first 3 months before response to ATG occurs; (2) stimulating residual HSC through the synergistic effect of GCSF with elevated circulating levels of endogenous growth factors, such as, stem cell factor and thrombopoietin; and (3) theoretically promoting mobilization of HSC. Three prospective randomized trials, however, showed no improvement in response, overall survival, or serious infections in the GCSF arms [26–28]. A further prospective randomized study compared four different IST regimens: horse ATG alone; horse ATG + CSA; horse ATG + CSA with GMCSF and rHu EPO; and rabbit ATG (Fresenius) with CSA, GMCSF, and rHu EPO. The response rates for each regimen were 58, 79, 73, and 53%, respectively. Compared to horse ATG and CSA alone, the addition of GMCSF to horse ATG and CSA had no impact on response, overall survival, the number of early deaths, or speed of hematologic recovery [29]. A major concern with prolonged use (and high doses) of GCSF in SAA is the emergence of monosomy 7 clones [30], either through selective pressure in the presence of a severely reduced stem cell pool or direct stimulation of preexisting small monosomy 7 clones [31]. For all these reasons, the routine use of GCSF with IST is no longer recommended.

ATG and CSA Combined With Additional Immunosuppressive Agents

Other approaches have been to combine ATG + CSA with other immunosuppressive drugs, specifically mycophenolate mofetil (MMF) or sirolimus. However, two prospective trials showed that adding either of these two drugs to standard combination of ATG and CSA showed no improvement in the response rate or reduction in the risk of relapse after ATG and CSA [32,33].

Use of Rabbit ATG Instead of Horse ATG in First Line IST

Historically, horse ATG has been used for the first course of IST and rabbit ATG usually reserved for a second course following relapse or failure to respond to a first course of ATG. The withdrawal of lymphoglobuline horse ATG in 2007 prompted three prospective studies comparing horse versus rabbit ATG with CSA as first line therapy for AA. It was anticipated that rabbit ATG would result in a higher response rate compared to horse ATG, because (1) it can induce response in a proportion of patients who have failed to respond to horse ATG; (2) it produces a more profound degree of lymphopenia and is more cytotoxic than horse ATG; and (3) it is more effective at preventing and treating acute renal allograft rejection episodes. Refer to Table 6.1 for a summary of prospective and retrospective studies. The first and largest prospective randomized study from NIH of 120 patients

showed an unexpected and significantly higher response rate at 6 months with horse ATG compared to rabbit ATG (68% vs. 37%, respectively) and overall survival [15]. Similar results were reported from a Phase II multicenter EBMT trial of 35 patients treated with horse ATG with CSA compared to 105 age- and disease severity matched patients receiving rabbit ATG with CSA from the EBMT registry, although no difference in response at 12 months [34]. In another Phase II study of 20 patients treated with rabbit ATG there was no significant difference in response compared historical controls treated with horse ATG (ATGAM), 45% versus 58%, respectively [35]. Data from retrospective studies have shown either worse outcome or similar outcome for rabbit- versus horse ATG [36–39]. A large retrospective Spanish study of rabbit ATG ($n = 169$) compared to horse ATG ($n = 62$) showed no significant differences in response at 3 or 12 months between the two products (63 % vs. 66% at 3 months; 84% vs. 76% at 12 months), respectively. However, the two groups were not matched for disease severity, as there were significantly more patients with NSAA in the rabbit ATG group compared to the horse ATG group (50% vs. 29%, respectively, $p < 0.01$), which is likely to have impacted on response rates reported with rabbit ATG, as less severe disease predicts for better response to ATG [36]. A retrospective study from China reported 292 patients treated with first line rabbit ATG and CSA with response rate of 60% and 68% at 6 and 12 months, respectively [40]. However, there was no control group treated with horse ATG for comparison.

Both the EBMT Phase II study and the retrospective Spanish study reported a higher number of infective deaths with rabbit ATG compared to horse ATG, which may reflect the more profound neutropenia and CD4+ lymphopenia seen after rabbit ATG. A further consequence of this more prolonged CD4 lymphopenia may be a longer time to achieve CR compared to horse ATG [14]. These observations may also reflect differences in ethnicity with possible genetic factors

TABLE 6.1 Studies Comparing Rabbit ATG With Horse ATG as First Line IST

Studies	Study designs	Country	No. of patients (R vs. H)	Age (range)	Horse ATG prep used	Rabbit ATG dose (mg/kg × 5)	Response at 6 mo [R vs. H (p)]	Response at 12 mo [R vs. H (p)]	Comments
Scheinberg (2011)	Prospective, randomized	USA	60 versus 60	31 ± 2.6	ATGAM	3.5	37% versus 68% (S)	NK	OS 55% versus 85% (S), R versus H
Marsh (2012)	Prospective Phase II	Europe (EBMT)	35 versus 105	36 (17–75)	Lympho	3.75	37% versus NK	60% versus 67%[a] (NS)	OS 68% versus 86% (S), R versus H More infective deaths with RATG
Afable (2011)	Prospective Phase II	USA	29 versus 42	55 (19–80)	ATGAM	3.5	45% versus 58% (NS)	50% versus 58% (NS)	
Vallejo (2015)	Retrospective	Spain	169 versus 62	44 (2–84)	Lympho and ATGAM	2.5–4	NK	84% versus 76% (NS)	More NSAA in RATG group (50% versus 29%); more infective deaths with RATG
Shin (2013)	Retrospective	South Korea	53 versus 46	37 (15–66)	Lympho	2.5	45% versus 39% (NS)	51% versus 48% (NS)	More CR at 6 mo but not at 12 or 18 mo with RATG
Atta (2010)	Retrospective	Brazil	29 versus 42	21 (4–63)	Lympho	2.5	35% versus 60% (S)	NK	OS 55% versus 78% (S), R versus H

CR, Complete response; H, horse; Lympho, lymphoglobuline; mo, months; NK, not known; NS, not statistically significant; NSAA, nonsevere aplastic anemia; OS, overall survival; prep, preparation; R, rabbit; S, statistically significant.

[a]In this study, later response was not determined at specific time point but instead as the best later response.

predisposing to response to ATG and variations in the dose of rabbit ATG used (Table 6.1). The latter effect is being examined in an ongoing prospective randomized study of the Asia Pacific Blood and Marrow Transplant (APBMT) group, comparing two different doses (2.5 vs. 3.5 mg/kg/day $\times$ 5) of rabbit ATG initiated by the Asia Pacific group (NCT01844635). Lastly, response to rabbit ATG may take longer than with horse ATG.

As rabbit ATG results in a more profound and prolonged PB CD4+ lymphopenia than horse ATG, does this increase the risk of infections and disease associated with CMV or EBV? Reactivation of EBV and CMV occurs commonly after ATG/CSA treatment, especially EBV reactivation in 87% of patients, with CMV reactivation less commonly in 33%. Rabbit ATG is associated with higher mean peak EBV levels and longer duration of PCR positivity compared to horse ATG, but despite this, EBV and CMV diseases are extremely rare in this setting and almost always self-limiting so treatment is rarely indicated [41]. There have been very rare reports of EBV lymphoproliferative disease, mostly from Japan, following treatment with either rabbit or horse ATG in combination with CSA [42,43].

As a consequence of these studies, and especially from the prospective studies, comparing rabbit- with horse ATG, horse ATG is currently recommended as the preferred animal source of ATG for first line IST [9,44].

Variations in CSA Dose and Duration

CSA is usually commenced at the start of ATG, at a dose of 5 mg/kg/day, although some centers use higher doses. There is also variability in the target CSA trough blood levels between centers (e.g., 100–200, 150–250 µg/L, and higher ranges), and recent data suggest that a lower trough level at around 100 µg/L may improve response by increasing proliferation of Tregs [45]. CSA should be continued while the blood count continues to rise. A slow tapering of the drug (25 mg every 2–3 months) can be started after at least a further

12 months of therapy, to help reduce the risk of later relapse [46,47]. However, a study from NIH showed that prolonged CSA therapy delayed, rather than reduced, the relapse risk [48].

Addition of Eltrombopag to ATG and CSA

The most recent approach has been to combine ATG and CSA with the novel thrombopoietin receptor agonist, eltrombopag. This followed results of a Phase II and extension study using eltrombopag as a single agent in refractory SAA, resulting in a 40–44% response rate [49,50]. The unexpected finding, of not only platelet response but also, in some patients bi- and trilineage hematologic responses, indicated that the drug may be acting as a direct stem-cell stimulant. Following later withdrawal of the drug in patients with a "robust" response (defined by Hb > 10, platelets > 50, and neutrophils > 1×10^9 L^{-1}), sustained responses were observed, leading the authors to propose that when a critical number of HSC regenerate after eltrombopag, hemopoiesis may be sustained without the need for continued drug exposure [51].

However, an abnormal cytogenetic clone was detected at a median of only 3 months in 19% of patients, with a high incidence of abnormalities of chromosome 7, either -7 or del(7q). There are currently two European multicenter prospective randomized trials, that will assess the use of eltrombopag with first line IST; one is enrolling patients with SAA and VSAA, the so-called "RACE" EBMT study (NCT02099747), which is comparing horse ATG and CSA with and without eltrombopag [21]. The second study will be for moderate AA, comparing the combination of CSA and eltrombopag with CSA and placebo, and sponsored by the University of Ulm, Germany. For all disease severity subtypes of AA, preliminary results from a single center, Phase II trial using ATG, CSA, and eltrombopag for first line therapy have recently been reported. Patients were subdivided into three cohorts according to the eltrombopag-dosing regimen of 150 mg/day (cohort 1: from day +14–3months, $n = 30$; cohort 2: day +14–6 months, $n = 31$; cohort 3: day +1–6 months, $n = 27$). The rationale for the different cohorts was the potential concern about clonal transformation with eltrombopag. Remarkably, overall response (OR) at 6 months was 85% and CR 54%, compared to the expected CR after ATG and CSA of 10%. Furthermore, for cohort 3, OR was 92% and CR 54%. A more rapid recovery of blood counts was observed compared to historical data. With a median follow up of only 15 months, there were sevem patients who acquired abnormal cytogenetic clones (loss or partial loss of chromosome 7 ($n = 4$), and one each for transient del(13q), complex: t(3;3)(q21;q26), −7) and +6 with +15 [20]. (See also Chapter 13). Longer follow up, serial monitoring of patients for somatic mutations [52], and results from the ongoing EBMT "RACE" prospective randomized study are crucial to evaluate further the risk of clonal transformation.

Conclusions

Until recently, the only change to IST regimen that had improved hematologic response was the addition of CSA to ATG. Other changes listed earlier (addition of GCSF, other immune suppressing agents, and use of rabbit ATG instead of horse ATG) have had no impact on response rates. This prompted the realization that the ceiling had probably been reached in terms of what further improvement in response could be achieved by the addition of other immunosuppressive drugs to the standard combination of ATG and CSA. [53]. Hence an alternative approach has recently been reported, using a drug that has a novel mode of action, for example, eltrombopag demonstrating trilineage responses in refractory SAA, and which has stimulated ongoing clinical trials of this drug in combination with horse ATG and CSA.

THE USE OF ALEMTUZUMAB IN AA

The humanized monoclonal antibody alemtuzumab recognizes the CD52 antigen and is expressed on T, B, NK, and dendritic cells but not on hemopoietic progenitors. It works by both antibody-dependent cellular cytotoxicity and complement-mediated lysis. Alemtuzumab was initially used in the treatment of lymphoid malignancies but has also been widely used in the prevention and treatment acute renal allograft rejection, prevention of graft versus host disease (GVHD) and graft rejection following allogeneic HSCT, and treatment of other autoimmune disorders, such as, multiple sclerosis, vasculitis, and autoimmune cytopenias [54].

A Phase II study of 13 patients with untreated AA was performed using alemtuzumab given as a 5-day course of consecutive doses of 3, 10, 30, 30, and 30 mg subcutaneously, with a lower dose for NSAA patients. Response was observed in nine (69%) of patients, with five CRs and four partial responses. Relapses were frequent but retreatment with alemtuzumab induced further response in a number of patients [55].

A large study from NIH examined three separate protocols for alemtuzumab in SAA used as a single agent (without CSA) [56]. Untreated AA patients were randomized as part of the prospective study comparing horse- versus rabbit ATG with CSA [15], the third arm of which was alemtuzumab. Out of 16 patients randomized to receive alemtuzumab, only 19% responded. In view of the low response to alemtuzumab and 3 early deaths, this arm of the randomized trial was discontinued. Therefore, alemtuzumab is not recommended as first line treatment for AA. In contrast, better responses were observed in refractory- and relapsed AA. Patients with refractory SAA were randomized to receive either alemtuzumab alone or the combination of rabbit ATG with CSA, and 37% of 27 patients responded to alemtuzumab compared to 33% with rabbit ATG and CSA. The best response to alemtuzumab was seen in a single arm study of 25 patients treated for relapsed SAA, with 56% response, which is comparable to response using ATG for relapsed SAA [56].

Two smaller separate studies examined the use of lower doses of alemtuzumab in combination with CSA in untreated SAA. The first compared two doses of alemtuzumab (60 mg and 90 mg) in 17 SAA patients. Response was seen in 35% of patients and surprisingly all who received the lower dose of alemtuzumab [57]. In the second study, 14 patients received 50 mg alemtuzumab and with a 57% OR [58].

From published studies thus far, the initial fears of a high risk of serious viral infections with alemtuzumab in SAA have not been substantiated but patients should receive prophylactic antibiotic, antifungal, and pneumocystis prophylaxis and antiviral prophylaxis for herpes simplex/zoster but not anti-CMV prophylaxis as this is myelosuppressive. Patients must also only receive irradiated blood products. Single agent alemtuzumab represents an alternative immunosuppressive agent for refractory- or relapsed AA, and may be particularly useful for patients with impaired renal function where use of CSA alone or in combination with ATG would be contraindicated [55].

TREATMENT OF NSAA

Prior to consideration of IST for patients with NSAA, careful assessment of the patient is required to exclude possible constitutional AA, which is more likely to present with milder pancytopenia and hypocellular myelodysplastic syndrome (MDS) [59,60]. Emerging diagnostic tests include peripheral blood telomere length analysis and next generation sequencing (NGS) gene panels for the known constitutional BMF disorders. Hypocellular MDS is often difficult to distinguish from NSAA on bone marrow morphological examination [59]. Furthermore, an abnormal cytogenetic clone does not always

signify hypocellular MDS instead of AA [61]. Further prospective studies of somatic mutations are needed to determine whether they might be useful as part of new diagnostic criteria for hypocellular MDS.

Having excluded constitutional NSAA, the decision to treat acquired NSAA with IST is straightforward if the patient is transfusion dependent. If the patient does not require transfusion support but has blood counts sufficiently low to impact on an individual's lifestyle, then it is reasonable to commence treatment [9]. Should this be with standard IST for SAA, that is the combination of ATG with CSA, or does oral CSA alone provide sufficient immune suppression to achieve hematologic response? This question was addressed in a prospective randomized EBMT study for NSAA comparing the combination of ATG with CSA with alone [19]. Compared to CSA alone, ATG with CSA resulted in an improved response at 6 months (74% vs. 46%, respectively). Although OS was similar between the two groups (93% vs. 91%), failure-free survival was significantly better using ATG with CSA (80%) compared to CSA alone (51%). In the CSA group, 25% of patients required a course of ATG within 6 months due to disease progression compared to only 6% in the ATG group. Lastly, patients treated with the combination of ATG and CSA achieved higher blood counts and needed fewer transfusions compared to those treated with CSA alone.

PREDICTIVE FACTORS FOR RESPONSE TO ATG

The ultimate goal at time of diagnosis of AA would be an individualized treatment approach based on an established predictive score for response of nonresponse to IST. Predictive factors for response to ATG include less severe disease; young age; absolute reticulocyte and lymphocyte counts of ≥ 25 and $\geq 1 \times 10^9 \, L^{-1}$, respectively [62]; the following chromosomal abnormalities, trisomy 8, del(13q), and LOH for chromosome 6p [63–65]; and discussed further in Section "Clonal Transformation to MDS/AML After IST." Short telomeres in children, but not in adults, also predict response to IST. Other independent predictive factors in children are the presence of a PNH clone, male sex, and a shorter time interval from diagnosis to treatment [66–68]. In several series, the presence of a PNH clone is also predictive for response in adults in several, but not all, studies [67–69]. This may reflect variation in the cut off level for defining the size of a PNH clone on flow cytometry. However, the presence of an acquired somatic mutation of *PIGA* or *BCOR/BCORL1* genes, does predict for good responses and outcomes after IST, whereas the presence of an unfavorable somatic mutation (*DNMT3A* and *ASXL1* predominantly), is associated with poor survival and response after IST [70]. The impact of *DNMT3A* or *ASXL1* on survival was even more significant in younger patients aged <60 years, indicating that early unrelated donor (UD) HSCT might be considered in preference to first line IST, in the absence of a MSD. Despite all this, there remains an unmet need for more robust predictive factors for response to IST at time of diagnosis of AA. A new approach has been explored by our group, using multidimensional mass cytometry to identify the detailed immunologic signature that predicts for response to IST at time of diagnosis of AA [7,8]. Validation of this novel observation is now required in large prospective studies.

REPEAT COURSES OF ATG FOR NONRESPONSE AND RELAPSE

Refractory AA is defined as lack of response with persistence of severe pancytopenia at 6 months after one course of IST. At this stage of the disease, before planning further treatment, it is important to reassess the patient with repeat bone marrow and cytogenetics to exclude clonal transformation to MDS/acute myeloid leukemia (AML), and to ensure that a possible constitutional BMF disorder has been excluded by using the latest molecular testing that may be available (with measurement of telomere length and NGS to detect somatic mutations and known constitutional BMF mutations). Referral to a center with specific expertise in AA should be considered to help expedite these specialist investigations in addition to providing expertise in treatment-management decisions with full discussions of all available therapeutic options [71].

The options for refractory AA include matched UD HSCT, eltrombopag, second course of IST with ATG, or alternative donor HSCT (haploidentical or cord blood) in the absence of a matched donor, as summarized in Fig. 6.3. UD HSCT is indicated after nonresponse to a single course of IST, although up-front UD HSCT may be considered in children who lack a MSD, and where a suitably matched unrelated donor (MUD) is available.

A repeat course of ATG may be given for failure to respond or relapse after a first course of ATG, and if the patient is ineligible for UD HSCT. There is a choice of animal ATG preparation for a second course, rabbit ATG or horse ATG may be given, although rechallenge with the same animal preparation may be associated with more immediate and late (serum sickness) side effects. The response rate depends on the indication: when using horse ATG again, the response rate is 60% and 35%, for relapse and nonresponse, respectively [72]. Similar results have been reported in a retrospective study from Brazil using rabbit ATG for both the first and second course of IST, with 22% response at 6 months for refractory AA, compared to 60% for relapsed AA, albeit only small numbers of patients treated for relapsed AA in this study [73].

However, patients treated with multiple courses of IST are at increased risk of later clonal transformation to MDS/AML [74].

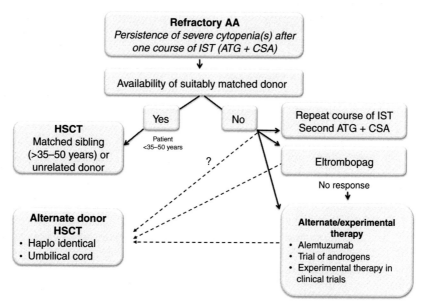

FIGURE 6.3 **Treatment of adult refractory SAA.** *Source: Modified with permission from Killick SB, Bown N, Cavenagh J, Dokal I, Foukaneli T, Hill A, Hillmen P, Ireland R, Kulasekararaj A, Mufti G, Snowden JA, Marsh JCW. Guidelines for the diagnosis and management of adult aplastic anaemia. Br J Haematol 2016;172(2):187–207 [9].*

CLONAL TRANSFORMATION TO MDS/AML AFTER IST

The most serious issue following IST with ATG is clonal transformation to MDS/AML, with a cumulative incidence of up to 20% at 10 years, most frequently with monosomy 7 [74–77]. Abnormal cytogenetic clones, most commonly abnormalities of chromosome 7, have been detected much earlier at a median of only 3 months following treatment of refractory SAA with eltrombopag [50].

However, some cytogenetic abnormalities, such as trisomy 8 (+8), del(13q), and loss of LOH involving the short arm of chromosome 6 (LOH6p), are associated with good response to IST and very low risk of transformation to MDS/AML [63–65]. Bone marrow HPCs from MDS patients with +8 show increased expression of WT1 antigen. This induces a specific T-cell response to WT1 peptides, leading to suppression of non-(+8) HPC through a bystander effect by activated CD8 T cells. In contrast, +8 HPCs survive this immune attack due to increased expression of antiapoptotic proteins survivin, cyclin D1, and increased proliferation due to increased expression of c-myc [78]. Another example of clones may escape the immune attack within the bone marrow environment and proliferate and attain a survival advantage over normal HSCs, is del(13q). In a study from Japan, all patients with del(13q) had PNH clones, and the del(13q) cells were only present in the nonmutant GPI− AP+ cells. Following IST, expansion of del(13q) clone occurred more frequently than a decrease in the clone size. So in this situation, both the del(13q) HSCs and the *PIGA*-mutant HSCs underwent preferential expansion and contributed to hematologic recovery [64]. Copy number neutral LOH6p was the most common acquired genetic event detected by SNP-A karyotyping in 13% of Japanese patients with AA [65]. LOH commonly affected the HLA locus with loss of expression of class I HLA-A molecules, particularly certain HLA-A*02:01, A*02:06, A*03:01, and B*40:02. In the presence of LOH6p, HSCs have lost the target for immune attack by CTLs and hence escape the immune attack, resulting in a growth advantage and clonal expansion over unaffected HSCs.

Other risk factors for MDS/AML and emergence of monosomy 7 are older age, short telomere length, nonresponse and multiple courses of IST, prolonged and high doses of GCSF, and the presence of a somatic mutation commonly mutated in myeloid malignancies [30,67,74,79]. Acquired somatic mutations (excluding *PIGA*) are present in 20–25% of patients with AA [70,76,77], and in the presence of a somatic mutation (as a whole), the risk of later MDS/AML is 38% compared to only 6% in the absence of a somatic mutation [76]. The same study found an association with *ASXL1 and DNMT3A* with monosomy 7. The possible mechanisms for emergence of monosomy 7 include stimulation of low level, previously undetectable clones by drugs, such as, GCSF or eltrombopag, or increased selective pressure on the few remaining HSCs, which then have a selective advantage, resulting in genomic instability and consequent malignant transformation. Telomere loss may be an important early event prior to emergence of monosomy 7 clones [79] (Fig. 6.4). Further understanding of the significance of somatic mutations in AA, particularly low level clones, will be crucial. In addition, serial characterization of the evolving clonal architecture and mutation hierarchy at the genomic level after IST as part of the monitoring for clonal transformation is necessary and this will be assessed prospectively

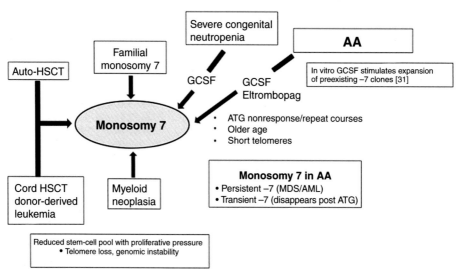

FIGURE 6.4   **Emergence of monosomy 7 in stressed hemato(myelo)poiesis.** Monosomy 7 clones may emerge in different scenarios where there is stressed hematopoiesis, for example in the setting of hematopoietic stem cell transplantation (HSCT), both autologous HSCT associated with low infused stem cell dose and monosomy 7 is commonly detected in donor-derived leukemia postallogeneic cord HSCT. Long-term administration of GCSF carries an increased risk of myelodysplastic syndrome (MDS)/acute myeloid leukemia (AML) with monosomy 7 in severe congenital neutropenia and AA, and has also been reported after use of eltrombopag for refractory SAA. Following ATG, the emergence of monosomy 7 is associated with a high risk of malignant transformation to MDS/AML and poor outcome. Rarely, after IST, monosomy 7 may be transient and disappear following hematologic response.

as part of a research study alongside the EBMT RACE clinical trial comparing ATG and CSA with or without eltrombopag.

FUTURE DIRECTIONS

Although around two-thirds of patients with AA will respond to IST, the recovery after IST is unstable and patients remain at high risk of clonal transformation and relapse. Progression to hemolytic PNH in around 10% of patients may indicate life-long anticomplement therapy. Only around 25% of patients will achieve normal or near-normal blood counts [25]. The NIH reported that an absolute platelet count and reticulocyte count of $\geq 50 \times 10^9$ L^{-1} is associated with improved long-term survival, and that the risk of monosomy 7was predicted by the 3 month platelet count [80]. But for complete responders there are no data to indicate whether they are at lower risk of acquiring somatic mutations compared to partial responders.

There is an important interaction between the immune response in AA and the emergence of abnormal clones, and ongoing and future research to correlate the molecular signature with the immunologic signature to understand the mechanism and determine the risk of malignant transformation following IST for AA will be essential. The current preliminary evaluation of the novel agent eltrombopag in combination with IST indicates a major improvement in overall and CR compared to historical data and prospective randomized trials are in progress. However, careful and long-term monitoring for malignant transformation to MDS/AML is vital in view of the reported high risk when eltrombopag is used as a single agent to treat refractory AA.

The emerging use of mass cytometry and NGS, to establish immune and molecular signatures for response to IST and later malignant transformation at time of diagnosis, potentially offers the future opportunity for patient-specific treatment. By understanding further the immune repertoire in those patients with SAA who have somatic mutations and how it differs from patients without mutations, may be predictive of transformation to MDS/AML. Recent EBMT data indicate that survival after MUD HSCT in adult patients is similar to MSD HSCT, although a higher rate of GVHD was observed after MUD HSCT, using ATG-based conditioning regimens. However, the significant improvement in both acute and chronic GVHD using alemtuzumab-based conditioning raises the future possibility of up-front UD HSCT for adults who lack a MSD, as a means of cure of AA and avoiding the risk of MDS/AML following IST. This may also extend the option of HSCT to older patients. The potential ability to define the molecular and immunologic signature of a patient at diagnosis that predicts not only response (or nonresponse) to IST but also the risk of malignant transformation would enable an individualized approach to therapy, thereby avoiding potential toxicity of ineffective IST and directing early HSCT for patients who either are predicted to be nonresponders to IST and/or have a high risk of later malignant transformation [81–83].

References

[1] Young NS, Bacigalupo A, Marsh JC. Aplastic anemia: pathophysiology and treatment. Biol Blood Marrow Transplant 2010;16(1 Suppl.):S119–25.

[2] Kordasti S, Marsh J, Al-Khan S, et al. Functional characterization of CD4+ T-cells in aplastic anemia. Blood 2012;119:2033–43.

[3] Solomou EE, Rezvani K, Mielke S, et al. Deficient CD4+ CD25+ FOXP3+ T regulatory cells in acquired aplastic anemia. Blood 2007;110:1603–6.

[4] de Latour RP, Visconte V, Takaku T, et al. Th17 immune responses contribute to the pathophysiology of aplastic anemia. Blood 2010;116:4175–84.

[5] Shi J, Lu S, Li X, et al. Intrinsic impairment of CD4+CD25+ regulatory T cells in acquired aplastic anemia. Blood 2012;120:1624–32.

[6] Sakaguchi S, Yamaguchi T, Nomura T, Ono M. Regulatory T cells and immune tolerance. Cell 2008;133:775–87.

[7] Kordasti S, Seidl T, Ellis R, et al. High resolution mass cytometry (CyTOF) in aplastic anaemia (AA) can

identify an aberrant Treg subset with pro-inflammatory properties, predicting poor response to immunosuppressive therapy. Blood 2014;124:1600.

[8] Kordasti S, Constantini B, Seidl T, et al. Deep-phenotyping of Tregs identifies an immune signature for idiopathic aplastic anemia and predicts response to treatment. Blood 2016:blood-2016-03-703702.

[9] Killick SB, Bown N, Cavenagh J, et al. British Society for Standards in Haematology. Guidelines for the diagnosis and management of adult aplastic anaemia. Br J Haematol 2016;172:187–207.

[10] Sureda A, Bader P, Cesaro S, et al. Indications for allo- and auto-SCT for haematological diseases, solid tumours and immune disorders: current practice in Europe. Bone Marrow Transplant 2015;50(8):1037–56.

[11] Tichelli A, Socie G, Henry-Amar M, et al. Effectiveness of immunosuppressive therapy in older patients with aplastic anemia. European Group for Blood and Marrow Transplantation Severe Aplastic Anaemia Working Party. Ann Intern Med 1999;130:193–201.

[12] Tichelli A, Marsh JC. Treatment of aplastic anaemia in elderly patients aged >60 years. Bone Marrow Transplant 2013;48:180–2.

[13] Kulasekararaj A, Young NS, Marsh JCW, Hillmen P. Aplastic anaemia and paroxysmal nocturnal haemoglobinuria. In: Hoffbrand V, Higgs DR, Keeling DM, Mehta AB, editors. Postgraduate haematology. 7th ed. Oxford: Wiley-Blackwell; 2014.

[14] Feng X, Scheinberg P, Biancotto A, et al. In vivo effects of horse and rabbit antithymocyte globulin in patients with severe aplastic anemia. Haematologica 2014;99:1433–40.

[15] Scheinberg P, Nunez O, Weinstein B, et al. Horse versus rabbit antithymocyte globulin in acquired aplastic anemia. New Engl J Med 2011;365:430–8.

[16] Speck B, Gluckman E, Haak HL, van Rood JJ. Treatment of aplastic anaemia by antilymphocyte globulin with or without allogeneic bone-marrow infusions. Lancet 1977;2:1145–8.

[17] Champlin R, Ho W, Gale RP. Antihtymocyte globulin treatment in patients with aplastic anemia. New Engl J Med 1983;308:113–8.

[18] Marsh JCW, Zomas A, Hows JM, Chapple M, Gordon-Smith EC. Avascular necrosis after treatment of aplastic anaemia with antilymphocyte globulin and high dose methyl prednisolone. Br J Haematol 1993;84:731–5.

[19] Marsh J, Schrezenmeier H, Marin P, et al. Prospective randomized multicenter study comparing cyclosporin alone versus the combination of antithymocyte globulin and cyclosporin for treatment of patients with nonsevere aplastic anemia: a report from the European Blood and Marrow Transplant (EBMT) Severe Aplastic Anaemia Working Party. Blood 1999;93:2191–5.

[20] Townsley DM, Dumitriu B, Scheinberg P, et al. Eltrombopag added to standard immunosuppression for aplastic anemia accelerates count recovery and increases response rates. Blood 2015;126:23.

[21] Risitano AM, et al. The RACE study: a SAAWP prospective randomized multicenter study comparing horse antithymocyte globulin (hATG) + cyclosporine A (CsA) with or without eltrombopag as front-line therapy for severe aplastic anaemia patients (WP007). Bone Marrow Transplant 2015;50(S1):S99.

[22] Bacigalupo A, Chaple M, Hows J, et al. Treatment of aplastic anaemia (AA) with antilymphocyte globulin (ALG) and methylprednisolone (MPred) with or without androgens: a randomized trial from the EBMT SAA working party. Br J Haematol 1993;83:145–51.

[23] Pasquini R, Carreras J, Pasquini MC, et al. HLA-matched sibling hematopoietic stem cell transplantation for fanconi anemia: comparison of irradiation and nonirradiation containing conditioning regimens. Biol Blood Marrow Transplant 2008;14(10):1141–7.

[24] Frickhofen N, Kaltwasser JP, Schrezenmeier H, Raghavachar A, Vogt HG, Herrmann F, Freund M, Meusers P, Salama A, Heimpel H. Treatment of aplastic anemia with antilymphocyte globulin and methylprednisolone with or without cyclosporine. The German Aplastic Anemia Study Group. New Engl J Med 1991;324:1297–3045.

[25] Frickhofen N, Heimpel H, Kaltwasser JP, Schrezenmeier H. German Aplastic Anemia Study Group. Antithymocyte globulin with or without cyclosporin A: 11-year follow-up of a randomized trial comparing treatments of aplastic anemia. Blood 2003;101:1236–42.

[26] Gluckman E, Rokicka-Milewska R, Hann I, European Group for Blood and Marrow Transplantation Working Party for Severe Aplastic Anemia. et al. Results and follow-up of a phase III randomized study of recombinant human-granulocyte stimulating factor as support for immunosuppressive therapy in patients with severe aplastic anaemia. Br J Haematol 2002;11:1075–82.

[27] Tichelli A, Schrezenmeie H, Socie G, et al. A randomized controlled study in patients with newly diagnosed severe aplastic anemia receiving antithymocyte globulin (ATG), cyclosporine, with or without G-CSF: a study of the SAA Working Party of the European Group for Blood and Marrow Transplantation. Blood 2011;117:4434–41.

[28] Teramura M, Kimura A, Iwase S, et al. Treatment of severe aplastic anemia with antithymocyte globulin and cyclosporin A with or without G-CSF in adults: a multicenter randomized study in Japan. Blood 2007;110:1756–61.

[29] Zheng Y, Liu Y, Chu Y. Immunosuppressive therapy for acquired severe aplastic anemia (SAA): a prospective

comparison of four different regimens. Exp Hematol 2006;34:826–31.

[30] Kojima S, Ohara A, Tsuchida M, Japan Childhood Aplastic Anemia Study Group. et al. Risk factors for evolution of acquired aplastic anemia into myelodysplastic syndrome and acute myeloid leukemia after immunosuppressive therapy in children. Blood 2002;100:786–90.

[31] Sloand EM, Yong AS, Ramkissoon S, et al. Granulocyte colony-stimulating factor preferentially stimulates proliferation of monosomy 7 cells bearing the isoform IV receptor. Proc Natl Acad Sci USA 2006;103:14483–8.

[32] Scheinberg P, Nunez O, Wu C, Young NS. Treatment of severe aplastic anaemia with combined immunosuppression: anti-thymocyte globulin, ciclosporin and mycophenolate mofetil. Br J Haematol 2006;133:606–11.

[33] Scheinberg P, Wu CO, Nunez O, et al. Treatment of severe aplastic anemia with a combination of horse antithymocyte globulin and cyclosporine, with or without sirolimus: a prospective randomized study. Haematologica 2009;94:348–54.

[34] Marsh JC, Bacigalupo A, Schrezenmeier H, European Blood and Marrow Transplant (EBMT) Group Severe Aplastic Anaemia (SAA) Working Party. et al. Prospective study of rabbit antithymocyte globulin and ciclosporin for aplastic anemia from the EBMT Severe Aplastic Anemia Working Party. Blood 2012;119:5391–6.

[35] Afable MG 2nd, Shaik M, Sugimoto Y, et al. Efficacy of rabbit anti-thymocyte globulin in severe aplastic anemia. Haematologica 2011;96(9):1269–75.

[36] Vallejo C, Montesinos P, Polo M, Bone Marrow Failure Spanish Study Group (Pethema-GETH). et al. Rabbit antithymocyte globulin versus horse antithymocyte globulin for treatment of acquired aplastic anemia: a retrospective analysis. Ann Hematol 2015;94:947–54.

[37] Atta EH, Dias DS, Marra VL, de Azevedo AM. Comparison between horse and rabbit antithymocyte globulin as first-line treatment for patients with severe aplastic anemia: a single-center retrospective study. Ann Hematol 2010;89:851–9.

[38] Shin SH, Yoon JH, Yahng SA, et al. The efficacy of rabbit antithymocyte globulin with cyclosporine in comparison to horse antithymocyte globulin as a first-line treatment in adult patients with severe aplastic anemia: a single-center retrospective study. Ann Hematol 2013;92:817–24.

[39] European Blood and Marrow Transplant Group, Severe Aplastic Anaemia Working Party. Rabbit ATG for aplastic anaemia treatment: a backward step? Lancet 2011;378:1831–3.

[40] Zhang L, Jing L, Zhou K, et al. Rabbit antithymocyte globulin as first-line therapy for severe aplastic anemia. Exp Hematol 2015;43:286–94.

[41] Scheinberg P, Fischer SH, Li L, et al. Distinct EBV and CMV reactivation patterns following antibody-based immunosuppressive regimens in patients with severe aplastic anemia. Blood 2007;109:3219–24.

[42] Nakanishi R, Ishida M, Hodohara K, et al. Occurrence of Epstein-Barr virus-associated plasmacytic lymphoproliferative disorder after antithymocyte globulin therapy for aplastic anemia: a case report with review of the literature. Int J Clin Exp Pathol 2014;7:1748–54.

[43] Takahashi T, Maruyama Y, Saitoh M, Itoh H, Yoshimoto M, Tsujisaki M. Fatal Epstein-Barr virus reactivation in an acquired aplastic anemia patient treated with rabbit antithymocyte globulin and cyclosporine A. Case Rep Hematol 2015;2015:926874.

[44] Scheinberg P, Young NS. How I treat acquired aplastic anemia. Blood 2012;120:1185–96.

[45] Bleyzac N, Philippe M, Bertrand A, Bertrand Y. Confounding effect of cyclosporine dosing when comparing horse and rabbit antithymocyte globulin in patients with severe aplastic anemia. Haematologica 2015;100(5):e211–2.

[46] Dufour C, Svahn J, Bacigalupo A. Severe Aplastic Anaemia-Working Party of the EBMT. Front-line immunosuppressive therapy of acquired aplastic anaemia. Bone Marrow Transplant 2013;48:174–7.

[47] Saracco P, Quarello P, Iori AP, Bone Marrow Failure Study Group of the AIEOP (Italian Association of Paediatric Haematology Oncology). et al. Cyclosporin A response and dependence in children with acquired aplastic anaemia: a multicentre retrospective study with long-term observation follow-up. Br J Haematol 2008;140:197–205.

[48] Scheinberg P, Rios O, Scheinberg P, Weinstein B, Wu CO, Young NS. Prolonged cyclosporine administration after antithymocyte globulin delays but does not prevent relapse in severe aplastic anemia. Am J Hematol 2014;89:571–4.

[49] Olnes MJ, Scheinberg P, Calvo KR, et al. Eltrombopag and improved hematopoiesis in refractory aplastic anemia. New Engl J Med 2012;367:11–9.

[50] Desmond R, Townsley DM, Dumitriu B, et al. Eltrombopag restores tri-lineage hematopoiesis in refractory severe aplastic anemia which can be sustained on discontinuation of drug. Blood 2014;123:1818–25.

[51] Desmond R, Townsley DM, Dunbar C, Young NS. Eltrombopag in aplastic anemia. Semin Hematol 2015;52:31–7.

[52] Marsh JC, Mufti GJ. Eltrombopag: a stem cell cookie? Blood 2014;123:1774–5.

[53] Passweg JR, Tichelli A. Immunosuppressive treatment for aplastic anemia: are we hitting the ceiling? Haematologica 2009;94:310–2.

[54] Young ME, Potter V, Kulasekararaj AG, Mufti GJ, Marsh JC. Allogeneic stem cell transplantation using

alemtuzumab-containing regimens in severe aplastic anemia. Curr Opin Hematol 2013;20:515–20.

[55] Risitano AM, Schrezenmeier H. Alternative immunosuppression in patients failing immunosuppression with ATG who are not transplant candidates: Campath (Alemtuzumab). Bone Marrow Transplant 2013;48:186–90.

[56] Scheinberg P, Nunez O, Weinstein B, Scheinberg P, Wu CO, Young NS. Activity of alemtuzumab monotherapy in treatment-naive, relapsed, and refractory severe acquired aplastic anemia. Blood 2012;119:345–54.

[57] Kim H, Min YJ, Baek JH, Shin SJ, Lee EH, Noh EK, Kim MY, Park JH. A pilot dose-escalating study of alemtuzumab plus cyclosporine for patients with bone marrow failure syndrome. Leuk Res 2009;33:222–31.

[58] Gómez-Almaguer D, Jaime-Pérez JC, Garza-Rodríguez V, et al. Subcutaneous alemtuzumab plus cyclosporine for the treatment of aplastic anemia. Ann Hematol 2010;89:299–303.

[59] Alter BP. Hematology diagnosis, genetics, and management of inherited bone marrow failure syndromes. Hematology Am Soc Hematol Educ Program 2007;29–39.

[60] Bennett JM, Orazi A. Diagnostic criteria to distinguish hypocellular acute myeloid leukemia from hypocellular myelodysplastic syndromes and aplastic anemia: recommendations for a standardized approach. Haematologica 2009;94:264–8.

[61] Gupta V, Brooker C, Tooze JA, et al. Clinical relevance of cytogenetics abnormalities in adult patients with acquired aplastic anaemia. Br J Haematol 2006;134:95–9.

[62] Scheinberg P, Wu CO, Nunez O, Young NS. Predicting response to immunosuppressive therapy and survival in severe aplastic anaemia. Br J Haematol 2009;144:206–16.

[63] Maciejewski JP, Risitano A, Sloand EM, Nunez O, Young NS. Distinct clinical outcomes for cytogenetic abnormalities evolving from aplastic anemia. Blood 2002;99:3129–35.

[64] Hosokawa K, Katagiri T, Sugimori N, et al. Favorable outcome of patients who have 13q deletion: a suggestion for revision of the WHO 'MDS-U' designation. Haematologica 2012;97:1845–9.

[65] Katagiri T, Sato-Otsubo A, Kashiwase K, et al. Frequent loss of HLA alleles associated with copy number-neutral 6pLOH in acquired aplastic anemia. Blood 2011;118:6601–9.

[66] Narita A, Muramatsu H, Sekiya Y, Japan Childhood Aplastic Anemia Study Group. et al. Paroxysmal nocturnal hemoglobinuria and telomere length predicts response to immunosuppressive therapy in pediatric aplastic anemia. Haematologica 2015;100:1546–52.

[67] Scheinberg P, Cooper JN, Sloand EM, Wu CO, Calado RT, Young NS. Association of telomere length of peripheral blood leukocytes with hematopoietic relapse, malignant transformation, and survival in severe aplastic anemia. JAMA 2010;304:1358–64.

[68] Sugimori C, Chuhjo T, Feng X, et al. Minor population of CD55– CD59– blood cells predicts response to immunosuppressive therapy and prognosis in patients with aplastic anemia. Blood 2006;107:1308–14.

[69] Kulagin A, Lisukov I, Ivanova M, et al. Prognostic value of paroxysmal nocturnal haemoglobinuria clone presence in aplastic anaemia patients treated with combined immunosuppression: results of two-centre prospective study. Br J Haematol 2014;164:546–54.

[70] Yoshizato T, Dumitriu B, Hosokawa K, et al. Somatic mutations and clonal hematopoiesis in aplastic anaemia. New Engl J Med 2015;373:35.

[71] Marsh JC, Kulasekararaj A. Management of the refractory aplastic anemia patient: what are the options? Blood 2013;122:3561–7.

[72] Scheinberg P, Nunez O, Young NS. Retreatment with rabbit anti-thymocyte globulin and ciclosporin for patients with relapsed or refractory severe aplastic anaemia. Br J Haematol 2006;133:622–7.

[73] Clé DV, Atta EH, Dias DS, et al. Repeat course of rabbit antithymocyte globulin as salvage following initial therapy with rabbit antithymocyte globulin in acquired aplastic anemia. Haematologica 2015;100(9):e345–7.

[74] Socié G, Henry-Amar M, Bacigalupo A, et al. Malignant tumors occurring after treatment of aplastic anemia. European Bone Marrow Transplantation-Severe Aplastic Anaemia Working Party. New Engl J Med 1993;329:1152–7.

[75] Young NS, Calado RT, Scheinberg P. Current concepts in the pathophysiology and treatment of aplastic anaemia. Blood 2006;108:2509–19.

[76] Kulasekararaj AG, Jiang J, Smith AE, et al. Somatic mutations identify a subgroup of aplastic anemia patients who progress to myelodysplastic syndrome. Blood 2014;124:2698–704.

[77] Mufti GJ, Kulasekararaj AG, Marsh JC. Somatic mutations and clonal hematopoiesis in aplastic anemia. New Engl J Med 2015;373:1674–5.

[78] Sloand EM, Pfannes L, Chen G, et al. CD34 cells from patients with trisomy 8 myelodysplastic syndrome (MDS) express early apoptotic markers but avoid programmed cell death by up-regulation of antiapoptotic proteins. Blood 2007;109:2399–405.

[79] Dumitriu B, Feng X, Townsley DM, et al. Telomere attrition and candidate gene mutations preceding monosomy 7 in aplastic anemia. Blood 2015;125:706–9.

[80] (a) Rosenfeld S, Follman D, Nunez O, Young NS. Antithymocyte globulin and cyclosporine for severe aplastic anaemia. Association between hematologic response and long term outcome. JAMA 2003;289:1130.
(b) Bacigalupo A et al, for the Aplastic Anemia Working Party of the European Group for Blood and Marrow Transplantation (WPSAA-EBMT). Current outcome of HLA identical sibling vs. unrelated donor transplants

in severe aplastic anemia: an EBMT analysis. Haematologica 2015;100:696–702.

[81] Marsh JC, Gupta V, Lim Z, et al. Alemtuzumab with fludarabine and cyclophosphamide reduces chronic graft versus host disease after allogeneic stem cell transplantation for acquired aplastic anemia. Blood 2011;118:2351–7.

[82] Grimaldi F, Barber L, Perez-Abellan P, et al. King's College Hospital FCC conditioning for severe aplastic anemia induces tolerance with mixed T-cell chimerism and extremely low incidence of GVHD. Blood 2014;124:1594.

[83] Samarasinghe S, Iacobelli S, Knol C, et al. Impact of different in vivo T cell depletion strategies on outcomes following hematopoietic stem cell transplantation for idiopathic aplastic anaemia: A study on behalf of the EBMT SAA Working Party. Blood 2015;126(23):1210.

Identical Sibling Donor Transplantation

G. Aldawsari, H. Alzahrani, M.D. Aljurf

Adult Hematology and Bone Marrow Transplantation, Oncology Center, King Faisal
Specialist Hospital and Research Center, Riyadh, Saudi Arabia

INTRODUCTION

Allogeneic hematopoietic stem cell transplantation (HSCT) is considered the standard treatment option for young patients with severe and very severe aplastic anemia (AA). The improved outcome of identical sibling (SIB) transplantation in AA that was observed in the last two decades had encouraged many centers worldwide to consider identical SIB transplantation for middle-age and selected older age patients.

This chapter will discuss the issues related to identical SIB HSCT in AA including conditioning regimens, stem-cell source, immunosuppression after transplantation, and issues related to posttransplantation follow up.

INDICATION FOR IDENTICAL SIBLING DONOR TRANSPLANTATION

HSCT provides curative therapy for the majority of patients with AA.

A randomized prospective trial from Seattle showed a survival benefit of matched related donor HCST over standard-of-care (supportive transfusions and androgen treatment) [1].

Candidates for upfront allogeneic HLA identical SIB HCST include patients younger than 40–50 years of age and selected biologically fit older patients with predictors of poor response to immunosuppressive therapy. HSCT is also recommended as a second line treatment for older patients who have failed previous immunosuppressive therapy [2–4].

Older age alone is not a contraindication for allogeneic HSCT, but age has been restricted to patients younger than 40–50 years because of the concerns about the potential toxicity of conditioning regimen and higher risk of GVHD in older patients.

The cure rate of allogeneic HSCT for younger patients now approaches 90% due to advances in supportive care and standardization of conditioning regimen [2,4].

The advantage of allogeneic HSCT over immunosuppressive therapy is related to the marked reduction in the risk of relapse and development of late clonal disorders, such as myelodysplastic syndrome and paroxysmal nocturnal hemoglobinuria. The risk of acute- and chronic graft versus host disease (GVHD) remains a challenge after HSCT [5].

A report from the European Group for Blood and Marrow Transplantation (EBMT) of over 1500 patients from 1991 to 2002 confirmed that predictors of survival following HSCT includes matched SIB donor, recipient age of less than 16 years, early HSCT (time from diagnosis to HSCT of less than 83 days), and a nonradiation-based conditioning regimen. The current survival data from pediatric patients' age (<16 years) receiving HSCT from identical SIB donor is 91%, significantly better than survival for patients over 16 years of age (74%) [4].

Currently, most of the centers offer upfront allogeneic transplantation for severe AA patients with an available matched SIB donor up to the age of 50 years and selected patients older than 50 years who are otherwise in excellent health with disease features suggestive of low likelihood of response to immunosuppressive therapy, such as very severe AA [6].

CONDITIONING REGIMEN

In severe AA, the goal of transplantation is to achieve optimal successful engraftment without acute- or chronic GVHD.

Graft failure is a major concern in transplantation for AA, occurring more frequently than many other allogeneic HSCT indications. As AA is an autoimmune disease, the antihematopoietic immune activity in the host can reject the graft by the same mechanism that attacks the recipient's stem cell and causes aplasia. Furthermore, most of the conditioning regimens used for AA HSCT are nonmyeloablative leading to a relatively high risk of graft rejection [7].

The occasional observation of cure of the aplasia with autologous recovery after transplantation had promoted the development of high dose cyclophosphamide (Cy) administration without grafting, pioneered by Brodsky et al. [8,9].

In the early experience with matched-related transplants in 1970s, using single agent Cy for conditioning and methotrexate (MTX) alone for GVHD prophylaxis, the incidence of graft failure was up to 30% and this was more evident in the previously transfused patients [9,10].

However, the rate of graft failure had fallen over in the previous years with the significant changes in the conditioning regimen and GVHD prophylaxis [5].

In the initial effort to reduce graft failure, radiation was added to the conditioning regimen and resulted in lower rates of graft failure, as documented in several series [11]; including the European Bone Marrow Transplant Registry (EBMT) and International Bone Marrow Transplantation Registry (IBMTR) studies. However, radiation was associated with significant early and late toxicity, including secondary malignancies with no survival advantage [12–14].

The addition of antithymocyte globulin (ATG) in the conditioning regimen pioneered by Smith et al. in Boston and by the Seattle group and the subsequent nonrandomized studies had resulted in a lower incidence of GVHD and improved survival compared with historical controls who received Cy-only conditioning [15–18].

The only prospective randomized trial published [19] had randomized 134 patients to receive Cy alone or in combination with ATG. All patient received HLA matched SIB transplantation. With a median follow up of 6 years, the 5-year survival was 74% for the Cy alone group and 80% for the Cy and ATG group, but this was not statistically significant ($P = 0.44$).

Therefore, the combination of Cy 50 mg/kg/day for 4 days and ATG conditioning is considered the standard conditioning regimen for younger patients with AA undergoing an HLA identical SIB transplantation with survival ranging from 65% to 95%, depending on the patient age [19]. This conditioning regimen was associated with a very low incidence of secondary malignancies [20] and preservation of fertility [21].

For older patients who are potential transplantation candidates, the best conditioning regimen is not known. Patients who are older than 40 years of age and who are medically fit enough for HSCT may receive a reduced intensity-conditioning regimen, using combination of Cy, fludarabine (FLU), and either ATG or alemtuzumab [22–23].

In a retrospective EBMT study [22], patients older than 30 years receiving FLU-based regimens were compared with a matched paired group of patients conditioned with Cy 200 mg/ kg over the same period of time (1998–2007). Patients conditioned with FLU had a higher probability of overall survival than control group ($P = 0.04$). The incidence of graft failure was reduced in patients receiving FLU (0% vs. 11%, $P = 0.09$), but no difference was observed in GVHD outcomes.

A recent UK study has compared the use of ATG with the use of alemtuzumab for in vivo T-cell depletion. One hundred patients received alemtuzumab and 55 patients, ATG-based regimens; the donor was a matched SIB in 56%, a matched unrelated donor (URD) in 39%, and other related or mismatched URD in 5% of patients. Engraftment failure occurred in 9% of the alemtuzumab group and 11% of the ATG group [24]. The overall survival was similar for SIB transplant using alemtuzumab or ATG (91% vs. 85%, respectively; $P = 0.562$).

SYNGENEIC STEM CELL TRANSPLANTATION IN APLASTIC ANEMIA

The availability of a genetically identical twin as a stem-cell donor for patients with AA is rarely encountered. This option is very appealing as it carries the least reported toxicities of all types of stem-cell transplantation. Although there are reports of its use without conditioning regimen and without GVHD prophylaxis further minimizing toxicities of the procedure of stem-cell transplantation, this seems to carry a high rate of graft failure and retransplantation. Several case reports and case series have been reported in the literature on the use of syngeneic stem cell transplantation in AA [25,26]. Recent analyses of the largest retrospective case series of severe AA and pediatric diseases working parties of the EBMT about syngeneic transplantation in AA have reported the outcome of 88 patients who received 113 transplants. Conditioning regimen was given in three-quarters of patients and 50% of patients received posttransplant immunosuppression therapy. The trend in the recent years is more toward using peripheral blood stem cells (PBSC) than bone marrow (BM) as a stem source. Overall survival at 10 years in this report was 93%. Graft failure was seen in one-third of patients and was associated with the avoidance of conditioning regimen, the use of BM as stem-cell source and lack of posttransplant immunosuppression therapy [27].

THE SOURCE OF THE STEM CELLS

Unmanipulated BM should be used as a stem-cell source for all patients with AA, as the use of PBSC is associated with increased risk of chronic GVHD.

Survival advantage for BM as a stem-cell source was confirmed in all age groups [28].

Suitable BM cell dose is recommended, as the results of transplantation are highly dependent on number of nucleated cells infused. It is recommended that at least 3×10^8 mononuclear cells/kg or 2×10^6 CD 34+ cells/kg of the recipient should be given.

The successful use of PBSC has been reported in AA. However, in a retrospective EBMT study, despite the faster engraftment with PBSC, it was associated with increased incidence of chronic GVHD and a significantly lower 2-year survival in patients <20 years from 85% to 73%, and in patients >20 years from 64% to 52% [29,30].

In 2012, EBMT published retrospective analysis of 1886 patients with AA who received a first matched SIB transplant between 1999 and 2009, with BM (n = 1163) or peripheral blood (PB) (n = 723) as a stem-cell source; PB grafts from HLA identical SIBs were associated with a greater risk of grade II–IV acute GVHD compared with BM grafts (17% vs. 11%, $P < 0.001$) and a greater risk of chronic GVHD (22% vs. 11%, $P < 0.001$) [29].

Therefore, PBSC should not be used in patients with AA as a stem-cell source and its use as an alternative stem-cell source should only be considered if BM harvest is not feasible or if the donor is not willing to donate BM.

There is accumulating evidence that GCSF-primed BM (G-BM) is associated with faster engraftment than unmanipulated BM and with lower incidence of GVHD in comparison with PBSC. This was looked at in several retrospective studies of which, six studies compared G-BM with BM and three studies compared G-BM with PB from related donor transplant in patients with hematologic malignancies [31–37].

In another retrospective CIBMTR study, three different stem-cell sources were compared in patients transplanted for AA from 1997–2003, namely BM (n = 547), PB (n = 134), and G-BM (n = 78).

The median times for neutrophil recovery were 15, 13, and 20 days after G-BM-, PB-, and BM transplantation, respectively. In multivariate analysis, the likelihood of achieving neutrophil recovery at 30 days after transplantation was similar in the three groups. Platelet recovery after transplantation of G-BM was slower compared to BM ($P = 0.015$) and PB ($P < 0.001$). Median times to platelet recovery were 26, 19, and 31 days after G- BM-, PB-, and BM transplantation, respectively.

At day 100, the cumulative incidence of grade III–IV acute GVHD were 6, 12, and 6 % after G-BM-, PB-, and BM transplantation, respectively.

Grade II–IV acute- and chronic GVHD were higher after PBSC as compared with G-BM and BM [38,39].

Therefore, G-BM should be considered in patients with AA when the expected yield of BM harvest is low, in the presence of major ABO incompatibility requiring red cell depletion, and when significant weight difference between donor and recipient is present. The routine use of G-BM should be restricted until further evidence is available.

POSTTRANSPLANTATION IMMUNOSUPPRESSION

Adequate posttransplantation immunosuppression is important not only for the prevention of GVHD but also to secure adequate suppression of the host immune system and prevention of graft rejection [6].

In the initial studies, MTX alone was used for GVHD prophylaxis and subsequently, cyclosporine (CsA) was introduced in 1980s and resulted in decreased transplant-related mortality (TRM), significantly reduced rejection rates, and improved survival [40]. In a prospective randomized trial comparing CsA + MTX with CsA alone, the TRM rate for patients given CsA/MTX or CsA alone were 3% and 15%, respectively.

The 5-year probability of survival was 94% in the CsA/MTX group and 78% for those in the CsA alone group [41].

The combination of CsA + MTX should be considered the standard posttransplant immunosuppression.

In the absence of a prospective, randomized comparison, no other regimen has shown clear superiority over the combination of calcineurin inhibitor (CsA or tacrolimus) and MTX. In a retrospective analysis of 949 patients reported to CIBMTR, including patients who had allogeneic BMT or PBSCT from HLA identical SIB or from URD for AA, the use of tacrolimus + MTX was associated with a lower risk of mortality among URD recipients and with slightly earlier neutrophil recovery among SIB recipients, with no

TABLE 7.1 Key Points Regarding Identical Sibling (SIB) Donor Transplantation in Aplastic Anemia (AA)

Indication for allogeneic HLA identical HSCT in severe AA:

- Patients younger than 40–50 years of age
- Older patients who have failed immunosuppressive therapy or with predictors of poor response to immunosuppressive therapy

Stem-cell source:

- BM is the recommended stem-cell source
- PB should not be used
- G-BM is the optimal stem-cell source to use in case of major ABO incompatibility

Conditioning regimen:

- Cy/ATG is the standard conditioning regimen in young patients
- FLU-based conditioning is a possible option for patients older than 40 years of age

Posttransplant immunosuppressive therapy:

- MTX/CsA is the standard for GVHD prophylaxis in younger patients

ATG, Antithymocyte globulin; BM, bone marow; CsA, cyclosporine; Cy, cyclophosphamide; FLU, fludarabine; G-BM, GCSF-primed bone marrow; GVHD, graft versus host disease; MTX, methotrexate; PB, peripheral blood.

statistically significant differences in other outcomes with the two regimens (CsA + MTX or tacrolimus + MTX) of prophylaxis [42].

There is only limited experience with ex vivo T-cell depletion in transplantation for AA and most of the data suggest a higher risk of graft failure [43]. Therefore, this cannot be recommended outside clinical trials.

For patient with renal impairment or other contraindications for the use of a calcineurin inhibitor for GVHD prophylaxis, the use of mycophenolate mofetil can be considered as an alternative. Two cases of SAA with significant renal impairment were reported in the literature and demonstrated that BMT is feasible with the use of mycophenolate mofetil for GVHD prophylaxis [44].

Table 7.1 summarizes of key points regarding identical SIB donor transplantation in AA.

POSTTRANSPLANT CARE

Chimerism

Chimerism is an indispensable tool to monitor the kinetics of engraftment and the ultimate fate of the graft. But whether chimerism analysis is useful to predict future graft rejection is less well known.

Engraftment and graft failure have been a major issue in HLA identical SIB transplants for acquired AA. Transient mixed chimerism is common in patients after marrow allografts for AA. In one study, 60% of patients had mixed chimerism in PB or BM after HSCT, and two-thirds of these eventually converted to complete donor-type hematopoietic cells while the remainder rejected their grafts [45]. In a larger study of 116 patients with AA transplanted from sex mismatched HLA identical SIBs, 45% had mixed chimeras detected in either blood or marrow. While patients with mixed chimerism had a higher incidence of graft rejection (14%) than those who were complete chimeras (9%), this was not statistically significant [46].

In another study on 94 SAA transplants, patients were classified as: (1) complete donor chimeras (43%); (2) transient mixed chimeras (16%); (3) stable mixed chimeras (20%); (4) progressive mixed chimeras (15%); and (5) early graft rejection (5%). This study showed that mixed chimerism occurs in a large proportion of patients and that the kinetics of sequential chimerism is a predictor of outcome [47].

Treatment of mixed chimerism has not been standardized; mixed chimerism associated with declining PB counts may be treated with low doses of donor lymphocyte infusion while maintaining GVHD prophylaxis with CsA, however, there is very limited published data on this approach.

In the aforementioned study, 9 patients with SAA had been treated with a total of 42 donor lymphocyte infusion for mixed chimerism: 6 patients achieved complete donor chimerism (100%) in BM cells and 4 in CD3+ cells; 2 patients died—1 of rejection and 1 of GVHD; and 7 patients survived 2–10 years after donor lymphocyte infusion.

Therefore, serial chimerism testing is needed after HSCT for all SAA patients after HSCT transplantation in SAA.

Survival

Overall survival for AA has increased over the last 30 years. In 1970s, survival rates of 40–60% were commonly seen, while the survival rates now range from 60% to 100% [48,49].

CIBMTR data for 1699 patients receiving HLA identical SIB transplantation for SAA between 1991 and 1997 showed a 5-year probability of survival (95% confidence interval) of 75 ± 3% for 874 patients ≤20 years of age, 68 ± 4% for 696 who were 21–39 years, and 35 ± 18% for 129 who were 40 years or older [50].

EBMT data for 1799 patients receiving HLA identical SIB transplantation between 1971 and 1998 confirmed CIBMTR data. The years of transplant and the age of the patient predicted outcome after HSCT. There has been striking improvement in the 5-year survival rates comparing patients transplanted before or after 1990 [51].

In 2012, CIBMTR reported the late effect of 1718 patients who had an allogeneic HSCT for acquired AA in 186 centers (1176 matched SIB donors and 542 URDs) in the period between 1995 and 2006. The median follow up was 70 months for matched SIB cases. The overall survival at 1, 2, and 5 years for the entire cohort was 76, 73, and 70%. Causes of death among the matched SIB HSCT group included: 32% secondary to infection, 22% organ failure, 14% graft rejection, 33% hemorrhage, and 10% death secondary to GVHD [52].

Fertility

Gonadal function in patients conditioned only with Cy often return to normal. Among 65 women with age between 13 and 25 years who received Cy-only conditioning, all had evidence of recovery of ovarian function, whereas among women aged 26–38 years, 37% developed primary ovarian failure [53–54]. Testicular function had returned to normal in most men aged 14–41 years who received Cy-only conditioning in one study [55].

Secondary Malignancy

Secondary malignancies following AA HSCT are traditionally uncommon. Analysis of transplant results in 700 patients with AA transplanted using total body irradiation-based conditioning showed an estimated incidence of secondary malignancies of 14% at 20 years [56]. However this incidence dropped to 1.4% at 10 years with regimen not containing radiation [57]. These findings were again confirmed the aforementioned CIBTMR study done in 2012 [52].

References

[1] Camitta BM, Thomas ED, Nathan DG, et al. Severe aplastic anemia: a prospective study of the effect of early marrow transplantation on acute mortality. Blood 1976;48:63–70.

[2] Armand P, Antin JH. Allogeneic stem cell transplantation for aplastic anemia. Biol Blood Marrow Transplant 2007;13:505–16.

[3] Locasciulli A, Oneto R, Bacigalupo A, et al. Outcome of patients with acquired aplastic anemia given first line bone marrow transplantation or immunosuppressive treatment in the last decade; a report from the European

Group for Blood and Marrow Transplantation (EBMT). Hematologica 2007;92:11–8.

[4] Bacigalupo A, Brand R, Oneto R, Bruno B, socie G, Passweg J, et al. Treatment of acquired severe aplastic anemia: bone marrow transplantation compared with immunotherapy—the European Group for Blood and Marrow Transplantation experience. Semin Hematol 2000;37:69–80.

[5] Amy E, DeZern, Eva C, Guinan. Aplastic anemia in adolescents and young adults. Acta Haematol 2014;132:331–9.

[6] AL-Jurf M, AL-Zahrani H, Van Lint MT, Passweg JR. Standard treatment of acquired SAA in adult patients 18–40 years old with an HLA-identical sibling donor. Bone Marrow Transplant 2013;48:178–9.

[7] Armand P, Antin JH. Allogeneic stem cell transplantation for aplastic anemia. Biol Blood Marrow Transplant 2007;13(5):505–16.

[8] Territo MC. Autologous bone marrow repopulation following high dose cyclophosphamide and allogeneic marrow transplantation in aplastic anemia. Br J Haematol 1977;36:305–12.

[9] Brodsky RA, Sensenbrenner LL, Smith BD, et al. Durable treatment-free remission after high dose cyclophosphamide therapy for previously untreated severe aplastic anemia. Ann Intern Med 2001;135:477–83.

[10] Story R, Longton G, Anasetti C, et al. Changing trends in marrow transplantation for aplastic anemia. Bone Marrow Transplant 1992;10(Suppl. 1):45–52.

[11] Gale RP, Ho W, Feig S, et al. Prevention of graft rejection following bone marrow transplantation. Blood 1980;57:9–12.

[12] McCann SR, Bacigalupo A, Gluckman E, et al. Graft rejection and second bone marrow transplants for acquired aplastic anaemia: a report from the Aplastic Anemia Working party of the European Bone Marrow Transplant Group. Bone Marrow Transplant 1994;13:233–7.

[13] Champlin RE, Horowitz MM, van Bekkum DW, et al. Graft failure following bone marrow transplantation for aplastic anemia: risk factors and treatment results. Blood 1989;73:606–13.

[14] Gluckman E, Horowitz MM, Champlin RE, et al. Bone marrow transplantation for severe aplastic anemia influence of conditioning and graft-versus-host-disease prophylaxis regimen on outcome. Blood 1992;79: 269–75.

[15] Deeg HJ, Socie G, Schoch G, et al. Malignancies after marrow transplantation for aplastic anemia and Franconia anemia: a joint Seattle and Paris analysis of results in 700 patients. Blood 1996;87:386–93.

[16] Smith BR, Guinan EC, Parkman R, et al. Effecacy of a cyclophosphamide-procarbazine-antithymocyte serum regimen for prevention of graft rejection following bone marrow transplantation for transfused patients with aplastic anemia. Transplantation 1985;39:671–3.

[17] Storb R, Etizioni R, Anasetti C, et al. Cyclophosphamide combined with antithymocyte globulin in preparation for allogeneic marrow transplants in patients with aplastic anemia. Blood 1994;84:941–9.

[18] Storb R, Weiden PL, Sullivan KM, et al. Second marrow transplants in patients with aplastic anemia rejecting the first graft: use of a conditioning regimen including cyclophosphamide and antithymocyte globulin. Blood 1987;70:116–21.

[19] Champlin RE, Perez WS, Passweg JR, Klein JP, Camitta BM, Gluckman E, et al. Bone marrow transplantation for severe aplastic anemia: a randomized controlled study of conditioning regimens. Blood 2007;109: 4582–5.

[20] Socie G, Henry-Amar M, Bacigalupo A, et al. Malignant tumors occurring after treatment of aplastic anemia. New Engl J Med 1993;329:1152–7.

[21] Hinterberger-Fischer M, Kier P, Kaths P, et al. Fertility, pregnancies and offspring complications after bone marrow transplantation. Bone Marrow Transplant 1991;7(1):5–9.

[22] Maury S, Bacigalupo A, Anderlini P, Aljurf M, March J, Socié G, et al. Improved outcome of patients older than 30 years receiving HLA-identical sibling hematopoietic stem cell transplantation for severe acquired aplastic anemia using fludarabine-based conditioning regimen: a comparison with conventional conditioning regimen. Haematologica 2009;94:1312–5.

[23] AL-Zahrani H, Nassar A, AL-Mohareb F, AL-Sharif F, Mohamed S, AL-Anazi K, et al. Fludarabine-based conditioning chemotherapy for allogeneic hematopoietic stem cell transplantation in acquired severe aplastic anemia. Biol Blood Marrow Transplant 2011;17:717–22.

[24] March JC, Pearce RM, Koh MB, British Society for Blood and Marrow Transplantation, Clinical Trias Committee. et al. Retrospective study of alemtuzumzb vs. ATG-based conditioning regimen without irradiation for unrelated and matched sibling donor transplants in acquired severe aplastic anemia: a study from the British Society for Blood and Marrow Transplantation. Bone Marrow Transplant 2014;49(1):42–8.

[25] Ghavamzadeh A, Alimonghaddam K, Ghaffari F, Derakhshandeh R, Jalali A, Jahani M. Twenty years of experince on stem cell transplantation in Iran. Iran Red Crescent Med J 2013;15(2):93–100.

[26] Ladeb S, Abdelkefi A, Torijman L, Ben Neji H, Lakhal A, Kaabi H, Ben Hamed L, Ennigrou S, Hmida S, Ben Othman T, Ben Abdeladhim A. Allogeneic hematopoietic stem cell transplantation for acquired aplastic anemia using cyclophosphamide and antithymocyte globulin: a single center experince. Bone Marrow Transplant 2009 [Epub 2009 Jul 27].

[27] Gerull S, Stern M, Apperley J, Beelen D, Brinch L, Bunjes D, Butler A, Ganser A, Ghavamzadeh A, Koh MB, Komarniger M, Maaertens J, Maschan A, Peters C, Rovira M, Sengelov H, Socie G, Tischer J, Oneto R, Passweg J, Marsh J. Syngeneic transplantation in aplastic anemia: pre-transplant conditioning and peripheral blood are associated with improved engraftment: an observational study on behave of the Severe Aplastic Anemia and Pediatric Diseases Working Parties of the European Group for Blood and Marrow Transplantation. Hematologica 2013;98(11):1804–9.

[28] Bacigalupo A, Socié G, Schrezenmeier H, Tichelli A, Locasciulli A, Fuehrer M, et al. Bone marrow versus peripheral blood as the stem cell source for sibling transplants in acquired aplastic anemia: survival advantage for bone marrow in all age group. Haematologica 2012;97:1142–8.

[29] March JC, Gupta V, Lim Z, Ho AY, Ireland RM, Hayden J, et al. Alemtuzumab with fludarabine and cyclophosphamide reduced chronic graft-versus-host disease after allogeneic stem cell transplantation for acquired aplastic anemia. Blood 2011;118:2351–7.

[30] Schrezenmeier H, Passweg JR, March JC, Bacigalupo A, Bredeson CN, Bullorsky E, et al. Worse outcome and more chronic GVHD with peripheral blood progenitor cells than bone marrow in HLA-matched sibling donor transplants for young patients with severe acquired aplastic anemia. Blood 2007;110: 1397–400.

[31] Deotare U, AL-Dawsari G, Couban S, Lipton JH. G-CSF-primed bone marrow as a source of stem cells for allografing: revisiting the concept. Bone Marrow Transplant 2015;1–7.

[32] Couban S, Messner HA, Andreou P, Egan B, Price S, Tinker L, et al. Bone marrow mobilized with granulocyte colony-stimulating factor in related allogeneic transplantation. Biol Blood Marrow Transplant 2000;6:422–42.

[33] Isola L, Seigliano E, Fruchtman S. Long-term follow-up after allogeneic granulocyte colony-stimulating factor—primed bone marrow transplantation. Biol Blood Marrow Transplant 2000;6:428–33.

[34] Ji SQ, Chen HR, Wang HX, Yan HM, Pan Sp, Xun CQ. Comparison of outcome of allogeneic bone marrow transplantation with and without granulocyte colonystimulating factor (lenograstim) donor-marrow priming in patients with chronic myelogenous leukemia. Biol Blood Marrow Transplant 2002;8:261–7.

[35] Serody JS, Sparks SD, Lin Y, Capel EJ, Bigelow SH, Kirby SL, et al. Comparison of granulocyte colony-stimulating factor (G-CSF)—mobilized peripheral blood progenitor cells and G-CSF—stimulated bone marrow as a source of stem cells in HLA-matched sibling transplantation. Biol Blood Marrow Transplant 2000;6:434–40.

[36] Morton J, Hutchins C, Durrant S. Granulocyte-colony-stimulating factor (G-CSF)-primed allogeneic bone marrow: significant less graft-versus-host-disease and comparable engraftment to G-CSF-mobilized peripheral blood stem cells. Blood 2001;98:3186–91.

[37] Elfenbein GJ, Sacktein R, Oblon DJ. Do G-CSF mobilized, peripheral blood-derived stem cells from healthy, HLA-identical donors really engraft more rapidly than do G-CSF primed, bone marrow-derived stem cells? Blood Cells Mol Dis 2004;32:106–11.

[38] Schrezenmeier H, Bredeson C, Bruno B, et al. Comparison of allogeneic bone marrow and peripheral blood stem cell transplantation for aplastic anemia: collaborative study of European Blood and Marrow Transplant Group (EBMT) and International Bone Marrow Transplant Registry (IBMTR). Blood 2003;102:267a.

[39] Chu R, Brazauskas R, Kan F, Bashey A, Bredeson C, Carmitta B, et al. Comparison of outcome after transplantation of G-CSF-stimulated bone marrow grafts versus bone marrow or peripheral blood grafts from HLA-matched sibling donors for patients with severe aplastic anemia. Biol Blood Marrow Transplant 2011;17:1018–24.

[40] Passweg JR, Socié G, Hinterberger W, et al. Bone marrow transplantation for severe aplastic anemia: Has outcome improved? Blood 1997;858–64.

[41] Locatelli F, Bruno B, Zecca M, Van-lint MT, McCann S, Arcese W, et al. Cyclosporin A and short-term methotrexate versus cyclosporin A as graft versus host disease prophylaxis in patients with severe aplastic anemia given allogeneic bone marrow transplantation from an HLA-identical sibling: results of GITMO/EBMT randomized trial. Blood 2000;96:1690–7.

[42] Inamoto Y, Mary ED, Flowers, Tao Wang, Alvaro Urbano-Ispizua, Michael T, Gale, Gupta Vikas, Betty K, Hamilton, Mohamed A, Kharfan-Dabaja, David I, Marks, Olle TH, Ringden, Gerard Socie, Melhem M, Solh, Gorgun Akpek, Mitchell S, Cairo, Nelson J, Chao, Robert J, Hayashi, Taiga Nishihori, Ran Reshef, Ayman Saad, Ami Shah, Takanori Teshima, Martin S, Tallman, Baldeep Wirk, Stephen R, Spellman, Mukta Arora, Paul J, Martin. Tacrolimus versus cyclosporine after hematopoietic cell transplantation for acquired aplastic anemia. Biol Blood Marrow Transplant 2015;1–7.

[43] Deeg HJ, Seidel K, Casper J, et al. Marrow transplantation from unrelated donors for patients with severe aplastic anemia who failed immunosuppressive therapy. Biol Blood Marrow Transplant 1999;5:243–52.

[44] Geerrie A, March J, Lipton JH, Messner H, Gupta V. Bone marrow transplantation for severe aplastic anemia with significant renal impairment. Bone Marrow Transplant 2007;39:311–3.

[45] Hill RS, Petersen FB, Storb R, et al. Mixed hematopoietic chimerism after allogeneic marrow transplantation

for severe aplastic anemia is associated with a higher risk of graft rejection and a lessened incidence of acute graft-versus-host disease. Blood 1986;67:811–6.

[46] Huss R, Deeg HJ, Gooley T, et al. Effect of mixed chimerism on graft-versus-host disease, disease recurrence, and survival after HLA-identical marrow transplantation for aplastic anemia or chronic myelogenous leukemia. Bone Marrow Transplant 1996;18:767–76.

[47] Lawler M, McCann SR, Marsh JC, Severe Aplastic Anaemia Working Party of the European Blood and Marrow Transplant Group. et al. Serial chimerism analysis indicates that mixed haematopoietic chimerism influences the probability of graft rejection and disease recurrence following allogeneic stem cell transplantation (SCT) for severe aplastic anemia (SAA): indication for routine assessment of chimerism post SCT for SAA. Br J Haematol 2009;144:933–45.

[48] Storb R, Thomas ED, Buckner CD, et al. Allogeneic marrow grafting for treatment of aplastic anemia. Blood 1974;43:157–80.

[49] Bortin MM, Rimm AA. Treatment of 144 patients with severe aplastic anemia using immunosuppression and allogeneic marrow transplantation: a report from International Bone Marrow Registry. Transplant Proc 1981;13:227–33.

[50] Horowitz MM. Current status of allogeneic bone marrow transplantation in acquired severe aplastic anemia. Semin Hematol 2000;37:30–42.

[51] Bacigalupo A, Brand R, Oneto R, et al. Treatment of acquired severe aplastic anemia: bone marrow transplantation compared with immunosuppressive therapy. The European Group for Blood and Marrow Transplantation experience. Semin Hematol 2000;37:69–80.

[52] Buchbinder D, Nugent DJ, Brazauskas R, Wang Z, Aljurf M, Cairo MS, Chow R, Duncan C, Eldjerou LK, Gupta V, Hale GA, Halter J, Hayes-Lattin BM, Hsu JW, Jacobshon DA, Kamble RT, Kasow KA, Lazarus HM, Mehta P, Myers KC, Parsons SK, Passweg JR, Pidala J, Reddy V, Carmen M, Savani BN, Seber A, Sorror ML, Steinberg A, Wood William A, Wall DA, Winiarski JH, Yu LC, Majhail NS. Late effects in hematopoietic cell transplant reciepients with acquired severe aplastic anemia: a report from the Late Effect Working Committee of the Center for International Blood and Marrow Transplant Research. Biol Blood Marrow Transplant 2012;18:1776–84.

[53] Hinterberger-Fischer M, Kier P, Kalhs P, et al. Fertility, pregnancies and offspring complication after bone marrow transplantation. Bone Marrow Transplant 1991;7:5–9.

[54] Schmidt H, Ehninger G, Dopfer R, Waller HD. Pregnancy after bone marrow transplantation for severe aplastic anemia. Bone Marrow Transplant 1987;2:329–32.

[55] Sanders JE. Seattle Marrow Transplant Team. The impact of marrow transplant preparative regimens on subsequent growth and development. Semin Hematol 1991;28:244–9.

[56] Deeg HJ, Socie G, Schoch G, et al. Malignancies after marrow transplantation for aplastic anemia and fanconi anemia: a joint Seattle and Paris analysis of result of 700 patients. Blood 1996;87:386–92.

[57] Sullivan KM. Comulative incidence of secondary solid malignant tumors in aplastic anemia patients given marrow graft after conditioning with chemotherapy alone. Blood 1992;79:289–92.

Unrelated Donor Transplants for Acquired Aplastic Anemia

A. Bacigalupo*, S. Sica*, M.T. Van Lint**,
C. Dufour†

*Istituto di Ematologia, Policlinico Universitario A. Gemelli, Universita' Cattolica
del Sacro Cuore, Roma, Italy; **Divisione di Ematologia e Trapianto di Midollo
Osseo, IRCCS AOU San Martino IST, Genova, Italy; †Hematology Unit, G. Gaslini
Children's Hospital; Unità di Ematologia Istituto Giannina Gaslini, Genova, Italy

ELIGIBILITY IN ACQUIRED SAA FOR UD TRANSPLANTATION

Failure to respond to one course of immunosuppressive therapy (IST) is reported in the European Group for Blood and Marrow Transplantation (EBMT) guidelines, as an indication for an unrelated donor (UD) transplant, in the absence of an HLA matched sibling [1]: the rate of nonresponse is variable, between 30% and 70% of patients receiving first line antithymocyte globulin (ATG) + cyclosporine (CsA) [1], although the percentage of refractory patients is bound to be reduced to less than 20%, with the introduction of eltrombopag in combination with ATG + CsA [2]. Pediatricians have attempted to use UD transplants upfront, in selected patients, and have compared the outcome with matched sibling donors (MSD) [3]: 2 year survival was in excess of 90% for 24 children undergoing a matched UD graft, similar to survival with MSD. In the same study the authors found a significant survival advantage for patients grafted with UD upfront (91%), as compared to UD transplants in children failing to respond to IST (74%). They also observed a significantly better event-free survival (92%) (events being death, relapse, nonresponse, need for transplant, and clonal events) over those receiving front line IST (40%) [3] thus implying a better quality of survivors, an issue of particular relevance in younger patients. These two studies may change treatment strategies in patients with severe aplastic anemia (SAA): on one hand there seems to be a significant improvement in response rates to IST, with the addition of eltrombopag, and on the other, upfront UD transplants show survival and event-free survival, in excess of 90%, at least in young patients.

Congenital and Acquired Bone Marrow Failure
http://dx.doi.org/10.1016/B978-0-12-804152-9.00008-7

However, for the time being, it seems reasonable to reserve upfront transplants for selected children with very severe aplastic anemia (vSAA) and a 10/10 matched UD; in all other instances, patients above the age of 18, one would follow current guidelines, that is, a course of IST first: indeed we do not have solid data of upfront UD transplants in adults.

UPPER AGE LIMIT FOR UD TRANSPLANTS

Age is known to be a negative predictor of transplant outcome: however in patients with leukemia, programs have been developed which allow transplants above the age of 70, with transplant-related mortality (TRM) of 10% [4]. This is not the case for patients with SAA and mortality is in the range of 40–50%, above the age of 40: in a Center for International Blood and Marrow Transplant Research (CIBMTR) analysis of MSD transplants, survival was 82, 72, and 53% for

patients aged 0–20, 21–40, and over 40 [5]. One study from the Seattle group analyzed 23 SAA patients over 40 years of age, and mortality was 37% [6]. We have recently looked at the EBMT data set, for patients grafted from either MSD or UD, between 2005 and 2009 (Fig. 8.1): survival is similar in patients aged 1–20 and 21–40 in both donor types, and significantly worse in patients above the age of 40. It is interesting that the 146 patients grafted from MSD have a mortality of 37%, exactly like the small group from Seattle [6], Taken altogether these reports suggest that an allogeneic transplant for SAA over the age of 40 exposes the patient to a risk of death up to 50%: we do not have published data showing that, things have changed in the last 2–3 years. This is strong evidence that one course of IS therapy is clearly indicated above the age of 40, and that transplantation should be reserved for patients not responding to IS therapy.

One could argue that results of transplantation in SAA are worse above the age of 40, precisely because many, if not all of these patients

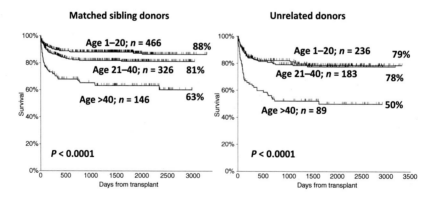

Transplants for acquired SAA; EBMT data 2005–09

FIGURE 8.1 Actuarial survival of patients transplanted between 2005 and 2009 from either HLA identical siblings (left) or unrelated donors (right): a strong age effect is seen in both groups, with patients over the age of 40 doing significantly worse, when compared to patients aged 1–20 or 21–40.

received a course of IS therapy first, and may have come to transplant with prolonged neutropenia and infections: in a separate analysis on EBMT data (unpublished) this does not seem to be the case. In fact the survival disadvantage in UD transplants, above the age of 40 also remains when looking only at patients grafted within 8 months from diagnosis (74% for younger patients vs. 54% for patients over 40 years), and also if grafted within 4 months from diagnosis (65% vs. 36%). Therefore SAA patients above the age of 40 years are fragile, for reasons which are not clear, unlike leukemia and myelodysplastic patients who can be grafted up and beyond the age of 70, with very low mortality [4].

OUTCOME OF PATIENTS ACTIVATING A UD SEARCH

One crucial question is whether finding a UD, after failing to respond to IS therapy, is beneficial. A French-Italian study looked at 295 patients with acquired SAA, who activated a UD search between 1994 and 2005 [7]: for 118 patients one or more matched UD were identified, but not for other 68 patients. A survival advantage could be shown for patients under the age of 17 (79% vs. 53%, $p < 0.01$) and for patients activating the UD search in the period 2000–2005 (74% vs. 47%, $p < 0.01$) [7]. Thus activating a UD search should be one of the first actions taken in a patient with SAA: we will be discussing age limits.

HLA MATCHED OR MISMATCHED DONORS

The four mandatory loci for HLA allele matching are A, B, C, DRB1, and UDs matched with high resolution typing at these four loci are referred to as 8/8 allele matched. Horan and coworkers have compared the outcome of 8/8 matched donors with mismatched donors (7/8 or 6/8) in nonmalignant disorders [8]:

mismatched UD transplants were significantly inferior to 8/8 matched grafts, the major problem being graft failure (GF). In a multicenter study in SAA patients, Deeg and coworkers also reported superior results with 8/8 matched UD as compared to <8/8 matched donors [9].

An EBMT analysis looked at 100 patients grafted with a homogeneous conditioning regimen [10]: among 75 unrelated transplants with full HLA typing, 46 were classified as HLA matched (reported as 8/8 or 10/10 allele matched) and 29 were mismatched, the donor being 1 or more allele mismatched with the recipient. The crude mortality of HLA matched and HLA mismatched transplants was 17% vs. 34% ($p = 0.1$), again confirming an effect of HLA matching in UD transplants for SAA. In patients grafted beyond 2 years with a mismatched UD, crude mortality was 53%.

The issue of permissive or nonpermissive mismatches has been studied in a Japanese paper on pediatric SAA [11]: the authors found that 7/8 matched UD had survival comparable to 8/8 matched donors, and only multiple mismatches were negative predictors of survival [11].

Therefore the aim of the UD search is to find an 8/8 high resolution HLA A, B, C, DRB1 matched donor; a 7/8 donor is possibly acceptable, particularly in the pediatric population. There may be some HLA mismatched which are more acceptable than others.

GRAFT REJECTION AND STEM CELL SOURCE

The issue of graft rejection, also referred to as GF, is a problem in patients with SAA, because of the nature of the disease, and possibly because conditioning regimens are typically nonmyeloablative, based on high dose cyclophosphamide (CY) alone [12]. In earlier days the addition of donor buffy coat was shown to reduce the incidence of rejection, but unfortunately increased the incidence of graft versus host disease (GvHD) [13]. With the increasing use of G-CSF mobilized peripheral blood (PB) stem cells in patients with leukemia, it was thought that changing the stem cell source from BM to PB would solve the problem of rejection, especially in patients undergoing a UD transplant. The CIBMTR has studied 296 SAA patients grafted from UD matched at HLA A, B, C, DRB1 [14]: although neutrophil and platelet recovery was faster after PB UD grafts, the incidence and GF was comparable in BM and PB transplants [14]. However overall mortality was significantly higher due to higher rates of GvHD in PB transplants [14]. Similar conclusions were reached in an EBMT study looking at 1163 BM and 723 PB transplants from sibling donors [15]: the proportion of patients recorded as nonengrafted in BM versus PB was, respectively, 5.5 and 7.2% (p = 0.1); the proportion of engrafted patients was 90.7 and 89.9% (p = 0.5), and patients with secondary GF were 1.3 and 1.4% (p = 0.2). Also in this study the survival after BM grafts was superior to PB grafts G [15].

Therefore there is no good reason to use PB grafts, and the CIBMTR study concludes that "BM is the preferred graft stem cell source for UD transplantation in SAA" [14].

CYCLOPHOSPHAMIDE AND THE CONDITIONING REGIMEN FOR UD TRANSPLANTS

High dose CY(200 mg/kg) has been the standard regimen for SAA grafted from sibling donors since the early 1970s [12] and remains standard of care for young patients undergoing a sibling transplant. However, CY200 mg/kg and ATG do not provide sufficient immunosuppression for sustained engraftment in SAA patients grafted from UD [16]: engraftment can be improved with high dose radiation (1200 cGy), but at the expense of severe toxicity [16] and a high risk of secondary tumors [17,18]. For these reasons Deeg and coworkers opened a study to optimize the dose of total body irradiation (TBI), to be given in combination with CY200 and ATG, for UD transplant in SAA [9]: 87 patients entered the study, which started with TBI at 3×200 cGy, to be escalated/de-escalated in steps of 200 cGy, dependent upon GF/toxicity [9]. The incidence of grade 3–4 toxicities was significantly reduced at lower doses of TBI (200 cGy), which also provided a low rate of GF (2%) with matched UD, and 11% with mismatch UD [9]. Best survival was achieved with a TBI dose of 200 cGy, especially in young patients (<20 years of age).

Having established the optimal dose of TBI, the CIBMTR embarked on a study looking at the optimal dose of CY to combine with fludarabine and TBI 200 cGy [19]: the doses of CY tested were 150, 100, 50, 0 mg/kg. The CY0 mg/kg dose was closed after three GFs out of three patients; The CY150 mg/kg was closed due to excess toxicity, leaving the 100 (CY100) and 50 mg/kg (CY50) doses open. A recent publication by Anderlini and coworkers reports the outcome of these two doses [19]: 79 patients entered this second phase of the study, and were grafted from UD after conditioning with fludarabine and TBI 200 cGy plus CY100 or 50 mg/kg. At day 100, 92% of patients were engrafted and alive in the CY50 mg/kg cohort (n = 38) and 85% in the 100 mg/kg cohort (n = 41). GF was seen in 7 and 15%, respectively, of the two cohorts. The proportion of patients with grade 3–4 toxicity by day 100 was 11 and 22%, respectively. Actuarial survival at 1 year was 97% for CY50 and 80% for CY100 groups. The crude overall mortality was 8% in the CY50 and 24% in the CY100 cohort [19].

There are several important messages that come from these CIBMTR studies: First, TBI should be reduced to a minimum, and 200 cGy seems currently the best choice. Second, CY is a crucial component of the conditioning regimen in SAA, and cannot be omitted. Third, CY150 mg/kg and thus also CY200 mg/kg should not be used because of excess toxicity. This leaves CY doses ranging between 50 and

100 mg/kg, probably with data favoring 50 mg/kg [19]. In keeping with these data most centers are using a combination of FLU CY ATG (FCA) with TBI 200 cGy [11,20,21].

AN UPDATE OF EBMT DATA ON UD TRANSPLANTS

We have updated the EBMT study on 100 SAA undergoing alternative donor transplants between 1998 and 2009, following preparative regimens FCA (FLU + CY + ATG) or FCA combined with low dose TBI, reported in 2010 [10]. With a median follow up of 8 years for surviving patients (range 1–15 years) there have been only 2 additional deaths, and the actuarial 5 year survival is 76% for FCA-TBI and 74% for FCA (Fig. 8.2) . It should be noted that these two conditioning regimens were offered to different patients: FCA was given primarily to pediatric patients (median age 13) and

FCA-TBI to adults (median age 27): the comparable survival is therefore noteworthy. Both regimens called for a low dose of CY, 300 mg/$m^2 \times 4$, total 1200 mg/m^2, which corresponds approximately to 40 mg/kg, close to the CY dose tested by CIBMTR [19]: because of the risk of GF (17%), the CY dose was increased to 30 mg/kg $\times$ 4 (total 120 mg/kg). We have grafted 13 patients in the Genova transplant Unit, with this modified FCA-TBI protocol (unpublished): median age was 38 years (17–54) and the interval from diagnosis to transplant 422 days, all of the patients having failed at least one course of immunosuppressive treatment. Rejection was seen in 1 patient on day 75 despite CY 120 mg/kg, and was fatal; the cumulative incidence of GF was 8% and the actuarial 5 year survival was 92% (Fig. 8.3). Very small number of patients, but median age now 38 years and excellent survival.

In conclusion the dose of CY to be used in the FCA regimen should range between 50 and

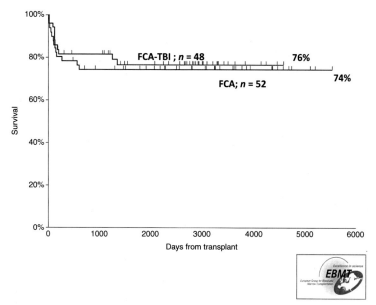

Update (2016) of 100 SAA patients grafted from alternative donors

FIGURE 8.2   Updated actuarial survival of 100 SAA patients grafted from alternative donors and reported in Ref. [10] .

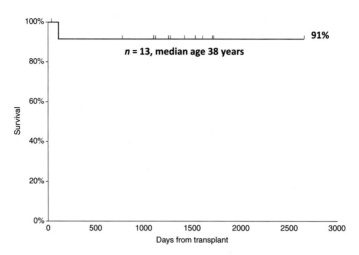

FIGURE 8.3 Actuarial survival of 13 consecutive patients who received the FCA-TBI regimen, with an escalated CY dose of 30 mg/kg × 4. FLU (30 mg/m² × 4), ATG (Thymoglobulin, Sanofi France, 3.75 mg/kg × 2) and TBI (200 cGy) doses remain the same as in Ref. [11].

120 mg/kg, and one would expect a higher rate of GF in the lower CY dose [22].

ALEMTUZUMAB INSTEAD OF ATG

One alternative to the use of ATG is alemtuzumab (CAMPATH, Sanofi, France). The British group has reported a low rate of GF (9%) and low risk of chronic GvHD (4%), with the FCC regimen (FLU 30 mg/m² × 4 + CY300 mg/m² × 4 + Campath 60 mg) in transplants for SAA [23]. The actuarial survival was 95% in a group of 50 patients, with a median age of 35 years, more than half of whom were UD grafts [23]. No patient achieved full donor chimerism, and this of course is associated with a low risk of chronic GvHD. Again we must call your attention to the fact that (1) this is a TBI-free regimen and (2) this is the low dose CY (approximately 40 mg/kg), both relevant factors for the short-term (toxicity) and long-term (infertility, second tumors?) outcome of patients with SAA. The British group has

also compared the FCC regimen with the FCA regimen, in 155 SAA patients, receiving either matched related or UD grafts [24]: GF occurred in 9% of the alemtuzumab group and 11% of the ATG group. Five-year survival was 90% for the alemtuzumab and 79% for the ATG groups ($p = 0.1$). For UD transplants, survival of patients was better when using alemtuzumab (88%) compared with ATG (57%), ($p = 0.02$), although smaller numbers of patients received ATG. Similar outcomes were seen for MSD transplants using alemtuzumab or ATG (91% vs. 85%, respectively). A lower risk of chronic GVHD was observed in the alemtuzumab group (11% vs. 26%, $p = 0.03$) [24]. Also in this study on multivariate analysis, the use of BM as a stem cell source was associated with better survival, and less acute/chronic GVHD.

Therefore it is possible to transplant acquired SAA with a TBI-free regimen: not all centers are familiar with the use of alemtuzumab, but the data are there and prospective studies comparing ATG and alemtuzumab are clearly warranted.

GRAFT VERSUS HOST DISEASE PROPHYLAXIS

ATG and alemtuzumab are key parts of the conditioning regimen, and exhert a dual role, on host T cells, thus favoring engraftment, and on donor T cells, by quencing GvHD. However for optimal prevention, pharmacologic immunosuppression is required with a calcineurin inhibitor (either CsA or tacrolimus, TAC) with or without methotrexate (MTX). CsA is used alone in alemtuzumab-based platforms; the combination of CsA + MTX is usually used in association with ATG [25]. A recent study has compared the combination of CsA + MTX with TAC + MTX [26]. The study included sibling and UDs: no firm conclusion was drawn from this retrospective analysis, but in UD grafts, the combination of TAC + MTX provided a reduction of mortality ($p = 0.008$) although GvHD risk was the same compared to CsA + MTX [26]. One additional way to prevent acute and chronic GvHD is high dose posttransplant CY (PT-CY) in combination with CsA and mycophenolate, successfully tested in SAA grafted from family haploidentical donors [27]; this combination could be tested also in the UD setting, in a prospective trial.

IMPROVEMENT OF UD TRANSPLANTS WITH TIME AND SUPPORTIVE CARE

The outcome of UD transplants has significantly improved in the last decade: in a CIBMTR analysis 3 year survival improved from 32% before 1998, to 61% between 1999 and 2005, to 75% after 2005 [14]. In an EBMT analysis, 3 year survival improved from 66% to 83% before or after 2004 [10]. A Japanese study in children shows current survival in the range of 90%, and same outcome is reported for children grafted in a British report [28].

Optimal allele HLA typing has contributed to improved survival, but supportive care has also changed significantly over the past decade, with careful monitoring of viral, bacterial and fungal infections, and preemptive therapy. The use of ATG is a risk factor for EBV infections [29], which can be fatal: we are currently giving a small dose of rituximab (200 mg total dose) on day +5 in SAA patients grafted from a UD, as we have shown this intervention to be well tolerated and to abrogate, almost completely, the risk of EBV disease posttransplant [30].

CONCLUSIONS

Results of UD transplants for SAA have significantly improved over the past few decades, as a consequence of several factors: better selection of donors with HLA allele matching, improved supportive care, and different conditioning regimens, better designed to allow sustained engraftment with low toxicity. The combination of FLU and CY appears now widely used, with doses of CY which should range between 50 and 100 mg/kg. Low dose TBI (2 Gy) is also widely used, and has been shown to be well tolerated and effective in allowing sustained engraftment. In the young patient, under the age of 40, survival after a well matched 8/8 UD transplant can be expected in the range of 80%, perhaps closer to 90% in the pediatric population. Patients above the age of 40 continue to have a significant mortality of 50%: until we do not have reliable data showing that this problem has been solved, patients above the age of 40 should always be treated with one course of immunosupopression (ATG + CsA). Children and patients under the age of 40 may be offered an upfront UD graft, when there is an nstitutional protocol ad hoc: upfront UD grafts are not standard of care in SAA.

Actuarial 10 year survival has increased in the last decade from 58% to 71%, for all patients reported to the EBMT registry and treated with IST or bone marrow transplantation (BMT) (Fig. 8.4): we anticipate that survival will further

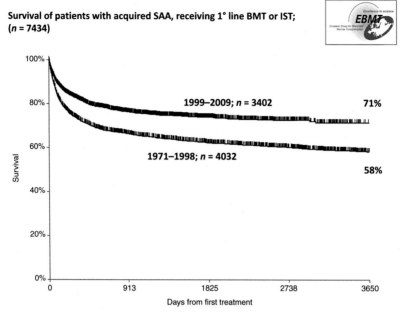

FIGURE 8.4 Actuarial survival of 7434 patients with acquired SAA treated with either transplantation (BMT) or immunosuppressive therapy (IST), reported to the registry of the EBMT.

increase, perhaps of an additional 10%, with the introduction of eltrombopag in upfront IST, and the improved outcome of alternative donor transplants. Prospective clinical trials remain the best therapeutic option for patients with acquired aplastic anemia.

References

[1] Dufour C, Svahn J, Bacigalupo A. Severe Aplastic Anemia–Working Party of the EBMT. Front-line immunosuppressive treatment of acquired aplastic anemia. Bone Marrow Transplant 2013;48:174–7.

[2] Townsley D, Dumitriu B, Scheinberg P, Desmond R, Feng X, Rios O, Weinstein B, Winkler T, Valdez J, Winkler T, Desierto M, Leuva H, Wu C, Calvo K, Larochelle A, Dunbar C, Young NS. Eltrombopag added to standard immunosuppression for aplastic anemia, accelerates count recovery and increases response rates. Blood 2015;126. LBA-2.

[3] Dufour C, Veys P, Carraro E, Bhatnagar N, Pillon M, Wynn R, Gibson B, Vora AJ, Steward CG, Ewins AM, Hough RE, de la Fuente J, Velangi M, Amrolia PJ, Skinner R, Bacigalupo A, Risitano AM, Socie G, Peffault de Latour R, Passweg J, Rovo A, Tichelli A, Schrezenmeier

H, Hochsmann B, Bader P, van Biezen A, Aljurf MD, Kulesekararaj A, Marsh JC, Samarasinghe S. Similar outcome of upfront-unrelated and matched sibling stem cell transplantation in idiopathic paediatric aplastic anaemia. A study on behalf of the UK Paediatric BMT Working Party, Paediatric Diseases Working Party and Severe Aplastic Anaemia Working Party of EBMT. Br J Haematol 2015;171:585–94.

[4] Blaise D, Fürst S, Crocchiolo R, El-Cheikh J, Granata A, Harbi S, Bouabdallah R, Devillier R, Bramanti S, Lemarie C, Picard C, Chabannon C, Weiller PJ, Faucher C, Mohty B, Vey N, Castagna L. Haploidentical T cell-replete transplantation with post-transplantation cyclophosphamide for patients in or above the sixth decade of age compared with allogeneic hematopoietic stem cell transplantation from an human leukocyte antigen-matched related or unrelated donor. Biol Blood Marrow Transplant 2016;22(1):119–24.

[5] Gupta V, Eapen M, Brazauskas R, Carreras J, Aljurf M, Gale RP, Hale GA, Ilhan O, Passweg JR, Ringdén O, Sabloff M, Schrezenmeier H, Socié G, Marsh JC. Impact of age on outcomes after bone marrow transplantation for acquired aplastic anemia using HLA-matched sibling donors. Haematologica 2010;95(12):2119–25.

[6] Sangiolo D, Storb R, Deeg HJ, Flowers ME, Martin PJ, Sandmaier BM, Kiem HP, Nash RA, Doney K, Leisenring WM, Georges GE. Outcome of allogeneic

hematopoietic cell transplantation from HLA-identical siblings for severe aplastic anemia in patients over 40 years of age. Biol Blood Marrow Transplant 2010;16(10):1411–8.

[7] Maury S, Balère-Appert ML, Pollichieni S, Oneto R, Yakoub-Agha I, Locatelli F, Dalle JH, Lanino E, Fischer A, Pession A, Huynh A, Barberi W, Mohty M, Risitano A, Milpied N, Socié G, Bacigalupo A, Marsh J, Passweg JR. Severe Aplastic Anemia Working Party of the European Group for Blood and Marrow Transplantation (EBMT). Outcome of patients activating an unrelated donor search for severe acquired aplastic anemia. Am J Hematol 2013;88(10):868–73.

[8] Horan J, Wang T, Haagenson M, Spellman SR, Dehn J, Eapen M, Frangoul H, Gupta V, Hale GA, Hurley CK, Marino S, Oudshoorn M, Reddy V, Shaw P, Lee SJ, Woolfrey A. Evaluation of HLA matching in unrelated hematopoietic stem cell transplantation for nonmalignant disorders. Blood 2012;120(14):2918–24.

[9] Deeg HJ, O'Donnell M, Tolar J, Agarwal R, Harris RE, Feig SA, Territo MC, Collins RH, McSweeney PA, Copelan EA, Khan SP, Woolfrey A, Storer B. Optimization of conditioning for marrow transplantation from unrelated donors for patients with aplastic anemia after failure of immunosuppressive therapy. Blood 2006;108(5):1485–91.

[10] Bacigalupo A, Socie' G, Lanino E, Prete A, Locatelli F, Locasciulli A, Cesaro S, Shimoni A, Marsh J, Brune M, Van Lint MT, Oneto R, Passweg J. Severe Aplastic Anemia Working Party of the European Group for Blood and Marrow Transplantation. Fludarabine, cyclophosphamide, antithymocyte globulin, with or without low dose total body irradiation, for alternative donor transplants, in acquired severe aplastic anemia: a retrospective study from the EBMT-SAA working party. Haematologica 2010;95(6):976–82.

[11] (a) Yagasaki H, Kojima S, Yabe H, Kato K, Kigasawa H, Sakamaki H, Tsuchida M, Kato S, Kawase T, Morishima Y, Kodera Y. Japan Marrow Donor Program. Acceptable HLA-mismatching in unrelated donor bone marrow transplantation for patients with acquired severe aplastic anemia. Blood 2011;118(11):3186–90. (b) Storb R, Thomas ED, Buckner CD, Clift RA, Johnson FL, Fefer A, Glucksberg H, Giblett ER, Lerner KG, Neiman P. Allogeneic marrow grafting for treatment of aplastic anemia. Blood 1974;43(2):157–80.

[12] Storb R, Thomas ED, Buckner CD, Clift RA, Johnson FL, Fefer A, Glucksberg H, Giblett ER, Lerner KG, Neiman P. Allogeneic marrow grafting for treatment of aplastic anemia. Blood 1974;43(2):157–80.

[13] Anasetti C, Storb R, Longton G, Witherspoon R, Doney K, Sullivan KM, Thomas ED. Donor buffy coat cell infusion after marrow transplantation for aplastic anemia. Blood 1988;72(3):1099–100.

[14] Eapen M, Le Rademacher J, Antin JH, Champlin RE, Carreras J, Fay J, Passweg JR, Tolar J, Horowitz MM, Marsh JC, Deeg HJ. Effect of stem cell source on outcomes after unrelated donor transplantation in severe aplastic anemia. Blood 2011;118:2618–21.

[15] Bacigalupo A, Socie' G, Schrezenmeier H, Tichelli A, Locasciulli A, Fuhrer M, Risitano A, Dufour C, Passweg J, Oneto R, Aljurf M, Flynn C, Mialou V, Hamladji RM, Marsh J. Bone marrow versus peripheral blood sibling transplants in acquired aplastic anemia: survival advantage for marrow in all age groups. Haematologica 2012;97(8):1142–8.

[16] Deeg HJ, Anasetti C, Petersdorf E, Storb R, Doney K, Hansen JA, Sanders J, Sullivan KM, Appelbaum FR. Cyclophosphamide plus ATG conditioning is insufficient for sustained hematopoietic reconstitution in patients with severe aplastic anemia transplanted with marrow from HLA-A, B, DRB matched unrelated donors. Blood 1994;83:3417–8.

[17] Storb R, Anasetti C, Appelbaum F, Bensinger W, Buckner CD, Clift R, Deeg HJ, Doney K, Hansen J, Loughran T. Marrow transplantation for severe aplastic anemia and thalassemia major. Semin Hematol 1991;28:235–9.

[18] Socié G, Henry-Amar M, Bacigalupo A, et al. Malignant tumors occurring after treatment of aplastic anemia. N Engl J Med 1993;329:1152–7.

[19] Tolar J, Deeg HJ, Arai S, Horwitz M, Antin JH, McCarty JM, Adams RH, Ewell M, Leifer ES, Gersten ID, Carter SL, Horowitz MM, Nakamura R, Pulsipher MA, Difronzo NL, Confer DL, Eapen M, Anderlini P. Fludarabine-based conditioning for marrow transplantation from unrelated donors in severe aplastic anemia: early results of a cyclophosphamide dose deescalation study show life-threatening adverse events at predefined cyclophosphamide dose levels. Biol Blood Marrow Transplant 2012;18(7):1007–11.

[20] Anderlini P, Wu J, Gersten I, Ewell M, Tolar J, Antin JH, Adams R, Arai S, Eames G, Horwitz ME, McCarty J, Nakamura R, Pulsipher MA, Rowley S, Leifer E, Carter SL, DiFronzo NL, Horowitz MM, Confer D, Deeg HJ, Eapen M. Cyclophosphamide conditioning in patients with severe aplastic anaemia given unrelated marrow transplantation: a phase 1-2 dose de-escalation study. Lancet Haematol 2015;2(9):e367–75.

[21] Kojima S, Matsuyama T, Kato S, Kigasawa H, Kobayashi R, Kikuta A, Sakamaki H, Ikuta K, Tsuchida M, Hoshi Y, Morishima Y, Kodera Y. Outcome of 154 patients with severe aplastic anemia who received transplants from unrelated donors: the Japan Marrow Donor Program. Blood 2002;100:799–803.

[22] Okuda S, Terasako K, Oshima K, Sato M, Nakasone H, Kako S, Yamazaki R, Tanaka Y, Tanihara A, Higuchi T, Nishida J, Kanda Y. Fludarabine, cyclophosphamide, anti-thymocyteglobulin, and low-dose total body

irradiation conditioning enables 1-HLA-locus-mismatched hematopoietic stem cell transplantation for very severe aplastic anemia without affecting ovarian function. Am J Hematol 2009;84(3):167–9.

[23] McGuinn C, Geyer MB, Jin Z, Garvin JH, Satwani P, Bradley MB, Bhatia M, George D, Duffy D, Morris E, van de Ven C, Schwartz J, Baxter-Lowe LA, Cairo MS. Pilot trial of risk-adapted cyclophosphamide intensity based conditioning and HLA matched sibling and unrelated cord blood stem cell transplantation in newly diagnosed pediatric and adolescent recipients with acquired severe aplastic anemia. Pediatr Blood Cancer 2014;61(7):1289–94.

[24] Marsh JC, Gupta V, Lim Z, Ho AY, Ireland RM, Hayden J, Potter V, Koh MB, Islam MS, Russell N, Marks DI, Mufti GJ, Pagliuca A. Alemtuzumab with fludarabine and cyclophosphamide reduces chronic graft versus host disease after allogeneic stem cell transplantation for acquired aplastic anemia. Blood 2011;118(8):2351–7.

[25] Marsh JC, Pearce RM, Koh MB, Lim Z, Pagliuca A, Mufti GJ, Perry J, Snowden JA, Vora AJ, Wynn RT, Russell N, Gibson B, Gilleece M, Milligan D, Veys P, Samarasinghe S, McMullin M, Kirkland K, Cook G. British Society for Blood and Marrow Transplantation, Clinical Trials Committee. Retrospective study of alemtuzumab vs ATG-based conditioning without irradiation for unrelated and matched sibling donor transplants in acquired severe aplastic anemia: a study from the British Society for Blood and Marrow Transplantation. Bone Marrow Transplant 2014;49(1):42–8.

[26] Locatelli F, Bruno B, Zecca M, Van-Lint MT, McCann S, Arcese W, Dallorso S, Di Bartolomeo P, Fagioli F, Locasciulli A, Lawler M. Bacigalupo A: cyclosporin A and short-term methotrexate versus cyclosporin A as graft versus host disease prophylaxis in patients with severe aplastic anemia given allogeneic bone marrow transplantation from an HLA-identical sibling: results of a GITMO/EBMT randomized trial. Blood 2000;96(5):1690–7.

[27] Inamoto Y, Flowers ME, Wang T, Urbano-Ispizua A, Hemmer MT, Cutler CS, Couriel DR, Alousi AM, Antin JH, Gale RP, Gupta V, Hamilton BK, Kharfan-Dabaja MA, Marks DI, Ringdén OT, Socié G, Solh MM, Akpek G, Cairo MS, Chao NJ, Hayashi RJ, Nishihori T, Reshef R, Saad A, Shah A, Teshima T, Tallman MS, Wirk B, Spellman SR, Arora M, Martin PJ. Tacrolimus versus cyclosporine after hematopoietic cell transplantation for acquired aplastic anemia. Biol Blood Marrow Transplant 2015;21(10):1776–82.

[28] Yagasaki H, Takahashi Y, Hama A, Kudo K, Nishio N, Muramatsu H, Tanaka M, Yoshida N, Matsumoto K, Watanabe N, Kato K, Horibe K, Kojima S. Comparison of matched-sibling donor BMT and unrelated donor BMT in children and adolescent with acquired severe aplastic anemia. Bone Marrow Transplant 2010;45(10):1508–13.

[29] Samarasinghe S, Steward C, Hiwarkar P, Saif MA, Hough R, Webb D, Norton A, Lawson S, Qureshi A, Connor P, Carey P, Skinner R, Vora A, Pelidis M, Gibson B, Stewart G, Keogh S, Goulden N, Bonney D, Stubbs M, Amrolia P, Rao K, Meyer S, Wynn R, Veys P. Excellent outcome of matched unrelated donor transplantation in paediatric aplastic anaemia following failure with immunosuppressive therapy: a United Kingdom multicentre retrospective experience. Br J Haematol 2012;157(3):339–46.

[30] Van Esser JW, van der Holt B, Meijer E, NiestersHG, Trenschel R, Thijsen SF, van Loon AM, Frassoni F, Bacigalupo A, Schaefer UW, Osterhaus AD, Gratama JW, Lowenberg B, Verdonck LF, Corneliss JJ. Epstein–Barr virus (EBV) reactivation is a frequent event after allogeneic stem cell transplantation (SCT) and qualitatively predicts EBV-lymphoproliferative disease following T cell depleted SCT. Blood 2001;98(4):972–8.

Umbilical Cord Blood Transplantation for Patients With Acquired and Inherited Bone Marrow Failure Syndromes on Behalf of Eurocord

E. Gluckman*, A. Ruggeri*,**, R. Peffault de Latour†

*Eurocord, Hospital Saint-Louis, Paris, France; **Hematology Department, Hospital Saint Antoine, Paris, France; †BMT Unit, French Reference Center for Aplastic Anemia and PNH, Saint-Louis Hospital, Paris, France

INTRODUCTION

Since the first human umbilical cord blood transplant (UCBT), performed in 1988 [1], cord blood banks (CBB) have been established worldwide for collection and cryopreservation of cord blood unit (CBU) for allogeneic hematopoietic stem cell transplant (HSCT). These advantages were first recognized in UCBT using related donors; subsequently, CBB established criteria for standardization of CBU collection, banking, processing, and cryopreservation for unrelated use in patients with hematologic malignant and nonmalignant diseases. Umbilical cord blood has now become one commonly used source of hematopoietic stem cells for allogeneic

HSCT. Today a global network of CBB and transplant centers has been established for a common inventory, an estimated 700,000 CBUs have been banked and more than 40,000 CBUs distributed worldwide for adults and children with severe hematologic diseases.

The first UCBT was performed in 1988 in a patient with Fanconi anemia (FA) [1]. This patient had a healthy HLA-identical sibling shown by prenatal testing to be unaffected by the disorder, and to have a normal karyotype. Her CBU was collected at birth, cryopreserved, and used after thawing for transplantation. The patient was conditioned by a procedure developed specifically for the treatment of FA patients who are extremely sensitive to the administration of alkylating agents like cyclophosphamide (Cy). Conditioning regimen consisted of low dose of Cy (20 mg/kg instead of 200 mg/kg) and 5 Gy total lymphoid irradiation. The frozen cells were hand-delivered from Indiana to Paris in a dry shipper that maintained the temperature at −175°C. The cells were thawed without further processing on day 0. Thawed cells were tested for viability and progenitor assays and results were similar to the counts recorded before freezing. First signs of engraftment appeared on day 22 with subsequent complete hematologic reconstitution and full donor chimerism. The patient had no GVHD and is currently more than 25 years after UCBT, healthy with a complete long-term hematologic and immunologic donor reconstitution.

This first success opened the way to a new field in the domain of HSCT as it showed that: (1) a single umbilical CBU contained enough hematopoietic stem cells to reconstitute definitely the host hematopoietic compartment; (2) a CBU could be collected at birth without any harm to the newborn infant, and (3) umbilical cord blood hematopoietic stem cells could be cryopreserved and transplanted in a myelo-ablated host after thawing without losing their repopulating capacity.

So far, several studies on UCBT from single center and registries have shown that the number of nucleated cells (TNC) is a critical factor for UCBT outcomes while some degree of HLA mismatches is acceptable [2–4] which is particularly true regarding patients with nonmalignant disorders.

Aplastic anemia is a heterogeneous disorder often called "bone marrow failure syndrome" (BMFS); it comprises idiopathic aplastic anemia (SAA)/paroxysmal nocturnal hemoglobinuria (PNH) and various hereditary disorders and the most frequent is FA. The indications of HSCT vary according to the type and the severity of the disease [5]. Idiopathic AA is considered as an autoimmune disease whereas hereditary disorders present specific genetic abnormalities

which might, as it is the case in FA, influence the choice for the conditioning regimen. In the absence of a related HLA-identical bone marrow donor, cord blood or haploidentical donor [6] might be a second choice.

CANDIDATES FOR CORD BLOOD TRANSPLANTATION FOR APLASTIC ANEMIA

Most related HLA-identical UCBT have been performed for inherited BMFS when the mother was pregnant of a healthy child and when there was a sibling who has an indication for HSCT. For this reason, family directed CBBs are not generally widespread explaining the small number of patients transplanted with this source of stem cells, as more often physicians use the bone marrow of the HLA-identical sibling [7].

In the absence of an HLA-identical sibling, the current standard treatment for SAA is the combination of horse antithymocyte globulin (ATG) and cyclosporine (CSA) [8]. After immunosuppressive therapy (IST), overall response is achieved in about two-thirds of patients, the cumulative incidence (CI) of relapse among responders is approximately 20–30%, and clonal evolution occurs in about 10–15% of cases. Relapsed patients are treated with repeated courses of IST leading to overall good results. Patients refractory to initial IST historically have a poor outcome with long-term survival rates of 20–30% and the overall management of those patients is challenging. HSCT from an unrelated donor represents an alternative salvage therapy. Better donor/recipient HLA matching has probably played a major role, but also significant changes in the conditioning regimen and improvement in supportive measures have occurred. Results of matched unrelated bone marrow (BM) transplants have improved to such an extent that treatment strategy may be affected. In children without a matched sibling donor, current guidelines recommend proceeding to

HSCT from a fully matched unrelated donor, after failing one course of IST. In adults, alternative donor transplant remains an option for second line treatment for patients failing 1 or 2 courses of IST [9–11].

Unfortunately, many patients, especially those from ethnic minority or less homogeneous populations, do not have a suitable unrelated donor. The percentage of such patients will vary between 5% and 40%, according to the patient's ethnic origin.

Eurocord has collected information of UCBT in European and non-European countries. On more than 11,000 UCBT reported to Eurocord, 121 UCBTs from a related cord blood donor and 346 UCBT for unrelated cord blood transplant were performed for BMFS. Table 9.1 shows the number of patients according to the type of donor and the etiology. Fig. 9.1 shows the probability of survival according to the diagnosis in unrelated and related UCBT. With a median follow-up time of 75 months overall survival (OS) after related UCBT was 84% for inherited BMFS and 73% for acquired BMFS. The probability of survival after unrelated UCBT was 47% for inherited BMFS and 44% for acquired BMFS. Fig. 9.2 shows the influence of cell dose on neutrophil recovery and OS in unrelated UCBT. These results are the basis of our recommendation to increase cell dose for unrelated UCBT for BMFS at more than 4×10^7 TNC/kg.

HLA-IDENTICAL SIBLING CORD BLOOD TRANSPLANT: EUROCORD RESULTS

Outcomes of 121 children with inherited or acquired BMF syndromes, who received UCBT from an HLA-identical related donor, were analyzed by Eurocord in collaboration with the Severe Aplastic Anemia Working Party of the European Group for Blood and Marrow Transplantation (SAA WP EBMT). Patients were transplanted in EBMT centers between 1988 and 2014. Ninety-five patients had an inherited and 26 an acquired BMF. Forty-eight patients had FA, 26 patients had Diamond–Blackfan anemia, and 26 had idiopathic SAA.

TABLE 9.1 Number of Patients With Acquired and Inherited Bone Marrow Failure Syndrome Transplanted With Cord Blood and Reported to Eurocord From 1988 to 2015

Acquired BMF			Inherited BMF		
	Related, $n = 26$	Unrelated, $n = 147$		Related, $n = 95$	Unrelated, $n = 199$
Idiopathic severe aplastic anemia	25	132	Fanconi anemia	48	119
Pure red cells aplasia	1	1	Diamond– Blackfan anemia	26	7
PNH		13	Schwachman–Diamond syndrome	2	4
Other acquired bone marrow failure syndrome		1	Dyskeratosis congenita	2	7
			Kostmann's syndrome	4	21
			Amegakariocytic thrombocytopenia	5	18
			Other inherited BMF	8	23

PNH, paroxysmal nocturnal hemoglobinuria.

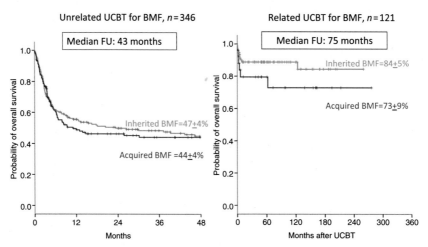

FIGURE 9.1 **Overall survival of unrelated and related cord blood transplant in patients with bone marrow failure syndrome.** Eurocord results.

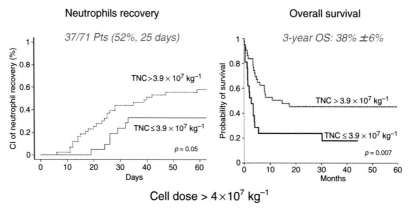

FIGURE 9.2 **Neutrophil recovery and overall survival according to cell dose in patients with bone marrow failure syndrome.** Eurocord results.

Eighty-eight patients received a single CBU and 33 received a combination of CB and BM from the same donor. Median age at UCBT was 6.7 (1–16) years (5.6 years for acquired and 6.9 years for inherited group). Median interval between diagnosis and UCBT was 49 months (21 for acquired and 55 for inherited group). Sixty-six patients (60%) received a reduced intensity (RIC) and 49 (40%) a myeloablative regimen (MAC). The most common protocols were Cy and fludarabine (54%) in RIC setting and Cy and

busulfan (46%) for patients receiving an MAC. Total body irradiation (TBI) was used in 8 patients (4 at a dose of 2 Gy). ATG was used in 46% of acquired and 37% of inherited BMF. GVHD prophylaxis consisted mainly of CSA in 53 patients (43%) and CSA and methotrexate in 23 patients (19%). Median number of infused TNC was 6.2 (1–25) $\times 10^7$ kg^{-1} for patients receiving a single CBU and 26 (5.3–41) $\times 10^7/kg$ for those receiving CB + BM. Median follow-up was 6 (range 0.6–27) years.

The CI of day 60 neutrophil recovery was $91 \pm 9\%$, with a median time of 22 days (29 days for acquired AA and 18 days for inherited AA). Eleven patients experienced primary graft failure 4 with acquired and 7 with inherited AA. Among them, 9 died and 2 are alive after a second HSCT using BM.

The 100-day CI of grade II–IV acute GVHD was $11 \pm 3\%$ and the 1-year CI of chronic GVHD was $12 + 4\%$. The CI of transplant-related mortality was $18 + 3\%$. Sixteen patients died: 9 due to PGF, 1 for cGVHD, and 6 of infections and other transplant-related causes.

OS was 91% at 6 years. At last follow-up OS was $73 + 9\%$ for acquired and $84 + 5\%$ for inherited BMF ($p = 0.09$) (Fig. 9.1). Information on long-term outcomes was available for 56 patients, of whom 3 patients were reported as experiencing secondary neoplasia.

These results show that in children with inherited or acquired BMF syndrome, UCBT from an HLA-identical sibling donor is associated with excellent long-term outcomes with particular low incidence of GVHD. In case of inherited BMF, collecting CBU at birth of a new sibling is recommended. If CBU does not contain enough cells addition of bone marrow from the same donor gives excellent results.

UNRELATED CORD BLOOD TRANSPLANTATION FOR BMFS

Results of Unrelated Cord Blood Transplants for Idiopathic Aplastic Anemia

To date, there are only a few reports on unrelated UCBT in patients with SAA. Primary reports showed poor outcome and high incidence of graft failure while few small series and case reports of successful UCBT for SAA have recently been reported [12–16]. A large cohort of 31 patients has been published by the Japanese group with a 2-year OS of 41%, suggesting that UCBT can be an alternative treatment for SAA patients who failed IST and have no suitable bone marrow donor [16].

Eurocord performed an analysis on 71 patients transplanted from 1996 to 2009 [17]. The median age was 13 years with 28 adults with SAA (9 with PNH) who received a single UCBT ($n = 57$; 79%) or double unit UCBT ($n = 14$; 19%). An RIC regimen was used in 68% of the patients. The CI of neutrophil recovery was $51 \pm 6\%$ at day 60, with significantly better engraftment seen in recipients of higher prefreezing TNC [median dose $3.9 \times 10^7 \text{ kg}^{-1}$; hazard ratio (HR), 1.5; $p = 0.05$]. The CI of platelet engraftment at day 180 was $37 \pm 7\%$, CI of grade II–IV acute GVHD was $20 \pm 5\%$, and CI of chronic GVHD at 3 years was $18 \pm 5\%$. With a median follow-up of 35 (range, 3–83) months, the probability of 3-year OS was $38 \pm 6\%$ (Fig. 9.2). Significantly improved OS was seen in recipients who received a high TNC of $>3.9 \times 10^7 \text{ kg}$ prefreezing (45%, compared with 18% for recipients of $<3.9 \times 10^7 \text{ TNC/kg}$; HR, 0.4; $p = 0.007$). These results highlight the critical role of cell dose for both engraftment and OS in patients with SAA undergoing UCBT.

Results of Unrelated Cord Blood Transplants for Fanconi Anemia

Eurocord analyzed results of UCBT in 93 FA patients [18]. The CI of neutrophil recovery at 60 days was $60 + 5\%$. In addition to high TNC dose, fludarabine-containing regimens (as in marrow recipients) were associated with better neutrophil engraftment. The incidence of acute and chronic GVHD was 32.5 and 16%, respectively. OS was $40 \pm 5\%$. In multivariate analysis factors associated with favorable outcome were use of fludarabine, high number of TNC, and negative recipient CMV serology. To date, there has been no formal comparison of outcomes in recipients of unrelated CB and marrow. However, results demonstrate that fludarabine is associated with better survival regardless of stem cell source in patients with

FA. This suggests that fludarabine, a potent immune suppressive agent, enhances engraftment without paying the price of extramedullary toxicity.

Unrelated UCBT is indicated in those FA patients, for whom an HLA-A, B, C, and DRB1 allele-matched unrelated volunteer donor cannot be identified, using a specific conditioning regimen disease-adapted.

Results of Cord Blood Transplant for Hereditary BMF Other than Fanconi Anemia

Sixty-four patients with hereditary BMFS were transplanted from related ($n = 20$) or unrelated UCBT ($n = 44$) and reported to Eurocord [19]. Diagnoses were Diamond–Blackfan anemia (21 patients), congenital amegakaryocytic thrombocytopenia (16 patients), dyskeratosis congenita (8 patients), Shwachman–Diamond syndrome (2 patients), severe congenital neutropenia (16 patients), and unclassified (1 patient). Among the 20 patients transplanted using a related UCBT, the median number of TNC infused was 5×10^7 kg. The CI of neutrophil recovery at 60 days was 95%. Two patients had grade II–IV aGVHD, while the 2-year CI of cGVHD was 11%. The 3-year OS rate was 95%.

In the group of unrelated UCBT, 86% had HLA-mismatched grafts and three received a double unit UCBT. The median number of TNC was 6.1×10^7 kg. The CI of neutrophil recovery at day 60 in this group was 55%. The 100-day CI of grade II–IV aGVHD was 24%, while the 2-year CI of cGVHD was 53%. OS at 3 years was 61%; better OS was associated with age less than 5 years ($p = 0.01$) and TNC $\geq 6.1 \times 10^7$ kg ($p = 0.05$).

These results suggest that in patients with hereditary BMFS, related UCBT is associated with excellent outcomes while increasing TNC dose and better HLA matching might provide better results in unrelated UCBT.

RECOMMENDATIONS FOR CORD BLOOD TRANSPLANTATION IN BMF

A prospective Phase II study on unrelated UCBT in acquired BMFS has been designed by the French Society for Stem Cell Transplantation and Eurocord-EBMT (APCORD Protocol). This protocol concerned all acquired BMF refractory to one course of IST in the absence of an HLA matched (related or unrelated) donor. The primary endpoint was OS at 1 year. Inclusion criteria are age: 3–55 years, acquired aplastic anemia (with SAA criteria) in treatment failure after IST without morphologic clonal evolution, and Karnofsky index >60%. One or two CBUs containing alone or together more than 4×10^7 cryopreserved TNC/kg with not more than two out of six HLA mismatches between the units and the patient, were recommended. The conditioning regimen consisted of fludarabine 30 mg/m^2 from D-6 to D-3, CY30 mg/kg from D-6 to D-3, ATG (thymoglobulin) 2.5 mg/kg from D-3 to D-2, TBI (2 Gray) on D-2. CSA is given alone as GVHD prophylaxis. One injection of anti-CD20 was recommended at day +5 (150 mg/m^2). All 26 patients have been included and results of the study will be submitted for publication.

According to the different studies reported, the criteria for donor choice are summarized in Table 9.2. A UCBT from an HLA-identical related donor is indicated as the HSCT using BM from related donors. When the cryopreserved CBU from the HLA-identical sibling does not contain enough cell dose, add-back of bone marrow cells at the time of transplant is feasible with excellent results and no increase in GVHD [20].

The place of UCBT from unrelated donors has to be considered in the frame of a clinical trial. The same conditioning regimen as for adult unrelated donor should be proposed and CBU with high number of cells and 0 or 1 HLA disparities need to be selected. Screening for donor-specific anti-HLA antibodies (DSA) should be

TABLE 9.2   Criteria of Cord Blood Donor Choice for Bone Marrow Failure Syndrome

1. *In case of absence of a full matched sibling donor*
2. *Possibility to perform an unrelated UCBT with high number of TNC*
    $>4 \times 10^7$ TNC/kg
3. Negative screening *for anti-HLA antibodies* against the cord blood unit
4. *HLA typing*:
    a. HLA-A and HLA-B typing at antigenic-level, HLA-DRB1 at allelic level
    b. Prefer class I than class II mismatch
    c. Avoid cord blood unit with more than 1 HLA mismatches
    d. If possible, include HLA-C typing, avoiding HLA-C mismatches when present[a]
5. If several cord blood units are available, choice of the best one should be also guided by
    a. Cord blood bank accreditation status and location
    b. ABO compatibility

TNC, total nucleated cell dose; UCBT, umbilical cord blood transplantation.
[a]*Especially when both HLA-C and HLA-DRB1 mismatches.*

performed systematically in patients with BMFS undergoing UCBT [21–23], given the increased risk of graft failure associated with DSA in unrelated UCBT. Results are particularly better in children and negative recipient cytomegalovirus serology.

FUTURE DIRECTIONS

Poor engraftment is the main complication of unrelated cord blood transplant for BMF. There are many ways to improve engraftment: choice of better units with high cell count and not more than 1 HLA mismatch, use of two CBUs, use of a combination of CB and BM from the same donor, or addition of CD34+ cells from a third party donor. Multiple investigators have explored CBU expansion strategies as a way to increase the cell dose and overcome the risk of graft failure. Delaney et al., using the notch ligand Delta 1, demonstrated expansion of short-term repopulating cells and an improvement in the time of neutrophil engraftment to 16 days [24]. The MD Anderson group used a coculture ex vivo with mesenchymal progenitor cells in one of the two UCB units in 31 patients, reporting a 30-fold expansion in CD34+ count, and median time to engraftment of 15 days [25], results comparing favorably to historic unma-

nipulated controls, who had a 24 day median engraftment. Some preliminary success has also been obtained using nicotinamide in combination with cytokines to ex vivo expand CD 133+ UCB in context of a double UCBT [26]. Other means to enhance the efficacy of UCBT are to increase the homing of cells within the hematopoietic microenvironment. Recent efforts include experimental studies using fucosylation of cells, inhibition of dipeptidylpeptidase 4 (DPP4, expressed as CD26 on the cell surface), and pretreatment of donor cells with a modified prostaglandin (PG) E molecule [27]. One approach is the upregulation of CXCR4 expression, which is expressed on CD34+ progenitor cells, to increase marrow homing. One of the two UCB units was incubated with PGE2; neutrophil engraftment improved by 3.5 days, and the PGE2 treated UCB provided long-term hematopoeisis in 10 of 12 patients. The use of an FDA approved orally active inhibitor of DPP4 on engraftment of single unit UCBT in adults was initiated in adults with high risk malignancies based on studies that demonstrated that DPP4 inhibition allows enhanced engraftment in animal studies [28], likely through effects on homing. However, inhibition of DPP4 can affect a number of cytokines produced by the BM nurturing environment, as well as homing of HSC. Thus, modifications of the timing and duration of the oral DPP4 inhibitor in the clinical setting may be more effective than reported. Homing and nurturing procedures may be relatively inexpensive to perform and more widely utilized without extensive ex vivo maneuvers or experience. Combinations of these procedures may result in greater improvement in engraftment capacity than any one procedure itself. An example of a potential combination treatment is that of either PGE-, cell fucosylation-, or DPP4 inhibitor treatment of donor cells, followed by the infusion of these cells into conditioned recipients that are given orally active DPP4 inhibitor prior to and following the CBU cell graft.

In conclusion, HLA-identical sibling cord blood transplantation if available is a good option in patients with BMFS especially in the group of patients with hereditary marrow failure where collection of cord blood for a healthy sibling can be anticipated before birth. If CBU does not contain enough cells addition of bone marrow from the same donor gives excellent results. Use of an unrelated cord blood transplant for BMFS is controversial because of the high incidence of graft failure and transplant-related mortality. However, unrelated cord blood transplantation using high number of cells (more than 5×10^7 TNC/kg) with not more than 0–1 HLA mismatches in selected patients (children or/and negative recipient cytomegalovirus serology) is of interest in refractory patients.

References

[1] Gluckman E, Broxmeyer HA, Auerbach AD, Friedman HS, Douglas GW, Devergie A, et al. Hematopoietic reconstitution in a patient with Fanconi's anemia by means of umbilical-cord blood from an HLA-identical sibling. New England J Med 1989;321(17):1174–8.

[2] Scaradavou A, Brunstein CG, Eapen M, Le-Rademacher J, Barker JN, Chao N, et al. Double unit grafts successfully extend the application of umbilical cord blood transplantation in adults with acute leukemia. Blood 2013;121(5):752–8.

[3] Gluckman E, Ruggeri A, Volt F, Cunha R, Boudjedir K, Rocha V. Milestones in umbilical cord blood transplantation. Br J Haematol 2011;154(4):441–7.

[4] Rocha V, Labopin M, Sanz G, Arcese W, Schwerdtfeger R, Bosi A, et al. Transplants of umbilical-cord blood or bone marrow from unrelated donors in adults with acute leukemia. New England J Med 2004;351(22):2276–85.

[5] Young NS, Bacigalupo A, Marsh JC. Aplastic anemia: pathophysiology and treatment. Biol Blood Marrow Transplant 2010;16(1 Suppl):S119–25.

[6] Dufour C, Veys P, Carraro E, Bhatnagar N, Pillon M, Wynn R, et al. Similar outcome of upfront-unrelated and matched sibling stem cell transplantation in idiopathic paediatric aplastic anaemia. A study on behalf of the UK Paediatric BMT Working Party, Paediatric Diseases Working Party and Severe Aplastic Anaemia Working Party of EBMT. Br J Haematol 2015;171(4):585–94.

[7] Gluckman E, Ruggeri A, Rocha V, Baudoux E, Boo M, Kurtzberg J, et al. Family-directed umbilical cord blood banking. Haematologica 2011;96(11):1700–7.

[8] Young NS. Current concepts in the pathophysiology and treatment of aplastic anemia. Hematology Am Soc Hematol Educ Prog 2013;2013:76–81.

[9] Bacigalupo A, Socie G, Hamladji RM, Aljurf M, Maschan A, Kyrcz-Krzemien S, et al. Current outcome of HLA identical sibling versus unrelated donor transplants in severe aplastic anemia: an EBMT analysis. Haematologica 2015;100(5):696–702.

[10] Peffault de Latour R, Porcher R, Dalle JH, Aljurf M, Korthof ET, Svahn J, et al. Allogeneic hematopoietic stem cell transplantation in Fanconi anemia: the European Group for Blood and Marrow Transplantation experience. Blood 2013;122(26):4279–86.

[11] Bacigalupo A, Marsh JC. Unrelated donor search and unrelated donor transplantation in the adult aplastic anaemia patient aged 18-40 years without an HLA-identical sibling and failing immunosuppression. Bone Marrow Transplant 2013;48(2):198–200.

[12] Lau FY, Wong R, Chui CH, Cheng G. Successful engraftment in two adult patients with severe aplastic anemia using nonmyeloablative conditioning followed by unrelated HLA-mismatched cord blood transplantation. J Hematother Stem Cell Res 2001;10(2):309–11.

[13] Mao P, Zhu Z, Wang H, Wang S, Mo W, Ying Y, et al. Sustained and stable hematopoietic donor-recipient mixed chimerism after unrelated cord blood transplantation for adult patients with severe aplastic anemia. Eur J Haematol 2005;75(5):430–5.

[14] Ohga S, Ichino K, Goto K, Hattori S, Nomura A, Takada H, et al. Unrelated donor cord blood transplantation for childhood severe aplastic anemia after a modified conditioning. Pediatr Transplant 2006;10(4):497–500.

[15] Chan KW, McDonald L, Lim D, Grimley MS, Grayson G, Wall DA. Unrelated cord blood transplantation in children with idiopathic severe aplastic anemia. Bone Marrow Transplant 2008;42(9):589–95.

[16] Yoshimi A, Kojima S, Taniguchi S, Hara J, Matsui T, Takahashi Y, et al. Unrelated cord blood transplantation for severe aplastic anemia. Biol Blood Marrow Transplant 2008;14(9):1057–63.

[17] Peffault de Latour R, Purtill D, Ruggeri A, Sanz G, Michel G, Gandemer V, et al. Influence of nucleated cell dose on overall survival of unrelated cord blood transplantation for patients with severe acquired aplastic anemia: a study by eurocord and the aplastic anemia working party of the European group for blood and marrow transplantation. Biol Blood Marrow Transplant 2011;17(1):78–85.

[18] Gluckman E, Rocha V, Ionescu I, Bierings M, Harris RE, Wagner J, et al. Results of unrelated cord blood transplant in Fanconi anemia patients: risk factor analysis for engraftment and survival. Biol Blood Marrow Transplant 2007;13(9):1073–82.

[19] Bizzetto R, Bonfim C, Rocha V, Socie G, Locatelli F, Chan K, et al. Outcomes after related and unrelated umbilical cord blood transplantation for hereditary bone marrow failure syndromes other than Fanconi anemia. Haematologica 2011;96(1):134–41.

[20] Tucunduva L, Volt F, Cunha R, Locatelli F, Zecca M, Yesilipek A, et al. Combined cord blood and bone marrow transplantation from the same human leucocyte antigen-identical sibling donor for children with malignant and non-malignant diseases. Br J Haematol 2015;169(1):103–10.

[21] Ruggeri A, Rocha V, Masson E, Labopin M, Cunha R, Absi L, et al. Impact of donor-specific anti-HLA antibodies on graft failure and survival after reduced intensity conditioning-unrelated cord blood transplantation: a Eurocord Societe Francophone d'Histocompatibilite et d'Immunogenetique (SFHI) and Societe Francaise de Greffe de Moelle et de Therapie Cellulaire (SFGM-TC) analysis. Haematologica 2013;98(7):1154–60.

[22] Takanashi M, Atsuta Y, Fujiwara K, Kodo H, Kai S, Sato H, et al. The impact of anti-HLA antibodies on unrelated cord blood transplantations. Blood 2010;116(15):2839–46.

[23] Cutler C, Kim HT, Sun L, Sese D, Glotzbecker B, Armand P, et al. Donor-specific anti-HLA antibodies predict outcome in double umbilical cord blood transplantation. Blood 2011;118(25):6691–7.

[24] Delaney C, Heimfeld S, Brashem-Stein C, Voorhies H, Manger RL, Bernstein ID. Notch-mediated expansion of human cord blood progenitor cells capable of rapid myeloid reconstitution. Nat Med 2010;16(2):232–6.

[25] de Lima M, McNiece I, Robinson SN, Munsell M, Eapen M, Horowitz M, et al. Cord-blood engraftment with ex vivo mesenchymal-cell coculture. New England J Med 2012;367(24):2305–15.

[26] Horwitz ME, Chao NJ, Rizzieri DA, Long GD, Sullivan KM, Gasparetto C, et al. Umbilical cord blood expansion with nicotinamide provides long-term multilineage engraftment. J Clin Investig 2014;124(7):3121–8.

[27] Robinson SN, Simmons PJ, Thomas MW, Brouard N, Javni JA, Trilok S, et al. Ex vivo fucosylation improves human cord blood engraftment in NOD-SCID IL-2Rgamma(null) mice. Exp Hematol 2012;40(6):445–56.

[28] Farag SS, Srivastava S, Messina-Graham S, Schwartz J, Robertson MJ, Abonour R, et al. In vivo DPP-4 inhibition to enhance engraftment of single-unit cord blood transplants in adults with hematological malignancies. Stem Cells Dev 2013;22(7):1007–15.

10

Haploidentical Transplantation

F. Ciceri*,**, M.T.L. Stanghellini*

*Hematology and BMT Unit, IRCCS San Raffaele Scientific
Institute, Milano, Italy; **University Vita-Salute San Raffaele,
IRCCS San Raffaele Scientific Institute, Milano, Italy

The great interest in transplantation from haploidentical donors (haplo-SCT) arises from several advantages: the immediate availability of a suitable one-haplotype mismatched donor for virtually all patients in the appropriate timing; transplantation is manageable according to patient needs; further access to donors for posttransplant immune interventions is easy. In addition, a haplo-SCT could rescue patients experiencing early graft failure (GF) from HLA-matched transplantation.

Early attempts of HSCT from haploidentical family donors were discouraging because of the development of refractory graft-versus-host disease (GvHD) and high transplant-related mortality (TRM). High rates of graft rejection (GR) and refractory GvHD were major limitations to the application of haploidentical HSCT [1]. In the absence of a fully HLA-matched donor, alternative donors, such as cord blood or haploidentical ones have been intensively investigated in the past 20 years in patients with poor-risk acute leukemias [2,3].

EX VIVO T-CELL DEPLETION

The first successful transplantation methodology in haplo-SCT has been based on T-cell-depleted grafts with high cell dose of CD34+ cells coming from family haploidentical donors [2,3]. Progress has been made in the optimization of conditioning regimens and graft selection to allow a stable hematopoietic engraftment across major HLA barriers, with promising disease-free survivals in adults with acute leukemias [2]. Unfortunately, large series from EBMR registry showed transplant-related deaths in a significant proportion of recipients. Leading causes of deaths reported are infections and interstitial pneumonia [3].

The ex vivo T-cell-depleted transplantation platform has been integrated by several complementary strategies aimed at improving posttransplant immune reconstitution; different cell therapy approaches have been established to a better control of graft rejection and GvHD [4–8]. A partial T-cell depletion can be provided by

Congenital and Acquired Bone Marrow Failure
http://dx.doi.org/10.1016/B978-0-12-804152-9.00010-5

alternative selections, such as CD3/CD19 negative selection or selective depletion of alloreactive T cells [9]; a T-cell content of $1 \times 10^5 \, kg^{-1}$ in the graft requires a posttransplant immune prophylaxis and translates into a definite risk of acute GvHD. A selective depletion of donor T cells bearing αβ+ T-cell receptor (TCR) has been provided by clinical grade technology and initially pivotally tested in pediatric and adults with malignancies [10–12]. This procedure depletes host-reactive donor T cells from hematopoietic stem cell allografts to prevent GvHD, reduces $CD19^+$ cells to decrease the risk of EBV-associated lymphoproliferative syndrome, and obtains a product enriched for stem cells and immune cells. In particular, high numbers of NK cells in T cell-depleted peripheral blood stem cells (PBSC) grafts facilitate engraftment and, together with γδ-TCR and monocytes, preserve antiinfective activity. This peculiar peripheral-blood (PB) graft manipulation, in the absence of any posttransplantation pharmacologic prophylaxis for GvHD, translates into a prompt recovery of γδ(+) T cells while αβ(+) T cells progressively ensued overtime sparing recipient from severe acute GvHD and chronic GvHD. The initial experience in children with nonmalignant diseases included several patients with severe aplastic anemia (SAA) and Fanconi anemia [10]; in this platform, two graft failures occurred in four patients with SAA, thus recommending the use of a maximal immune-suppressive preparation, including high dosages of cyclophosphamide or of total body or lymphoid irradiation. Functional and phenotypic characteristics of immune reconstitution after αβ(+) T cells and CD19(+) B-cells-depleted graft shows that γδ T lymphocytes γδ T cells are the predominant T-cell population in patients during the first weeks after transplantation, being mainly, albeit not only, derived from cells infused with the graft and expanding in vivo. Moreover, Vδ1 cells are expanded in patients experiencing cytomegalovirus reactivation and are more cytotoxic compared with those of children who did not experience reactivation.

UNMANIPULATED GRAFT HAPLO-SCT

Recently, the practice of haploidentical transplantation has benefited from the development of platforms of unmanipulated grafts. Pivotal experiences have been provided by Chinese centers based on bone marrow (BM) graft source and an extended pharmacologic GvHD prophylaxis. Huang et al. presented a large series of G-CSF primed BM and PB, antithymocyte globulin (ATG) in the conditioning, and a powerful posttransplantation GvHD prophylaxis [13–16]. The conditioning regimen prior to HSCT included busulfan, cyclophosphamide, and thymoglobulin. The recipients received CsA, mycophenolate mofetil, and short-term MTX for GvHD prophylaxis. All patients achieved 100% donor myeloid engraftment; the median time for myeloid engraftment was 12 days and for platelets was 18 days. Factors that promoted full engraftment in this trial were the intensified conditioning regimen which adds busulfan to CTX + ATG, the combination of G-BM and G-PB grafts that work synergistically to enhance engraftment, and the employment of a potent GvHD prophylactic regimen using CSA, MTX, and MMF. The overall survival (OS) was 64% with a median follow-up time of 746 days for surviving patients. However, reported chronic GvHD incidence was 56%. The clinical data of 18 children with SAA treated from 2010 to 2014 were recently reviewed by Zhang [17] with a median follow-up time of 2 years 23.5 months (range, 3–52 months). OS rate was 66.7%; however GvHD occurred in 15/18 patients of the HHCT group, including five cases with grade III or higher. Lu compared haplo-SCT to unrelated donor transplantation (UD-HSCT) [18]. Of a cohort of 50 SAA patients between 2012 and 2014, 26 patients underwent UD-HSCT and 24 patients haplo-SCT. OS rate was 91.3% with a median follow-up of 9 months. In this study, haplo-SCT developed a significant high incidence of aGvHD and cGvHD (37%). Wang et al. reviewed 17 children and adolescents with SAA after haplo-HSCT [19]. Neutrophil engraftment was achieved in all 17 patients in a median time of 16 days and platelet engraftment was achieved in 16 patients in a median time of 22 days. Acute GvHD was 30% and cGvHD was 21%. Secondary graft failure with autologous hematopoiesis recovery occurred in one patient. The overall survival was 70% at a median follow-up of 1 year.

Clinical trials based on post-BM transplantation cyclophosphamide targeting activated donor or host alloreactive T cells have been developed in Baltimore and Seattle [20–22]. Esteves et al. reported from Brazil the outcome of 16 patients who underwent haploidentical transplantation using a reduced-intensity conditioning regimen with posttransplant Cy [23]. Stem cell sources were BM ($N = 13$) or PBSCs ($N = 3$). The rate of neutrophil engraftment was 94% and of platelet engraftment was 75%. Two patients had secondary graft failure and were successfully salvaged with another transplant. Three patients developed acute GvHD and five patients have died with 1-year OS 67% (95% confidence interval: 36.5–86.4%). Clay et al. reported pilot findings of haplo-SCT using reduced-intensity conditioning with postgraft high-dose cyclophosphamide in eight patients with immune suppressive therapy (IST)-refractory SAA or as salvage treatment in patients who rejected a prior UD or cord blood transplant [24]. Six of eight patients engrafted; graft failure was associated with donor-directed HLA antibodies, despite intensive pre-HSCT desensitization with plasma exchange and rituximab. PT-Cy was highly effective in GvHD prevention as only one patient showed grade II skin GvHD. This initial trial documented the relevance of donor-directed HLA antibodies in the outcome of haplo-SCT in patients with SAA; routine screening for HLA antibodies is recommended pretransplant and,

if present an alternative donor lacking HLA antigens against which recipient HLA antibodies are directed should be used. PBSCs have also recently been used for reduced-intensity haplo-HSCT with postgraft CY in the setting of hematologic malignancy with no adverse impact on GvHD or survival compared with BM [25,26].

A calcineurin inhibitor-free GvHD prophylaxis based on rapamycin, mycophenolate mofetil (MMF), and anti-T lymphocyte globulin (ATG-Fresenius) was explored in the attempt to promote a fast posttransplant immune recovery with a preferential accumulation of regulatory T cells (Tregs) [27]. Rapamycin is an immunosuppressive drug that, in contrast to calcineurin inhibitors, promotes the generation of natural Tregs. In this initial study, GvHD prophylaxis with sirolimus-mycophenolate-ATG-F-rituximab promoted a rapid immune reconstitution skewed toward Tregs, allowing the infusion of unmanipulated haploidentical PBSC grafts in patients with advanced malignancies. More recently, the feasibility of PBSC grafts followed by a PTCy and sirolimus-based GvHD prophylaxis (Sir-PTCy) was investigated and low rates of GvHD and NRM as well as the favorable immune reconstitution profile were reported [28].

CONCLUSIONS

In conclusion, haploidentical SCT is a clinical option in patients with SAA failing at least one course of IST and lacking an HLA-matched related or unrelated donor.

Multiple platforms are currently explored and data are insufficient to propose an accepted standard of care. Ex vivo T-cell-depleted and unmanipulated grafts both represent available options potentially safe and effective to provide a stable donor hematopoietic engraftment with acceptable GvHD toxicities. Haploidentical transplants for refractory SAA should be offered by center with major experience with haplo-SCT procedures, and within clinical protocol designed specifically to address the primary objectives of a stable haploidentical hematopoietic engraftment at a low GvHD rate.

References

[1] Beatty PG, Clift RA, Mickelson EM, Nisperos BB, Flournoy N, Martin PJ, Sanders JE, Stewart P, Buckner CD, Storb R. Marrow transplantation from related donors other than HLA-identical siblings. N Engl J Med 1985;313(13):765–71.

[2] Aversa F, Tabilio A, Velardi A, Cunningham I, Terenzi A, Falzetti F, et al. Treatment of high risk acute leukemia with T-cell-depleted stem cells from related donors with one fully mismatched HLA haplotype. N Engl J Med 1998;339:1186–93.

[3] Ciceri F, Labopin M, Aversa F, Rowe JM, Bunjes D, Lewalle P, et al. Acute Leukemia Working Party (ALWP) of European Blood and Marrow Transplant (EBMT) Group. A survey of fully haploidentical hematopoietic stem cell transplantation in adults with high-risk acute leukemia: a risk factor analysis of outcomes for patients in remission at transplantation. Blood 2008;112:3574–81.

[4] Reisner Y, Hagin D, Martelli MF. Haploidentical hematopoietic transplantation: current status and future perspectives. Blood 2011;118(23):6006–17.

[5] Perruccio K, Tosti A, Burchielli E, Topini F, Ruggeri L, Carotti A, et al. Transferring functional immune responses to pathogens after haploidentical hematopoietic transplantation. Blood 2005;106:4397–406.

[6] Di Ianni M, Falzetti F, Carotti A, Terenzi A, Castellino F, Bonifacio E, et al. Tregs prevent GVHD and promote immune reconstitution in HLA-haploidentical transplantation. Blood 2011;117(14):3921–8.

[7] Ciceri F, Bonini C, Stanghellini MT, Bondanza A, Traversari C, Salomoni M, et al. Infusion of suicide-gene-engineered donor lymphocytes after family haploidentical haemopoietic stem-cell transplantation for leukaemia (the TK007 trial): a non-randomised phase I-II study. Lancet Oncol 2009;10(5):489–500.

[8] Lucarelli B, Merli P, Bertaina V, Locatelli F. Strategies to accelerate immune recovery after allogeneic hematopoietic stem cell transplantation. Expert Rev Clin Immunol 2016;12(3):343–58.

[9] Bethge WA, Haegele M, Faul C, Lang P, Schumm M, Bornhauser M, Handgretinger R, Kanz L. Haploidentical allogeneic hematopoietic cell transplantation in adults with reduced-intensity conditioning and CD3/CD19 depletion: fast engraftment and low toxicity. Exp Hematol 2006;34:1746–52.

[10] Bertaina A, Merli P, Rutella S, Pagliara D, Bernardo ME, Masetti R, Pende D, Falco M, Handgretinger R, Moretta F, Lucarelli B, Brescia LP, Li Pira G, Testi M, Cancrini C,

Kabbara N, Carsetti R, Finocchi A, Moretta A, Moretta L, Locatelli F. HLA-haploidentical stem cell transplantation after removal of αβ+ T and B cells in children with nonmalignant disorders. Blood 2014;124(5):822–6.

[11] Airoldi I, Bertaina A, Prigione I, Zorzoli A, Pagliara D, Cocco C, Meazza R, Loiacono F, Lucarelli B, Bernardo ME, Barbarito G, Pende D, Moretta A, Pistoia V, Moretta L, Locatelli F. δ T-cell reconstitution after HLA-haploidentical hematopoietic transplantation depleted of TCR-αβ+/CD19+ lymphocytes. Blood 2015;125(15):2349–58.

[12] Tumino M, Mainardi C, Pillon M, Calore E, Gazzola MV, Destro R, Strano A, Varotto S, Gregucci F, Basso G, Messina C. Haploidentical TCR A/B and B-cell depleted hematopoietic SCT in pediatric SAA and aspergillosis. Bone Marrow Transplant 2014;49(6):847–9.

[13] Huang XJ, Liu DH, Liu KY, Xu LP, Chen H, Han W, et al. Haploidentical hematopoietic stem cell transplantation without in vitro T-cell depletion for the treatment of hematological malignancies. Bone Marrow Transplant 2006;38(4):291–7.

[14] Liu D, Huang X, Liu K, Xu L, Chen H, Han W, et al. Haploidentical hematopoietic stem cell transplantation without in vitro T cell depletion for treatment of hematological malignancies in children. Biol Blood Marrow Transplant 2008;14(4):469–77.

[15] Huang XJ, Liu DH, Liu KY, Xu LP, Chen H, Han W, et al. Treatment of acute leukemia with unmanipulated HLA-mismatched/haploidentical blood and bone marrow transplantation. Biol Blood Marrow Transplant 2009;15(2):257–65.

[16] Xu LP, Liu KY, Liu DH, Han W, Chen H, Chen YH, Zhang XH, Wang Y, Wang FR, Wang JZ, Huang XJ. A novel protocol for haploidentical hematopoietic SCT without in vitro T-cell depletion in the treatment of severe acquired aplastic anemia. Bone Marrow Transplant 2012;47(12):1507–12.

[17] Zhang Y, Guo Z, Liu XD, He XP, Yang K, Chen P, Chen HR. Comparison of haploidentical hematopoietic stem cell transplantation and immunosuppressive therapy for the treatment of acquired severe aplastic anemia in pediatric patients. Am J Ther 2016.

[18] Gao L, Li Y, Zhang Y, Chen X, Gao L, Zhang C, Liu Y, Kong P, Wang Q, Su Y, Wang C, Wang S, Li B, Sun A, Du X, Zeng D, Li J, Liu H, Zhang X. Long-term outcome of HLA-haploidentical hematopoietic SCT without in vitro T-cell depletion for adult severe aplastic anemia after modified conditioning and supportive therapy. Bone Marrow Transplant 2014;49(4):519–24.

[19] Wang Z, Zheng X, Yan H, Li D, Wang H. Good outcome of haploidentical hematopoietic SCT as a salvage therapy in children and adolescents with acquired severe aplastic anemia. Bone Marrow Transplant 2014;49(12):1481–5.

[20] Im HJ, Koh KN, Choi ES, Jang S, Kwon SW, Park CJ, Chi HS, Seo JJ. Excellent outcome of haploidentical hematopoietic stem cell transplantation in children and adolescents with acquired severe aplastic anemia. Biol Blood Marrow Transplant 2013;19(5):754–9.

[21] Luznik L, O'Donnell P, Symons H, Chen AR, Leffell MS, Zahurak M, et al. HLA haploidentical bone marrow transplantation for hematologic malignancies using nonmyeloablative conditioning and high-dose, posttransplantation cyclophosphamide. Biol Blood Marrow Transplant 2008;14(6):641–50.

[22] Tuve S, Gayoso J, Scheid C, Radke J, Kiani A, Serrano D, et al. Haploidentical bone marrow transplantation with post-grafting cyclophosphamide: multicenter experience with an alternative salvage strategy. Leukemia 2011;25(5):880–3.

[23] Esteves I, Bonfim C, Pasquini R, Funke V, Pereira NF, Rocha V, Novis Y, Arrais C, Colturato V, de Souza MP, Torres M, Fernandes JF, Kerbauy FR, Ribeiro AA, Santos FP, Hamerschlak N. Haploidentical BMT and post-transplant Cy for severe aplastic anemia: a multicenter retrospective study. Bone Marrow Transplant 2015;50(5):685–9.

[24] Clay J, Kulasekararaj AG, Potter V, Grimaldi F, McLornan D, Raj K, de Lavallade H, Kenyon M, Pagliuca A, Mufti GJ, Marsh JC. Nonmyeloablative peripheral blood haploidentical stem cell transplantation for refractory severe aplastic anemia. Biol Blood Marrow Transplant 2014;20(11):1711–6.

[25] Castagna L, Crocchiolo R, Furst S, et al. Bone marrow compared with peripheral blood stem cells for haploidentical transplantation with a nonmyeloablative conditioning regimen and post-transplantation cyclophosphamide. Biol Blood Marrow Transplant 2014;20:724–72.

[26] Ciceri F, Lupo-Stanghellini MT, Korthof ET. Haploidentical transplantation in patients with acquired aplastic anemia. Bone Marrow Transplant 2013;48(2):183–5.

[27] Peccatori J, Forcina A, Clerici D, Crocchiolo R, Vago L, Stanghellini MT, Noviello M, Messina C, Crotta A, Assanelli A, Marktel S, Olek S, Mastaglio S, Giglio F, Crucitti L, Lorusso A, Guggiari E, Lunghi F, Carrabba M, Tassara M, Battaglia M, Ferraro A, Carbone MR, Oliveira G, Roncarolo MG, Rossini S, Bernardi M, Corti C, Marcatti M, Patriarca F, Zecca M, Locatelli F, Bordignon C, Fleischhauer K, Bondanza A, Bonini C, Ciceri F. Sirolimus-based graft-versus-host disease prophylaxis promotes the in vivo expansion of regulatory T cells and permits peripheral blood stem cell transplantation from haploidentical donors. Leukemia 2015;29(2):396–405.

[28] Cieri N, Greco R, Crucitti L, Morelli M, Giglio F, Levati G, Assanelli A, Carrabba MG, Bellio L, Milani R, Lorentino F, Stanghellini MT, De Freitas T, Marktel S, Bernardi M, Corti C, Vago L, Bonini C, Ciceri F, Peccatori J. Post-transplantation cyclophosphamide and sirolimus after haploidentical hematopoietic stem cell transplantation using a treosulfan-based myeloablative conditioning and peripheral blood stem cells. Biol Blood Marrow Transplant 2015;21(8):1506–14.

Management of Acquired Aplastic Anemia in Children

C. Dufour*, S. Samarasinghe**, M. Miano*

*Hematology Unit, G. Gaslini Children's Hospital, Genova, Italy; **Department of Haematology, Great Hormond Street Hospital, London, United Kingdom

DIAGNOSIS AND CLINICAL CHARACTERISTICS

Acquired aplastic anemia (AA) is a rare disease whose incidence in the western general population is 2–3 case/million [1]. Incidence is up to threefold higher in the Far East. Incidence is likely to be lower in childhood than in adult population. The diagnosis of AA in pediatric age is particularly challenging because of the rarity of the disease and the consistent risk of misdiagnosis with inherited bone marrow failure syndromes (IBMFSs) that occur more frequently in childhood than in adults. This may generate relevant treatment and follow-up implications. For this reason, in addition to the diagnostic algorithm available for the adults [2], it is wise to rely also on age adapted diagnostic packages [3] when dealing with a potential case of AA in childhood. If clinical and laboratory criteria suggest a diagnosis of AA some elements should be carefully looked at to identify underlying congenital defect. They include positive family history for anemia, for cytopenia, for hematologic malignancy and the presence of physical abnormalities (short stature, facial dismorphisms, cleft palate, cardiac or renal/genitourinary tract malformations, skeleton, nails, teeth, skin, and eyes abnormalities). These findings should encourage physicians to investigate patients with AA for possible IBMFS. However, even in the absence of family history or physical abnormalities, some diagnostic exams are highly recommended or even mandatory to rule out cryptic IBMFS. This is the case of the chromosomal fragility test for excluding Fanconi Anemia and the measurement of telomere length (TL) in leukocytes. TL shortening although occurring in subsets of patients with IBMFS including Fanconi Anemia, Diamond–Blackfan Anemia, and Schwachman–Diamond syndrome [4,5] is the classical hallmark of Dyskeratosis Congenita (DKC). Typically, TL below first centile for age points to a likely diagnosis of telomeropathy and indicates mutation study in one of the genes involved in the telomere maintenance. TL between first and tenth centile may suggest masked forms of telomeropathies, with no or limited somatic involvement (particularly if coexisting with a positive history for telomere-related problems like marrow failure, leukemia, lung, liver, skin diseases, and immune dysfunction) that might be due to lesions of TERC and TERT genes [6]. As some of these disorders may also be associated with immunodeficiency, peripheral blood immune-phenotyping and immunoglobulin serum level should also be part of the diagnostic screening. Table 11.1 shows the suggested work-up for children with AA.

Obviously, the advent of the next generation sequencing (NGS) and of the whole exome sequencing is likely to facilitate the identifications of cryptic forms of IBMFS thus providing an effective tool for the differential diagnosis with acquired AA.

Exposure to medications and history of infections should also be looked for in patient's history as drugs, hepatitis viruses, EBV, CMV, parvovirus, HHV6, HSV, HIV, adenovirus, and Varicella-Zoster may cause marrow hypoplasia and pancytopenia. As a PNH clone may be present in up to 41% of newly diagnosed children with AA [7] and may become clinically relevant throughout the course of AA, PNH clone analysis should be part of the diagnostic work-up bearing in mind that the small amount of leukocytes at diagnosis may induce falsely negative result which imposes retesting after neutrophil recovery. HLA typing should also be performed at diagnosis to enable a rapid finding of registry donor to be used for front-line transplant in children who lack an HLA identical family

TABLE 11.1 Diagnostic Work-Up of Aplastic Anemia in Children

Mandatory tests
- Full blood count with differential count
- Reticulocyte count
- Peripheral blood film
- Liver function tests
- Liver virus tests (antibodies and DNA/RNA)
- Bone marrow aspirate for morphology, cytogenetics, immunophenotype, Pearl's staining (for intracytoplasmic iron)
- Bone marrow trephine biopsy with immunostaining for CD34 and CD117
- Flow cytometry for PNH clones
- Autoantibody screening (antinucleus and anti-DNA for SLE detection)
- Vitamin B12 and folate serum levels
- Fibrinogen and serum ferritin (detection of HLH)
- Stool pancreatic elastase, serum pancreatic lipase (for identification of Shwachman–Diamond syndrome)
- Serum bilirubin and LDH
- Chest X-ray
- Abdomen US scan and echocardiography (for liver, spleen, lymph node enlargement, and malformations)

Mandatory tests for differential diagnosis with constitutional marrow failure syndromes
- Chromosomal fragility tests (MMC or DEB). Gold standard for the diagnosis of Fanconi Anemia
- Telomere length measurement (in alternative a TERC mutation analysis —for detection of hidden forms of DKC and TERT mutation analysis—for those who do not respond to IST)

Ancillary tests for the diagnosis of AA
- Search for mycobacteria infection (atypical mycobacteria more frequently than TB mycobacteria)
- Marrow progenitor assay (not available in all centers)

Ancillary tests for differential diagnosis with constitutional marrow failure syndromes
- TNF2, NHP2, NOP10, DKC1, and cMPL mutation analysis
- Shwachman–Diamond syndrome mutation analysis
- MRI of vertebral column

Modified from "Barone A, Lucarelli A, Onofrillo D, et al. Diagnosis and management of acquired aplastic anemia in childhood. Guidelines from the Marrow Failure Study Group of the Pediatric Haemato-Oncology Italian Association (AIEOP). Blood Cells Mol Dis 2015;55(1): 40-7.

donor. Moreover, HLA typing may also have prognostic meaning as subjects with HLA-DR2 and HLA-DRB1*15 are reported not only to have good probability to respond to Cyclosporin A (CsA) but also to become dependent on this treatment [8,9].

Anemia and thrombocytopenia are the most frequent symptoms at presentation of AA. Although less frequently neutropenia-related infections may occur at onset. A history of single lineage cytopenia or seronegative hepatitis may also be present and suggests a diagnosis of hepatitis-associated AA.

Clonal evolution of AA is considered the most cumbersome event. It is usually announced by worsening of blood counts (in responders to immunosuppression), by the appearance of dysplastic changes, and/or cytogenetic abnormalities in the bone marrow. Sometimes cytogenetic abnormalities, with the exclusion of monosomy 7, can be transient and need to be confirmed over follow-up [10]. New molecular diagnostic techniques, such as NGS have recently been applied [11] to investigate genes involved in clonal evolution of AA toward MDS. This may prove to be a very useful tool to survey on transformation of AA into MDS. A recent study showed that about 20% of AA patients carry somatic mutations of genes involved in epigenetic regulation of DNA transcription (TET2, DNMT3A, and BCOR) or in upregulation of immune response (ASXL1) that predicts the evolution to MDS. In particular in patients with shorter telomere and longer duration of the disease these mutations were associated with 40% risk of transformation [12,13].

SUPPORTIVE TREATMENT

Anemia should be treated with transfusions of packed red blood cells with the only aim to improve symptoms but not to maintain Hb above a particular level or to normalize anemia. Iron overload should be routinely monitored and chelation should be started when transfusion load is higher than 200 mL/kg or liver iron is >7 mg/g dry weight. If liver iron content is impossible to assess then ferritin, although fairly less specific, may be an acceptable surrogate marker of iron overload. Levels persistently above 1000 ng/L may indicate iron overload [10]. Thrombocytopenia should be treated with transfusion of platelets only when count drops below $10 \times 10^9 \, L^{-1}$ or if bleeding occurs. Some recommend prophylactic transfusions when platelets are below $30 \times 10^9 \, L^{-1}$ during antithymocyteglobuline (ATG) administration. All blood products should be leukodepleted to reduce the risk of sensitization and irradiated. Prophylaxis against pneumocystis pneumonia should be started in lymphopenic patients. Although no data from clinical studies in children are available, antifungal prophylaxis may be considered when neutrophil count is persistently $<0.2 \times 10^9 \, L^{-1}$. Antibiotic prophylaxis can also be considered an option with neutrophils $<0.2 \, L^{-1}$, between day +30 and +90 after combined immunosuppression therapy (IST) [3].

To maximize the chances of success of the treatment it is very important that all children diagnosed with AA should be referred to appropriate centers with expertise in AA and marrow failure disorders and offered an accurate diagnostic work-up and follow-up program.

GENERAL CONCEPTS FOR SPECIFIC TREATMENT

Timely specific treatment should consider the severity of the disease, the age, and the clinical status. Children with nonsevere AA might just be engaged in a surveillance program [14] especially if they are not transfusion-dependent and may eventually turn to a treatment regimen if blood count drops. On the contrary, all patients with severe or very severe disease should receive prompt treatment with the exceptions of those with severe clinical comorbidities who may be at serious risk not to tolerate aggressive therapies and should therefore be carefully clinically evaluated. In these cases, treatment options have to be discussed on the basis of the single patient's clinical status and needs.

OPTIONS FOR FIRST-LINE TREATMENT

Currently, at least three first line options are available for children diagnosed with AA. They are as follows: hematopoietic stem cell transplantation (HSCT) from matched sibling donor (MSD), combined IST, and HSCT from matched unrelated donor (MUD).

HSCT from MSD

Standard treatment for children who have an MSD is HSCT that provides a cure in about 90% of patients [15,16]. The European Blood and Marrow Transplantation (EBMT) Society recommends a conditioning regimen including cyclophosphamide (Cy) (200 mg/kg) given in 4 days, and rATG (7.5 mg/kg) [17]. The use of ATG may be considered optional as a prospective study comparing patients receiving Cy with or without ATG showed similar results [18]. An alternative option is fludarabine (150 mg/m^2) associated with lower Cy doses with the aim to prevent potential fertility impairment. Graft-versus-host disease (GvHD) prophylaxis with methotrexate and CsA was superior to CsA alone [19]. Acute GvHD rate varies between 10% and 20%. Chronic GvHD rate ranges between 8% and 20% in the younger patients and looks far lower than that observed in adults (30–40%) [15]. Risk factors for chronic GvHD are previous acute GvHD [20], marrow cell dose [21] and the use of peripheral blood stem cells (PBSC) [22]. Noteworthy, at variance from malignant diseases, the graft-versus-leukemia (GvL) effect of GvHD does not provide any benefit in AA. A significantly better outcome and a lower occurrence of GvHD has been shown in different studies [15,23] with the use of bone marrow that for this reason should be utilized as a preferred source of cells. To counterbalance the negative effect of GvHD, alemtuzumab has been incorporated in a conditioning regimen with fludarabine 120 mg/ m^2 and Cy1200 mg/m^2 used also in pediatric patients [24] and proved to generate a lower incidence of chronic GvHD. In addition, alemtuzumab permitted to avoid posttransplant methotrexate thus reducing toxic effects. The use of alentuzumab prior to and after transplant increased the incidence of graft failure but this negative effect was emended implying this drug only before the graft.

Graft rejection (GR) is another important issue in AA transplants. Low cell dose, donor recipient gender mismatching, and mixed chimerism are the most important risk factors. After the transplant, patients should receive Cyclosporin (CsA) at full dose for at least 9 months before de-escalating during the next 3–6 months. Rejection may occur many years after transplant [25,26] and this suggests the need for long-term frequent assessment of chimerism particularly during CsA reduction.

Immunosuppressive Therapy

For those children who lack an identical donor in the family and cannot be transplanted from an HLA MUD within 2–3 months since diagnosis, combined immunosuppression still represents a reasonable front-line option because of the high survival rate, which peaks to over 90% [16] and enables in case of failure a good rescue treatment with second-line unrelated donor (UD) HSCT.

ATG and CsA have been since many years the standard regimen in first line IST because they proved to be superior to ATG alone [27]. The addition of further immunosuppressive agents like mychophenolate or sirolimus to standard regimens, aiming to improve the response, was tested in prospective studies and failed to show outcome improvement [28,29]. Horse ATG (hATG) (ATGAM) at the dose of 40 mg/kg/day for 4 days proved to be the most effective agent, and superior to rabbit ATG (rATG) in two large prospective controlled trials. In the National Institutes of Health (NIH) study [30] the 3-year probability of survival was 96% with horse versus 66% with rATG and that was also associated with a higher incidence of early deaths. The EBMT study compared patients prospectively treated with hATG to matched historical controls who received rATG and confirmed these differences: hATG patients had a 2-year OS of 86% versus 68% of rATG patients. A higher incidence of fatal infection was observed in the rabbit group, although later (after day +100) than in the NIH study [31]. Interestingly, in a retrospective Japanese study, patients receiving rATG had similar response rates to those treated with hATG. However, 2-year and 10-year OS were superior in the hATG group. All these results had an important impact on the management of hATG supply [32]. In fact hATG was withdrawn from the market in Europe after 2007 leaving rATG, the former second-line choice in patients who failed horse ATG, as the only available ATG. The results of the aforementioned prospective trials contributed to make hATG again available in many European countries.

CsA should be started at day +1 at the dose of 5 mg/kg/day, divided into two daily doses, continued for 12 months and then slowly tapered off over another 12 months. CsA serum levels should be maintained around 200 ng/mL. Steroids (prednisone 2 mg/kg/day) should also be associated for at least 14 days to prevent serum sickness. The duration of CsA administration is a critical point in the management of responders to immunosuppression. Controversial results have been obtained in different studies in pediatric and adult populations. In a retrospective Italian study in children [33] a slow tapering of CsA (0.3–0.7 mg/kg/month) was correlated to a lower relapse incidence when compared to a rapid drop (>0.8 mg/kg). An NIH study [34] on adult patients using higher doses did not show differences in terms of relapse between patients who stopped CsA after 6 months and those who continued and tapered it off over the following 18 months. However given that side effects due to CsA treatment were not reported as a major problem in children [33], as a general rule, 1-year treatment with full dose CsA followed by a slow tapering off over the next 12 months can be considered a reasonable schedule.

In a prospective randomized study conducted by the EBMT [35] G-CSF was shown to increase the count of neutrophils and reduce both infections and days of hospitalization although it did not affect response rate, EFS and OS. G-CSF demonstrated to predict response, since patients who under G-CSF experienced a neutrophil count $\geq 0.5 \times 10^9$ L^{-1} on day + 30 had a higher chance to respond to IST. However, many concerns have risen toward the use of this growth factor mainly because of an increased risk of clonal diseases that was reported in a retrospective analysis [35], but that however was not confirmed in prospective trials [36,37]. Overall based on these concerns, G-CSF could be used during the first 30 days of treatment and beyond in case of infective episodes in neutropenic patients.

Most of responses to IST are seen during the first 3–4 months after ATG, with few patients experiencing a slower response within 6 months or even afterward [38]. Failure of AA patients to respond to IST may also be explained by the presence of cryptic IBMFS, a differential diagnosis that can be reconsidered in this circumstance. However in case of no response within 3–4 months from IST start, patients can have good chances of rescue if an HSCT from HLA matched UD is done quickly before invasive infections or excess sensitization due to transfusions occur. Hence there is the need to start the donor research early at diagnosis to have the donor available 3–4 months after IST start.

HSCT from MUD

In the last few years, the outcome of this procedure has remarkably improved thanks to better supportive therapies and donor selection. A recent study from the European Blood and Marrow Transplantation (EBMT) Society [17] showed that HLA MUD HSCT gives a probability of EFS of 92% that compares with the 87% achieved by MSD HSCT and is far superior to the 40% provided by IST (Fig. 11.1). Noteworthy, the interval between diagnosis and neutrophil engraftment was not significantly longer in MUD (0.39 years) over MSD HSCT (0.31 years; $p = 0.93$). Chronic GvHD rate was not low (19%), but it was limited to skin in all cases and subsided with local steroids. Viral reactivation was frequent whereas viral disease was not. Posttransplant median hospital stay was 42 days and the transfusion requirement was quite low with a median number of red cells and platelets transfusions of 5 and 19 units, respectively. Whole blood chimerism was 100% and T-cell chimerism 96.5%. The conditioning regimen used in the UK serie included fludarabune 150 mg/m², cyclophosphamide 120 mg/m², and Campath 0.9–1 mg/kg (FCC). GvHD prophylaxis was carried out with CsA alone. Of relevance, the outcome of MUD HSCT upfront is superior (OS and EFS 95%) to that of the same treatment performed as rescue option after failure of IST (OS and EFS 74%; $p = 0.02$). This, in addition to the fact that HSCT provides a better hematopoietic reconstitution [39] and lowers the risk of secondary malignancies in comparison to IST [15,16] is another factor in favor of this treatment option.

A recent study showed that a longer TL of the donor (and not in the recipient) is associated with better post-HSCT survival thus pointing to telomeres length measurement as a new criteria for donor selection aimed to further improve the outcome of these transplants [40].

For all these reasons, if the clinician evaluates as highly likely that the transplant be done within 2–3 months since diagnosis, MUD HSCT can be reasonably considered as a front-line option in children diagnosed with AA if no family donor is found. This is an additional reason for an early HLA typing and donor research start at diagnosis if no HLA matched donor is available in the family.

OPTIONS FOR SECOND LINE TREATMENTS

Patients in need of further treatment are those who fail front-line IST or, more rarely, upfront HSCT. In case of failure of IST front line, HSCT from MUD proves to be a successful option with EFS of about 80% reported in children [16] and of about 70% in adolescents [15]. By using a reduced intensity conditioning regimen this transplant provided an excellent outcome with OS and failure free survival (FSS) above 90% in children who failed front-line IST [41]. Of note although data are derived from a small cohort, encouraging results [18] were obtained with one antigen mismatched UD HSCT that might therefore also be considered when a matched donor is not available.

In patients who rejected a first transplant, second HSCT proved to be a good rescue option in about 60% of cases [42]. Children who fail upfront IST and lack a UD are eligible to undergo second-line immunosuppressive treatments. Studies including pediatric populations showed that for those who relapsed after an initial course of rATG a second course of the same ATG offers a probability of response of 65%. For those who instead did never respond to the first cycle of rATG, a second round of the same drug provided a lower chance of response of about 30%. Survival in responders was 90% whereas it dropped to 60–65% in nonresponders to second IST [43].

Also alentuzumab has been used as second-line treatment for patients, including children, who failed a first course of IST. Also with this

(A)

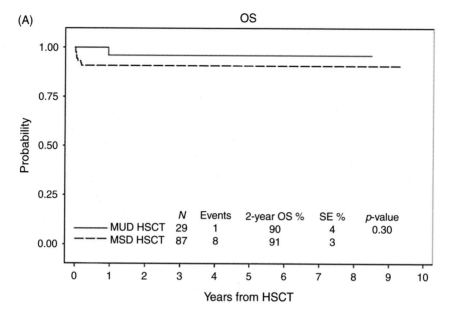

(B)

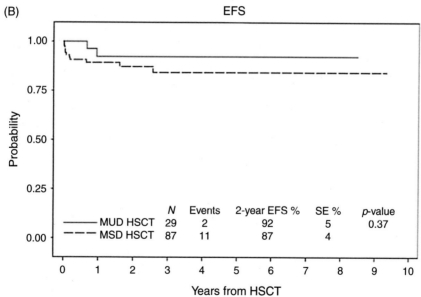

FIGURE 11.1   OS And EFS of 29 AA children treated with MUD HSCT upfront compared to matched historical control who received MSD HSCT. *From Dufour C, Veys P, Carraro E, et al. Similar outcome of upfront-unrelated and matched sibling stem cell transplantation in idiopathic paediatric aplastic anaemia. A study on behalf of the UK Paediatric BMT Working Party, Paediatric Diseases Working Party and Severe Aplastic Anaemia Working Party of EBMT. Br J Haematol. 2015;171(4):585-94*

agent response was higher in relapsing (56%) over refractory patients (37%) to upfront immunosuppression [44].

OPTIONS FOR THIRD LINE TREATMENTS

For those children who after failure of upfront IST do not have an available UD or for those in whom second line IST did not work, the following third line options are available: haploidentical HSCT, cord blood transplant, and other nontransplant treatment.

Haploidentical HSCT

Patients who failed immunosuppressive treatments and lack an unrelated donor might be considered eligible for haploidentical HSCT. Some experiences using different conditioning regimens have been published, but the limited number of patients and the absence of prospective studies render difficult to make recommendation. Both unmanipulated and ex vivo T-cell-depleted transplants have been performed. The first platform was used in patients conditioned with TBI and Cy or with Busulfan, Cy and ATG. This latter regimen was used in 19 consecutive patients who received a combination of bone marrow and PBSC from haploidentical family donors [45]. OS was 64% but a high rate of acute and chronic GvHD (42 and 56%, respectively) was reported.

A single center study on 17 children and adolescents with AA undergoing haploidentical HSCT showed an OS of 70%, one case of graft failure and again a high incidence of acute and chronic GvHD of 30 and 21%, respectively [46].

Manipulated grafts were administered to three children after a myeloablative regimen (fludarabine–Cy–ATG or low dose TBI–fludarabine–Cy) and positive selection of CD34+ cells [47]. Outcome was encouraging with 2/3

patients surviving with no GvHD. In vitro T-cell-depleted haploidentical PBSCs were infused to 12 children and adolescents conditioned with Flu, Cy, ATG ± TBI. After a follow-up of 14 months engraftment and OS were excellent (both 100%) but GvHD rate was still high (three out of nine evaluable patients) [48]. Removal of α+, β+ T and CD19+ B cells has been recently used before haploidentical transplantation in 23 children with nonmalignant disorders including 4 patients with SAA. Results of the whole cohort were promising but no conclusions are allowed in the setting of SAA [49].

Posttransplant cyclophosphamide was used in haploidentical HSCT to selectively deplete donor alloreactive T cells and reduce acute GvHD. This approach was used in 16 SAA patients including children mostly refractory to IST, who received BM (13) or PB (3) after an RIC consisting of Cy, Flu, TBI (200–600 Cy), and Cy 50 mg/kg on days +3 and +4. One-year OS was 67%. Acute and chronic GvHD were reported in 3/16 and 4/16 patients, respectively [50].

Cord Blood Transplant

Umbilical cord blood transplantation (CBT) may be a treatment option for patients who lack a MS and a MUD. The risk of graft failure in advanced and sensitized AA patients is higher when using a potentially rather poor source of cell as cord blood [51]. Limited data are available so far on its use in children with AA. In a Eurocord study 71 children (median age 13 years) with SAA received either single or double unit of cord blood. Prevalent (69%) conditioning regimen was fludarabine-based. Full donor chimerism was achieved in 82% and overall survival was 38%. Cell dose was considered a critical factor on outcome and a total of $>3.9 \times 10^7 \text{ kg}^{-1}$ nucleated cells were associated with a better outcome [51]. To overcome cell dose limitations, double unit CBTs were performed in AA. Although 2-year OS was good

(80%), the incidence of acute GvHD was high (71%) [52]. A recent study from Eurocord and EBMT on congenital and acquired marrow failure syndromes, including AA patients, showed a 5-year probability of survival of 73 ±9% [53].

Other Nontransplant Treatments

A prospective trial in which Cy 200 mg/kg for 4 days without HSCT rescue was used [54,55] showed a good response rate but also a very high incidence of neutropenia-related infections. A recent study from the NIH [56] tested moderate dose of Cy (120 mg/kg) along with low dose CSA. The data safety monitoring board eventually recommended the termination of the accrual of this study due to low response rate and high rate of fungal infections.

There is strong evidence that Eltrombopag, an analog of the thrombopoietin receptor, stimulates hemapopoietic stem cells and can reconstitute good level of hematopoiesis in patients with aplastic marrow. In the original phase II trial, eltrombopag induced a sustained hematologic response in 44% of 25 adult patients refractory to IS without heavy toxicity [57]. In some case this response was tri-lineage and this raised the issue of using this drug in upfront therapy. However, it has to be noted that follow up of this study showed that 8 out 43 patients (18.6%) developed clonal cytogenetic abnormalities after 3–13 months of eltrombopag administration [58]. Data from a prospective trial investigating the role of eltrombopag in addition to standard IST showed an OS of 90% and a response rate of about 60% in the arm in which eltrombopag was given since day 1 [59]. A recent retrospective study on the use of recombinant human thrombopoietin (rhTPO) associated with IST in 40 AA patients >16 years showed an increase in both platelets and red cells count. Even if no differences in terms of OS were shown compared with controls, its efficacy on both erythroid cells and megakaryocytes

suggests a potential role in reducing transfusion dependency [60].

Although reports sustain a possible role in some patients [61,62], androgens did not prove to be effective when tested in randomized studies in AA patients [63–65]. They might be considered as an alternative option in patients with refractory disease. Treatment should be continued for at least 3 months to show its benefit.

Other agents might have a role as alternative immunosuppressive option. Some reports suggest that levamisole can be a practicable and economical solution for patients with AA. Levamisole, used as an anthelminthic agent for many years, is also known to have an immunomoduolatory effect [66]. Its association with CsA induced a response rate of 52.9% at 6 months in adults and children with refractory disease [67]. A 5-year OS of 80.5% was observed in moderate AA [68] thus suggesting a potential role of the combination of this drug with CsA in the immunosuppressive strategy of moderate AA.

Porcine ATG was used as an alternative option in adults and children with AA in combination with cyclosporine and showed similar efficacy to rabbit ATG but at lower costs [69,70]. Age ≥45 years, very severe disease, and lower baseline absolute lymphocyte count ($<1 \times 10^9$/L) were unfavorable predictors of overall survival [71]. These findings need confirmation in prospective studies but this source of ATG may have a future as a less expensive form of immunosuppression in countries where hATG is not marketed.

Finally a scheme including Flu, Cy, and CsA has been reported to be successful in 7 adults and in one 14-year-old boy and can be either regarded as a third line option or as more upfront treatment in less developed countries where ATG costs are unaffordable [72].

Fig. 11.2 shows a modified treatment algorithm from the Marrow Failure Group of the Pediatric Haemato-Oncology Italian Association (AIEOP) for children with aplastic anemia.

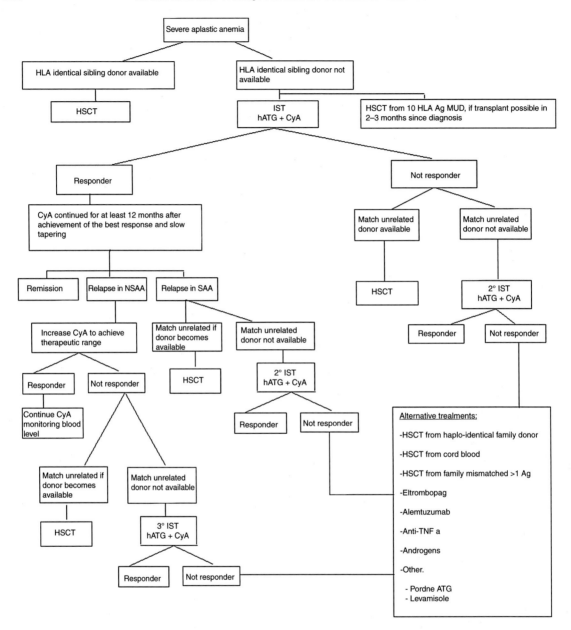

Legend:
HLA: human leucocyte antigen. HSCT: hematopoitic stem cell transplant. MUD: matched unrelated donor. IST: immunosoppressive therapy.
hATG: horse anti-thymocyte globulin. CyA: cyclosporine. NSAA: non-severe aplstic anemia. SAA: severe aplastic anemia.

FIGURE 11.2   **Treatment algorithm in children with AA.** *Modified from "Barone A, Lucarelli A, Onofrillo D, et al. Diagnosis and management of acquired aplastic anemia in childhood. Guidelines from the Marrow Failure Study Group of the Pediatric Haemato-Oncology Italian Association (AIEOP). Blood Cells Mol Dis. 2015;55(1):40-7.*

References

[1] Young NS. Bone Marrow Failure Syndromes. Philadelphia: W.B. Saunders Company; 1996.

[2] Rovó A, Tichelli A, Dufour C. SAA-WP EBMT. Diagnosis of acquired aplastic anemia. Bone Marrow Transplant 2013;48(2):162–7.

[3] Barone A, Lucarelli A, Onofrillo D, et al. Diagnosis and management of acquired aplastic anemia in childhood. Guidelines from the Marrow Failure Study Group of the Pediatric Haemato-Oncology Italian Association (AIEOP). Blood Cells Mol Dis 2015;55(1):40–7.

[4] Pavesi E, Avondo F, Aspesi A, et al. Analysis of telomeres in peripheral blood cells from patients with bone marrow failure. Pediatr Blood Cancer 2009;53(3):411–6.

[5] Alter BP, Giri N, Savage S, Rosenberg PS. . Haematologica 2015;(1):49–54.

[6] Townsley DM, Dumitriu B, Young NS. Bone marrow failure and the telomeropathies. Blood 2014;124(18):2775–83.

[7] Timeus F, Crescenzio N, Longoni D, et al. Paroxysmal nocturnal hemoglobinuria clones in children with acquired aplastic anemia: a multicentre study. PLoS One 2014;9(7):e101948.

[8] Kapustin SI, Popova TI, Lyschov AA, Togo AV, Abdulkadyrov KM, Blinov MN. HLA-DR2 frequency increase in severe aplastic anemia patients is mainly attributed to the prevalence of DR15 subtype. Pathol Oncol Res 1997;3(2):106–8.

[9] Sugimori C, Yamazaki H, Feng X, et al. Roles of DRB1 *1501 and DRB1 *1502 in the pathogenesis of aplastic anemia. Exp Hematol 2007;35(1):13–20.

[10] Marsh JC, Ball SE, Cavenagh J, et al. Guidelines for the diagnosis and management of aplastic anaemia. Br J Haematol 2009;147(1):43–70.

[11] Khincha PP, Savage SA. Genomic characterization of the inherited bone marrow failure syndromes. Semin Hematol 2013;50(4):333–47.

[12] Kulasekararaj AG, Jiang J, Smith AE, Mohamedali AM, Mian S, Gandhi S. Somatic mutation identify a subgroup of aplastic anemia patients that progress to myelodisplastic syndrome. Blood 2014;124(17):2698–704.

[13] Yoshizato T, Dumitriu B, Hosokawa K, et al. Somatic mutations and clonal hematopoiesis in aplastic anemia. N Engl J Med 2015;373(1):35–47.

[14] Kwon JH, Kim I, Lee YG, et al. Clinical course of non-severe aplastic anemia in adults. Int J Hematol 2010;91(5):770–5.

[15] Dufour C, Pillon M, Passweg J, et al. Outcome of aplastic anemia in adolescence. A survey of the Severe Aplastic Anemia Working Party of the European Group for Blood and Marrow Transplantation. Haematologica 2014;99(10):1574–81.

[16] Dufour C, Pillon M, Socie G, et al. Outcome of aplastic anaemia in children. A study by the severe aplastic anaemia and paediatric disease working parties of the European group blood and bone marrow transplant. Br J Haematol 2015;169(4):565–73.

[17] Dufour C, Veys P, Carraro E, et al. Similar outcome of upfront-unrelated and matched sibling stem cell transplantation in idiopathic paediatric aplastic anaemia. A study on behalf of the UK Paediatric BMT Working Party, Paediatric Diseases Working Party and Severe Aplastic Anaemia Working Party of EBMT. Br J Haematol 2015;171(4):585–94.

[18] Champlin RE, Perez WS, Passweg JR, et al. Bone marrow transplantation for severe aplastic anemia: a randomized controlled study of conditioning regimens. Blood 2007;109(10):4582–5.

[19] Locatelli F, Bruno B, Zecca M, et al. Cyclosporin A and short-term methotrexate versus cyclosporin A as graft versus host disease prophylaxis in patients with severe aplastic anemia given allogeneic bone marrow transplantation from an HLA identical sibling: results of a GITMO/EBMT randomized trial. Blood 2000;96:1690–1697.

[20] Storb R, Prentice RL, Sullivan KM Predictive factors in chronic graft-versus host disease in patients with aplastic anemia treated by marrow transplantation from HLA identical siblings. Ann Intern Med 1983;98:461–466.

[21] Kahl C, Leisenring W, Deeg HJ, et al. Cyclophosphamide and antithymocyte globulin as a conditioning regimen for allogeneic marrow transplantation in patients with aplastic anaemia: a long-term follow-up. Br J Haematol 2005;130(5):747–51.

[22] Eapen M, Le Rademacher J, Antin JH, et al. Effect of stem cell source on outcomes after adult unrelated donor transplantation in severe aplastic anemia. Blood 2011;118:2618–2621.

[23] Bacigalupo A, Socié G, Schrezenmeier H, et al. Bone marrow versus peripheral blood as the stem cell source for sibling transplants in acquired aplastic anemia: survival advantage for bone marrow in all age groups. Haematologica 2012;97(8):1142–8.

[24] Marsh JC, Gupta V, Lim Z, et al. Alemtuzumab with fludarabine and cyclophosphamide reduces chronic graft versus- host disease after allogeneic stem cell transplantation for acquired aplastic anemia. Blood 2011;118:2351–7.

[25] Dufour C, Dallorso S, Casarino L, et al. Late graft failure 8 years after first bone marrow transplantation for severe acquired aplastic anemia. Bone Marrow Transplant 1999;23(7):743–5.

[26] Eapen M, Davies SM, Ramsay NK. Late graft rejection and second infusion of bone marrow in children with aplastic anaemia. Br J Haematol 1999;104(1):186–8.

[27] Locasciulli A, Oneto R, Bacigalupo A, et al. Severe Aplastic Anemia Working Party of the European Blood and Marrow Transplant Group. Outcome of patients with acquired aplastic anemia given first line bone

marrow transplantation or immunosuppressive treatment in the last decade: a report from the European Group for Blood and Marrow Transplantation (EBMT). Haematologica 2007;92:11–8.

[28] Scheinberg P, Nunez O, Wu C, Young NS. Treatment of severe aplastic anaemia with combined immunosuppression: anti-thymocyte globulin, ciclosporin and mycophenolate mofetil. Br J Haematol 2006;133(6):606–11.

[29] Scheinberg P, Wu CO, Nunez O, et al. Treatment of severe aplastic anemia with a combination of horse antithymocyte globulin and cyclosporine, with or without sirolimus: a prospective randomized study. Haematologica 2009;94(3):348–54.

[30] Scheinberg P, Nunez O, Weinstein B, et al. Horse versus rabbit antithymocyte globulin in acquired aplastic anemia. N Engl J Med 2011;365:430–8.

[31] Marsh JC, Bacigalupo A, Schrezenmeier H, et al. Prospective study of rabbit antithymocyte globulin and cyclosporine for aplastic anemia from the EBMT Severe Aplastic Anaemia Working Party. Blood 2012;119(23):5391–6.

[32] Yamazaki H, Saito C, Sugimori N, et al. Thymoglobuline is as effective as lymphoglobuline in Japanese patients with aplastic anemia possessing increased glycosylphosphatidylinositol-anchored protein (GPI-AP) deficient cells [abstract]. Blood (ASH Annual Meeting Abstracts) 2011;118(21):1339.

[33] Saracco P, Quarello P, Iori AP, et al. Cyclosporin A response and dependence in children with acquired aplastic anaemia: a multicentre retrospective study with long-term observation follow-up. Br J Haematol 2008;140(2):197–205.

[34] Scheinberg P, Rios O, Scheinberg P, Weinstein B, Wu CO, Young NS. Prolonged cyclosporine administration after antithymocyte globulin delays but does not prevent relapse in severe aplastic anemia. Am J Hematol 2014;89(6):571–4.

[35] Socie G, Mary JY, Schrezenmeier H, et al. Granulocyte-stimulating factor and severe aplastic anemia: a survey by the European Group for Blood and Marrow Transplantation. Blood 2007;109:2794–6.

[36] Teramura M, Kimura A, Iwase S, et al. Treatment of severe aplastic anemia with antithymocyte globulin and cyclosporine A with or without G-CSF in adults: a multicenter randomized study in Japan. Blood 2007;110:1756–61.

[37] Gluckman E, Rokicka-Milewka R, Hann I, et al. Results and follow-up of a phase III randomized study of recombinant human- granulocyte stimulating factor as support for immunosuppressive therapy in patients with severe aplastic anaemia. Br J Haematol 2002;119:1075–82.

[38] Jeong DC, Chung NG, Cho B, et al. Long-term outcome after immunosuppressive therapy with horse or rabbit antithymocyte globulin and cyclosporine for severe aplastic anemia in children. Haematologica 2014;99(4):664–71.

[39] Podesta M, Piaggio G, Frassoni F, et al. The assessment of the hematopoietic reservoir after immunosuppressive therapy or bone marrow transplantation in severe aplastic anemia. Blood 1998;91(6):1959–65.

[40] Gadalla SM, Wang T, Haagenson M, et al. Association between donor leukocyte telomere length and survival after unrelated allogeneic hematopoietic cell transplantation for severe aplastic anemia. JAMA 2015;313(6):594–602.

[41] Samarasinghe S, Steward C, Hiwarkar P. Excellent outcome of matched unrelated donor transplantation in paediatric aplastic anaemia following failure with immunosuppressive therapy: a United Kingdom retrospective multicentre experience. Br J Haematol 2012;157(3):339–46.

[42] Cesaro S, de Latour RP, Tridello G, et al. Severe Aplastic Anaemia Working Party of the European Group for Blood and Marrow Transplantation. Second allogeneic stem cell transplant for aplastic anaemia: a retrospective study by the severe aplastic anaemia working party of the European society for blood and marrow transplantation. Br J Haematol 2015;171(4):606–14.

[43] Scheinberg P, Nunez O, Young NS. Re-treatment with rabbit antithymocyte globulin and ciclosporin for patients with relapsed or refractory severe aplastic anaemia. Br J Haematol 2006;133:622–7.

[44] Scheinberg P, Nunez O, Weinstein B, Scheinberg P, Wu CO, Young NS. Activity of alemtuzumab monotherapy in treatment-naive, relapsed, and refractory severe acquired aplastic anemia. Blood 2012;119(2):345–54.

[45] Xu LP, Liu KY, Liu DH, et al. A novel protocol for haploidentical hematopoietic HSCT without in vitro T-cell depletion in the treatment of severe acquired aplastic anemia. Bone Marrow Transplant 2012;47(12):1507–12.

[46] Wang Z, Zheng X, Yan H, Li D, Wang H. Good outcome of haploidentical hematopoietic SCT as a salvage therapy in children and adolescents with acquired severe aplastic anemia. Bone Marrow Transplant 2014;49(12):1481–5.

[47] Woodard P, Cunningham JM, Benaim E, et al. Effective donor lymphohematopoietic reconstitution after haploidentical CD34 + -selected hematopoietic stem cell transplantation in children with refractory severe aplastic anemia. Bone Marrow Transplant 2004;33:411–8.

[48] Im HJ, Koh KN, Choi ES, et al. Excellent outcome of haploidentical hematopoietic stem cell transplantation in children and adolescents with acquired severe aplastic anemia. Biol Blood Marrow Transplant 2013;19(5):754–9.

[49] Bertaina A, Merli P, Rutella S, et al. HLA-haploidentical stem cell transplantation after removal of αβ+ T and B cells in children with nonmalignant disorders. Blood 2014;124(5):822–6.

[50] Esteves I, Bonfim C, Pasquini R, et al. Haploidentical BMT and post-transplant Cy for severe aplastic anemia: a multicenter retrospective study. Bone Marrow Transplant 2015;50(5):685–9.

[51] Peffault de Latour R, Purtill D, Ruggeri A, et al. Influence of nucleated cell dose on overall survival of unrelated cord blood transplantation for patients with severe acquired aplastic anemia: a study by Eurocord and the Aplastic Anemia Working Party of the European Group for Blood and Marrow Transplantation. BBMT 2010;1:78–85.

[52] Ruggeri A, Peffault de Latour R, Rocha V, et al. Double cord blood transplantation in patients with high risk bone marrow failure syndromes. Br J Haematol 2008;143:404–8.

[53] Pagliuca S, Peffault de Latour R, Locatelli F. Outcomes of cord blood transplantation from an HLA-identical sibling for patients with inherited or acquired bone marrow failure disorders: a report from Eurocord, Cord Blood Committee (CBC-CTIWP) and Severe Aplastic Anemia Working Party (SAAWP) of the European Group of Blood and Bone Marrow Transplantation (EBMT). Abstract 153 Blood Abstracts: 57th Annual Meeting Abstracts.vol. 126, issue 23, December 3, 2015.

[54] Tisdale JF, Dunn DE, Geller N, et al. High-dose cyclophosphamide in severe aplastic anaemia: a randomised trial. Lancet 2000;356:1554–9.

[55] Tisdale JF, Maciejewski JP, Nunez O, Rosenfeld SJ, Young NS. Late complications following treatment for severe aplastic anemia (SAA) with high-dose cyclophosphamide (Cy): follow-up of a randomized trial. Blood 2002;100:4668–70.

[56] Scheinberg P, Townsley D, Dumitriu B, et al. Moderate dose cyclophosphamide for severe aplastic anemia has significant toxicity and does not prevent relapse and clonal evolution. Blood 2014;124(18):2820–3.

[57] Olnes MJ, Scheinberg P, Calvo KR, et al. Eltrombopag and improved hematopoiesis in refractory aplastic anemia. N Engl J Med 2012;367(1):11–9.

[58] Desmond R, Townsley DM, Dumitriu B, et al. Eltrombopag restores trilineage hematopoiesis in refractory severe aplastic anemia that can be sustained on discontinuation of drug. Blood 2014;123(12):1818–25.

[59] Zhao X, Feng X, Townsley DM. Persistent elevation of plasma thrombopoietin levels in severe aplastic anemia, even with hematologic recovery. Abstract 3609 Blood Abstracts: 57th Annual Meeting Abstracts. vol. 126, issue 23, December 3, 2015.

[60] Wang H, Dong Q, Fu R, et al. Recombinant human thrombopoietin treatment promotes hematopoiesis recovery in patients with severe aplastic anemia receiving immunosuppressive therapy. Biomed Res Int 2015;2015:597293.

[61] Gardner FH, Juneja HS. Androstane therapy to treat aplastic anaemia in adults: an uncontrolled pilot study. Br J Haematol 1987;65(3):295–300.

[62] French Cooperative Group for the Study of Aplastic and Refractory Anemias. Androgen therapy in aplastic anaemia: a comparative study of high and low-doses and of 4 different androgens. Scand J Haematol 1986;36(4):346–52.

[63] Kojima S, Hibi S, Kosaka Y, et al. Immunosuppressive therapy using antithymocyte globulin, cyclosporine,and danazol with or without human granulocyte colony-stimulating factor in children with acquired aplastic anemia. Blood 2000;96(6):2049–54.

[64] Doney K, Pepe M, Storb R, et al. Immunosuppressive Therapy of aplastic anemia: results of a prospective, randomized trial of antithymocyte globulin (ATG), methylprednisolone, and oxymetholone to ATG, very high-dose methylprednisolone, and oxymetholone. Blood 1992;79(10):2566–71.

[65] Champlin RE, Ho WG, Feig SA, Winston DJ, Lenarsky C, Gale RP. Do androgens enhance the response to antithymocyte globulin in patients with aplastic anemia? A prospective randomized trial. Blood 1985;66(1):184–8.

[66] Stevenson HC, Green I, Hamilton JM, Calabro BA, Parkinson DR. Levamisole: known effects on the immune system, clinical results, and future applications to the treatment of cancer. J Clin Oncol 1991;9:2052–66.

[67] Shao Y, Li X, Shi J, Ge M, et al. Cyclosporin combined with levamisole for refractory or relapsed severe aplastic anaemia. Br J Haematol 2013;162(4):552–5.

[68] Wang M, Li X, Shi J, et al. Outcome of a novel immunosuppressive strategy of cyclosporine, levamisole and danazol for severe aplastic anemia. Int J Hematol 2015;102(2):149–56.

[69] Chen M, Liu C, Zhuang J, et al. Long-term follow-up study of porcine anti-human thymocyte immunoglobulin therapy combined with cyclosporine for severe aplastic anemia. Eur J Haematol. 2016;96(3):291–6.

[70] Liu L, Ding L, Hao L, et al. Efficacy of porcine antihuman lymphocyte immunoglobulin compared to rabbit antithymocyte immunoglobulin as a first-line treatment against acquired severe aplastic anemia. Ann Hematol 2015;94(5):729–37.

[71] Wei J, Huang Z, Guo J, et al. Porcine antilymphocyte globulin (p-ALG) plus cyclosporine A (CsA) treatment in acquired severe aplastic anemia: a retrospective multicenter analysis. Ann Hematol 2015;94(6):955–62.

[72] Liu Q, Zhao X, Xu N, et al. Frontline therapy of severe aplastic anaemia with fludarabine, cyclophosphamide and ciclosporin. Br J Haematol 2015;171(3):427–30.

Treatment of Elderly Patients With Aplastic Anemia

A. Tichelli*, A. Rovó**, J.C.W. Marsh†

*Hematology, University Hospital of Basel, Basel, Switzerland; **Hematology, University Hospital of Bern, Bern, Switzerland; †Department of Haematological Medicine, King's College Hospital/King's College London, London, United Kingdom

Aplastic anemia (AA) is a rare disease occurring in all age groups but with two peak incidences from 10 to 20 years and over 60 years. Particularly in the elderly, it can often be difficult to distinguish between AA and hypocellular myelodysplastic syndrome (MDS), as this last entity is a relatively common bone-marrow disorder in advanced age, as is the finding of clonal hematopoiesis. This is demonstrated by the finding of acquired somatic mutations in genes typical for MDS/AML in around 10% of healthy individuals aged more than 45 years [1]. A careful review of the history of the patient and an examination of the blood film and bone-marrow investigations, including, aspirate, trephine, and cytogenetics/FISH by a specialist is of particular importance. The diagnostic approach in patients with AA is discussed in Chapter 3 "Diagnosis of acquired AA." The treatment decision-making process in an elderly patient with AA is based on disease-related factors but also, importantly, on his general health condition and functional ability. In the first part of this chapter, general issues on aging and the way to assess the ability of an elderly patient to undergo a treatment, and in the second part, treatment options of elderly patients with AA will be discussed.

AGING AND ITS CONSEQUENCES ON THE APPROACH TO TREATMENT

Aging is a physiologic process characterized by the progressive decline in one's ability to withstand stress due to functional, cognitive, and social limitations. Elderly population is usually defined by chronologic age. For clinical trials the age to define elderly patients is often set at 70 years. Indeed, patients who are physiologically old are usually older than 70 years of age. However, many individuals aged 70 years or older are still in a good health condition and are able to live an autonomous and active life. Increase in longevity and improvement in physical and cognitive health means that many elderly individuals are now functionally younger at a given chronologic age than adults of previous generations. Yet, today, the physiologic age may greatly differ from the chronologic age. For treatment decision in elderly patients, the physiologic age is crucial, however, its determination may be much more complex.

What can affect the physiologic age in the elderly? Chronic health conditions are generally more common among the elderly. Many of these conditions are not immediately life threatening. However, they may lead to a loss of functional autonomy, to the development of a geriatric syndrome, and be associated with a decrease in health and quality of life. With advancing age, even a healthy person presents age-related physiologic changes with a decline in functional reserves and consequently, a decreased adaptation to any sort of stressors [2]. While a healthy elderly individual still has acceptable functioning in a steady state, he might not be able to adapt adequately to any additional stress [3]. Comorbidity, which is defined as the coexistence of disorders in addition to a primary disease of interest [4], increases with advancing age. The most frequent chronic conditions in the elderly are heart diseases, peripheral vascular diseases, hypertension, stroke, diabetes mellitus, kidney disease, Parkinsonism, neurologic and musculoskeletal diseases, and malignancies. "Geriatric syndrome" is the term used to highlight the unique features of common health conditions observed in the elderly, such as, delirium, fall, incontinence, and frailty [5]. Additionally, elderly patients are particularly subject to polypharmacy, leading to a high rate of adverse drug reactions and increased risk of drug–drug interactions [6]. Due to age-related decline in function of diverse organ systems, absorption, distribution, metabolism, and elimination of a drug may be altered. As a consequence, the aging body can be more susceptible to drug side effects. Other factors, such as, nutritional status, social support, education, and environmental influences may contribute to age-related diseases.

In order to make a reliable treatment decision in an elderly patient, all these factors have to be integrated into a global patient assessment.

In younger adults the aim of a treatment is, whenever possible, to cure a severe disease. In this setting, prolongation of lifespan is usually closely related with increased healthspan. This may differ for the elderly. The cure of a disease with prolongation of overall survival often requires intensive treatments and consequently is associated with high toxicity and a number of complications. Therefore in older adults, extension of healthspan could represent a more valuable goal than prolongation of lifespan (Fig. 12.1). This corresponds to what is called in the geriatric language "compression of morbidity," meaning compressing years of disability into the end of life and can be used as a basis for assessing health gains and losses during the aging process [7]. From this perspective, the treatment aim of an elderly person should not only be based on disease characteristics, but also take into consideration functional autonomy, healthy physical and cognitive status, and active engagement with life.

Severe AA is a rare disease. If left untreated, many patients will die from bleeding or infection. The risk of bleedings and serious infections increases substantially in older adult patients. Comparatively, in chronic ITP patients with persistent low platelet counts, the risk for major and fatal hemorrhage increases from 0.4% in younger age, up to 13% per year in patients over 60 years of age [8]. Moreover, the age-related decline of the immune system is reflected in the increased susceptibility to infectious diseases [9]. Theoretically, the same treatment options are available for elderly patients with AA as for younger adults or children; they are immunosuppressive treatment and hematopoietic stem-cell transplantation (HSCT). However, as it will be discussed later, HSCT does not belong to the first-line treatment in elderly patients. In elderly AA patients, the treatment-decision process is more complicated and should be done stepwise.

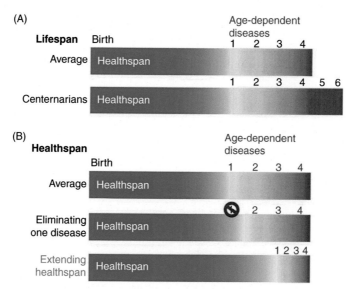

FIGURE 12.1 (A) Extending lifespan without extending healthspan; (B) Extending healthspan by preserving functional capacity and quality of life rather than only prolonging lifespan (presentation received by Ewald Collin).

First, one needs to evaluate whether the patient needs a treatment intervention; second, in those cases where treatment is needed, the patient's health condition has to be assessed to determine eligibility for any treatment; and third, according to the stage of the disease and the general health condition, optimal treatment modality has to be chosen. In geriatric medicine, the decision to treat always concerns two levels, function and disease. The functional ability of the patient determines the eligibility to be treated and the disease characteristics determine the therapeutic possibilities, along with the expressed wishes of the patient [10].

COMPREHENSIVE GERIATRIC ASSESSMENT

Traditional medical parameters are insufficient to properly address the issues needed for treatment decision of older patients. The success of a treatment in older patients does not only depend on the primary medical disease, but also on their general health condition and functional and cognitive status before the treatment. The comprehensive geriatric assessment (CGA) is an instrument assessing the global health condition of an elderly patient, which is needed for tailoring the treatment for fit, vulnerable, and frail patients (Table 12.1). Parameters of the CGA include determination of functional and cognitive status, comorbidity, presence or absence of a geriatric syndrome, nutritional assessment, and the social support of the patient [11].

The functional ability of an elderly patient can be determined at different levels: basic activities of daily living (ADL), instrumental activities of daily living (IADL), and advanced activities of daily living (AADL). The ADL are the tasks that a patient has to be able to complete in order to live autonomously in his own residence. It concerns activities, such as, bathing, showering, dressing, eating, swallowing, mobility, and personal and toilet hygiene. The IADL are activities that a patient needs to be able to perform in order to maintain an independent household. It concerns housework, taking medication, shopping

TABLE 12.1 Evaluation of an Older Person Based on a Comprehensive Geriatric Assessment (CGA)

Functional status	
ADL	Basic self-care skills needed to maintain independence in the home • Bathing • Showering • Dressing • Eating • Swallowing • Mobility at home • Personal and toilet hygiene
IADL	Skills required to maintain independence in the community • Housework • Taking medication • Shopping for groceries and clothing • Telephone and other technologies • Use of public transportation
Gait speed	Normal gait speed • Aged 70–79 years, 1.13 m/s (women), 1.26 m/s (men) • Aged 80–89 years, 0.94 m/s (women), 0.97 m/s (men)
Comorbidity	Number of comorbid conditions and comorbidity indices • For instance, Charlson age-comorbidity index
Emotional condition	Assessment of cognitive status and depression
Nutritional status	Under nourishment
Polypharmacy	Adverse drug reactions Risk of drug interactions
Geriatric syndrome	• Delirium, dementia • Depression • Falls • Incontinence • Frailty
Availability of family and social support	• Presence and effectiveness of the caregiver responsible for managing the daily life • Social resources available • Education • Environmental influences

ADL, Activities of daily living; IADL, instrumental activities of daily living.

for groceries and clothing, use of telephone and other technologies, and use of public transportation. AADL are measurements to assess societal, family recreational, and occupational tasks [12].

Gait speed is a quick, inexpensive, and reliable measure of functional capacity with well-documented predictive value for major health-related outcomes [13]. A normal gait speed for healthy women and men aged between 70 and 79 years is 1.13 m/s and 1.26 m/s, respectively. For women and men aged 80–89 years, it is 0.94 m/s and 0.97 m/s, respectively. A gait speed of less than 0.8 m/s is a predictor of poor clinical outcome and a gait speed of less than 0.60 m/s predicts further functional decline in those elderly patients [14].

Comorbidity is frequently considered as an index. Several scores have been evaluated to identify the comorbidities in cancer patients. The most used comorbidity scores in the oncologic setting are Adult Comorbidity Evaluation (ACE-7), Cumulative Index Rating Scales in Geriatrics (CIRS-G), and Charlson index. [15–17] For patients treated with HSCT, the Sorror score enables risk assessment before undertaking the transplantation [18]. The index has been shown to capture the prevalence and magnitude of severity of various organ impairments before HSCT and provides valuable prognostic information after HSCT. As age is a poor prognostic factor for HSCT in malignant disease, a composite measure has been established to allow integration of both comorbidity and age into the clinical decision [19]. No comorbidity score exists for treatment of AA with an immunosuppressive treatment. However, multimorbidity predicts negative outcome in the adult population. Heart disease, cerebrovascular disease, chronic obstructive pulmonary disease, and diabetes are recurrent risk factors. The Charlson age-comorbidity index is still the most extensively studied comorbidity index for predicting mortality of patients with various diseases, taking also into account the age of the patient. (http://farmacologiaclinica.info/scales/Charlson_Comorbidity/).

TREATMENT OF APLASTIC ANEMIA IN THE ELDERLY

Once the eligibility for treatment has been determined, the type of treatment for AA has to be defined. Supportive measures, including treatment of infections, and cell replacement with transfusions of red blood cells and platelets when indicated, are essential in all cases. Older age, even over the age of 80, is not per se a reason for withholding definitive treatment in patients with AA. Immunosuppression is the first-line treatment of choice for eligible older patients. For some patients with nonsevere AA and minimal need for supportive measures, transfusions alone can be considered as long as the situation remains stable. HSCT is not indicated as

a first-line treatment of patients with AA older than 60 years, unless with the use of a syngeneic donor [20].

Immunosuppression in the Elderly

In order to choose the optimal immunosuppression for elderly patients, we have to review some data on the treatment of younger adults. In younger adults, the combination of antithymocyte globulin (ATG) and cyclosporine (CSA) is considered the gold standard for immunosuppressive treatment [21]. As compared to ATG alone, more patients respond to the combination treatment, the response is usually faster and more complete, but there is no difference in respect of relapse after immunosuppression. The drawback of a treatment including ATG is the need for hospitalization of the patient and its acute and delayed toxicity.

Response to treatment and relapse of AA after immunosuppression are independent of age. However, immunosuppressive treatment is associated with increased mortality with advancing age, mainly due to an excess of infections and bleedings [22]. In a prospective randomized study comparing a treatment with ATG and CSA, with or without G-CSF, the overall survival at 6 years of patients older than 60 years was 56%, as compared to 71% for patients aged between 40 to 60 years, and 92% ± 4% for those aged 20–40 years (Fig. 12.2). In patients older than 60 years, death rate was 41%, as compared to 15% for patients younger than 60 years [23]. For patients above the age of 60, the main cause of death was infection (63%), followed by cardiovascular disease (12%), nonresponse (8%), and malignancies (8%). In patients younger than 60 years, for the same causes of death, mortality ratio was 45, 5, 30, and 5%, respectively (unpublished data from the G-CSF study [23]).

In younger adults with nonsevere AA, the combination of ATG and CSA is significantly better than CSA alone in respect of response rate and disease-free survival. Patients treated with

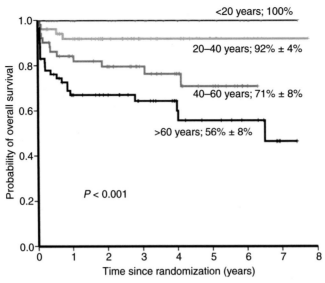

FIGURE 12.2   Overall survival of 192 patients with severe aplastic anemia *(SAA)* from the randomized G-CSF study, treated with horse antithymocyte globulin *(ATG)* and cyclosporine *(CSA)* with or without G-CSF. Here all patients (with and without G-CSF) are included and stratified according to age groups [23].

CSA alone more often need to be retreated with a second course of immunosuppression using ATG and CSA. However, there is no difference in the overall survival because CSA-refractory patients may respond to second-line treatment with ATG and CSA [24]. Patients who respond to first-line CSA alone will benefit because they do not need hospitalization to receive ATG treatment and are less exposed to treatment toxicity. For patients who fail to CSA alone, as they are not at a high risk of life-threatening infections and severe bleedings, a delay in response is acceptable and will not expose the patients to an additional risk. This, as a matter of fact, can be applied for decision making in the elderly patients and will be discussed later in the chapter.

The question whether reduced dose of ATG would be a less toxic, but still an efficient treatment option, in elderly AA patient is still under debate. A single center study, including 24 elderly patients with AA, aged between 61 and 78 years (median age 70 years) reported on 13 out of 24 patients who received an attenuated dose

of ATG (≤50% of the normal dosage) in combination with CSA. Three of the 13 patients with standard ATG dose and 10 of 13 patients with attenuated dose did not present any serious adverse events during treatment. Overall response at 2 years was 42%, with no difference between both, the standard and the attenuated ATG-treatment group [25]. These data suggest that attenuated ATG dose could be efficient and less toxic in the elderly. However, discordant results have been reported in a study on 14 elderly patients, aged between 62 and 74 years (median age 71 years) and treated with one-third of the standard dose of ATG. Only 1 of the 14 patients responded to this low a dose of ATG. In contrast to the previous reported study, the reduced dose ATG was not combined with CSA [26]. These latter results suggest that lower dose of ATG has a low efficacy in the treatment of older patients with AA. Differences in dose reduction of ATG (≤50% dose vs. 1/3 of the standard dose) could be an explanation of the discrepancy of the results between both studies. However, more likely, the higher response rate of the first study is due to the combination of ATG with CSA. This would mean that at a lower dose, ATG would simply be an innocent bystander. Thus, if ATG is needed, standard rather than reduced dose should be used.

Algorithm for First-Line Treatment of Elderly Aplastic Anemia Patients

The choice of first-line immunosuppressive treatment in elderly patients is based on the immediate risk of disease-related complications, the possible need for hospitalization because of the patient's clinical condition, as well as, the health condition needed to overcome treatment toxicity (Fig. 12.3). Patients with severe infection or with very high risk to develop severe infections (neutrophils < 0.2 G/L), and/or needing a hospitalization anyway may be considered for the combination of standard dose of ATG and CSA (Fig. 12.3). The prerequisite for such

intensive immunosuppression is a reasonable health condition assessed by a CGA. Treatment with CSA alone might be an alternative, less aggressive option, but would be associated with a higher probability of delayed response or even refractoriness. Any delay in the hematopoietic response may expose these high-risk AA patients to serious infections, with increased risk of death. The benefit from a fast and complete response to treatment may here exceed the risk of toxicity due to the more intense immunosuppression, provided the elderly patient has been adequately assessed for his functional ability and the comorbidity state.

For patients who are not at an immediate risk for severe infections and who can therefore be managed as outpatients with supportive care until response, a first-line treatment with CSA alone is recommended (Fig. 12.3). This group includes mainly patients with nonsevere AA and some patients with severe AA but neutrophil count above 0.5 G/L and without severe

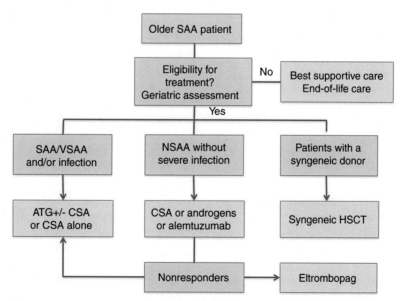

FIGURE 12.3 **Algorithm of treatment of aplastic anemia *(AA)* patients aged > 60 years.** ATG, Antithymocyte globulin; BMT, bone-marrow transplantation; CSA, cyclosporine; IS, immunosuppression; SAA, severe aplastic anemia; VSAA, very severe aplastic anemia. *Source: Modified from Tichelli A, Marsh JC. Treatment of aplastic anaemia in elderly patients aged > 60 years. Bone Marrow Transplant 2013;48(2):180–2 [20].*

infection. Indeed severe thrombocytopenia and anemia alone are not usually an indication for hospitalization. The use of alemtuzumab or androgens, particularly in men, can be considered in patients intolerant of CSA, for example, with renal impairment. However, androgens are associated with hepatotoxicity, congestive cardiac failure, prostatic enlargement, elevated blood lipids, mood changes, and other side effects and require careful monitoring of the patient. Patients not responding to CSA alone after 6 months, and still eligible for intensive immunosuppression, may be considered for combination therapy with ATG and CSA, alternative immunosuppression, or eltrombopag (Fig. 12.3).

Transplantation in the Elderly

Currently, allogeneic HSCT is the first-line treatment of choice for children and young adults below the age of 40 with an eligible matched-sibling donor [27]. Mortality risks increase in patients aged more than 40 years and are also higher when performance score is poor and the interval between diagnosis and transplantation is longer than 3 months [28]. This does not apply to patients with a syngeneic donor, in whom transplantation should be the first-line therapy, even in advanced age. Between 1979 and 2009, 88 patients received 113 syngeneic transplants. The 10-year overall survival was 93% with five transplant-related deaths. Syngeneic transplant has an excellent long-term outcome when conditioning is given [29]. The median age of the cohort with syngeneic donor was 21 years with an upper range of 68 years. The problem for syngeneic transplantation in older patients is, that the donor has the same old age as the recipient, and therefore has a higher probability of not being medically suitable for stem-cell donation. Experience with adult-related donors is available up to a donor age of 75 years. The physician assessing the donor's suitability should be aware of the higher risk due to the prevalence of many age-related health disorders [30]. Complications of hematopoietic stem-cell collection have been observed more in older age donors [31].

Increasingly, patients aged 40–50 years and even up to 60 years, if medically fit, are considered for transplantation from a matched sibling or unrelated donor [32,33]. There are few data on allogeneic HSCT of patients older than 60 years [28]. However, with newer approaches in transplantation, there is much debate concerning the upper age limit for HSCT in AA patients. Recent results report encouraging outcomes in patients older than 50 years with minimal comorbidities. Using a conditioning regimen with a standard high dose (200 mg/kg) of cyclophosphamide and ATG among 23 patients aged 43–68 years, an overall survival of 65% at 10 years was reported. Cardiac toxicity and fluid overload was the major concern in the elderly patients [34]. A large CIBMTR study of 1307 patients showed that older age at transplantation is a negative factor for various reasons, such as, prior immunosuppressive treatment, poor performance score, increased time between diagnosis and HSCT, and the use of peripheral blood instead of bone marrow [28]. Improved outcomes have been reported with the use of a fludarabine- or alemtuzumab-based conditioning regimen. In a cohort of 50 patients receiving transplants from matched sibling or matched unrelated donor and using a conditioning regimen with cyclophosphamide and alemtuzumab, 12 patients were 50 years or older (range 50–62) at the time of transplantation. Overall survival at 2 years was 71% as compared to 92% for younger patients. This study shows excellent outcomes among the small subgroup of patients older than 50 years [35]. More recently, encouraging outcomes with improved overall survival, particularly in patients with minimal comorbidity, clearly suggest that age itself can no longer remain a contraindication for HSCT. Future decisions making will rather be based on both age and comorbidities [35,36].

OPEN QUESTIONS IN THE TREATMENT OF ELDERLY PATIENTS WITH APLASTIC ANEMIA

Rabbit ATG is inferior to horse ATG as a first-line treatment for severe AA, with lower hematologic responses and survival [37]. Patients with rabbit ATG have a higher death rate, mainly due to prolonged immunosuppression and deaths due to severe infections [38]. There are no direct data comparing both forms of ATG in elderly patients. However, from the 34 patients treated with rabbit ATG, 11 died (32%), 7/24 (29%) in the 21–60 age group, and 4/7 (57%) in the group of patients >60 years. It is likely that, as patients treated with rabbit ATG have longer immunosuppression and more deaths from severe infections, the risk of adverse complication and deaths in elderly patients will be higher when using rabbit ATG. Furthermore, a relative low salvage rate is obtained with repeated immunosuppression with rabbit ATG [39]. It is therefore reasonable to consider, whenever possible the use of horse ATG as first-line immunosuppression also in the elderly.

Eltrombopag, an oral thrombopoietin mimetic, promotes megakaryopoiesis and release of platelets from mature megakaryocytes. Treatment with eltrombopag is efficacious in the treatment of AA and has been associated with multilineage clinical response in some patients with refractory disease [40]. Clinical trials are ongoing to define other settings of bone-marrow failure. The aim is to optimize the response to immunosuppressive treatment, and to prevent short-term complications, such as, infections and bleeding including an adjunct with good tolerability to the standard immunosuppression [41]. In August 2014, the US Food and Drug Administration approved the use of eltrombopag in patients with severe AA who fail to respond adequately to immunosuppressive therapy. There are no direct data on the use of eltrombopag in elderly AA patients. In a randomized study of treatment of thrombocytopenia in patients with advanced MDS or acute myeloid leukemia, eltrombopag had an acceptable safety profile, despite the advanced median age of 73 years (range 29–88 years) of the patients. Treatment with eltrombopag did not result in any unexpected safety problems. Patients with eltrombopag had fewer severe hemorrhages than the placebo group and there was no worsening of leukemia [42]. Due to its relative low toxicity profile, the simplicity of handling the drug, and the preliminary efficacy data, the addition of eltrombopag to immunosuppressive treatment in elderly patients with AA appears promising. Surprising results have been observed in a single center study including Japanese adults with immune thrombocytopenia. Older age (≥70 years) and normal or decreased cellularity were both significantly associated with a better response to eltrombopag. The beneficial effect of age is in contradiction to previous findings on patients including mainly Caucasian populations. Factors, such as changes in the pharmacokinetics, might modulate the effects of eltrombopag in older Japanese patients, as East Asians show higher bioavailability to eltrombopag, by a yet unknown mechanism [43].

In MDS, large retrospective studies demonstrated the feasibility of allogeneic HSCT in elderly patients [44]. In the near future a number of questions on HSCT in elderly patients with AA will arise. Unsolved questions are the optimal timing for HSCT, the selection of the donor (matched-sibling donor, matched-unrelated donor, cord blood, and haploidentical donor), the type of conditioning, as well as the way to select the recipient in respect of the disease and the patient's health condition.

CONCLUSIONS

Patients aged 60 or older with AA have a poor prognosis if untreated, and may die from complications of infections or bleedings. Much progress has been obtained with adequate red blood cells and platelets transfusion and infectious support. However, in many cases a definitive treatment is necessary. The higher age is not, per se, a contradiction in the treatment of a patient. The functional ability, comorbidity, nutritional status, and social support rather than the chronologic age of the patient, determine whether a patient should be treated and what type of treatment should be proposed. A stepwise strategy evaluating a patient's eligibility for intensive treatment and timing must be determined for each individual case. With the exception of the rare situation of patients having a syngeneic donor, currently HSCT is not part of the first-line treatment in elderly patients; immunosuppression is. Depending on the urgency to obtain a response (severe neutropenia or active infections), treatment should start with either the combination of ATG and CSA or CSA alone. In refractory patients, who are still eligible for ongoing management, second-line immunosuppression or HSCT are treatment options. The role of new drugs, such as eltrombopag, in the treatment of AA in the elderly has still to be defined. Finally, in geriatric medicine an important message is that the decision to treat always concerns two levels, function and disease. The functional ability of the patient determines the eligibility to be treated and the disease determines the therapeutic possibilities [10].

Acknowledgments

We thank Collin Ewald for the idea and design of Fig. 12.1.

References

[1] Steensma DP, Bejar R, Jaiswal S, et al. Clonal hematopoiesis of indeterminate potential and its distinction from myelodysplastic syndromes. Blood 2015;126(1):9–16.

[2] Boss GR, Seegmiller JE. Age-related physiological changes and their clinical significance. West J Med 1981;135(6):434–40.

[3] Fulop T, Larbi A, Witkowski JM, et al. Aging, frailty and age-related diseases. Biogerontology 2010;11(5):547–63.

[4] Feinstein AR. The pre-therapeutic classification of co-morbidity in chronic disease. J Chronic Dis 1970;23(7):455–68.

[5] Inouye SK, Studenski S, Tinetti ME, Kuchel GA. Geriatric syndromes: clinical, research, and policy implications of a core geriatric concept. J Am Geriatr Soc 2007;55(5):780–91.

[6] Craftman AG, Johnell K, Fastbom J, Westerbotn M, von Strauss E. Time trends in 20 years of medication use in older adults: findings from three elderly cohorts in Stockholm, Sweden. Arch Gerontol Geriatr 2016;63:28–35.

[7] Cannon ML. What is aging? Dis Mon 2015;61(11):454–9.

[8] Cohen YC, Djulbegovic B, Shamai-Lubovitz O, Mozes B. The bleeding risk and natural history of idiopathic thrombocytopenic purpura in patients with persistent low platelet counts. Arch Intern Med 2000;160(11):1630–8.

[9] Derhovanessian E, Solana R, Larbi A, Pawelec G. Immunity, ageing and cancer. Immun Ageing 2008;5:11.

[10] Stähelin HB. Besonderheiten der geriatrie. In: Zöllner N, Gresser U, Hehlmann R, editors. Innere Medizin. Berlin Heidelberg: Springer; 1991. p. 657–68.

[11] Brunello A, Sandri R, Extermann M. Multidimensional geriatric evaluation for older cancer patients as a clinical and research tool. Cancer Treat Rev 2009;35(6):487–92.

[12] Rosen SL, Reuben DB. Geriatric assessment tools. Mt Sinai J Med 2011;78(4):489–97.

[13] Peel NM, Kuys SS, Klein K. Gait speed as a measure in geriatric assessment in clinical settings: a systematic review. J Gerontol A 2013;68(1):39–46.

[14] Abellan van Kan G, Rolland Y, Andrieu S, et al. Gait speed at usual pace as a predictor of adverse outcomes in community-dwelling older people an International Academy on Nutrition and Aging (IANA) Task Force. J Nutr Health Aging 2009;13(10):881–9.

[15] Charlson M, Szatrowski TP, Peterson J, Gold J. Validation of a combined comorbidity index. J Clin Epidemiol 1994;47(11):1245–51.

[16] Miller MD, Paradis CF, Houck PR, et al. Rating chronic medical illness burden in geropsychiatric practice and research: application of the Cumulative Illness Rating Scale. Psychiatry Res 1992;41(3):237–48.

[17] Piccirillo JF, Tierney RM, Costas I, Grove L, Spitznagel EL Jr. Prognostic importance of comorbidity in a hospital-based cancer registry. JAMA 2004;291(20):2441–7.

[18] Sorror ML, Maris MB, Storb R, et al. Hematopoietic cell transplantation (HCT)-specific comorbidity index: a new tool for risk assessment before allogeneic HCT. Blood 2005;106(8):2912–9.

[19] Sorror ML, Storb RF, Sandmaier BM, et al. Comorbidity-age index: a clinical measure of biologic age before allogeneic hematopoietic cell transplantation. J Clin Oncol 2014;32(29):3249–56.

[20] Tichelli A, Marsh JC. Treatment of aplastic anaemia in elderly patients aged >60 years. Bone Marrow Transplant 2013;48(2):180–2.

[21] Frickhofen N, Kaltwasser JP, Schrezenmeier H, et al. Treatment of aplastic anemia with antilymphocyte globulin and methylprednisolone with or without cyclosporine. The German Aplastic Anemia Study Group. N Engl J Med 1991;324(19):1297–304.

[22] Tichelli A, Socie G, Henry-Amar M, et al. Effectiveness of immunosuppressive therapy in older patients with aplastic anemia. European Group for Blood and Marrow Transplantation Severe Aplastic Anaemia Working Party. Ann Intern Med 1999;130(3):193–201.

[23] Tichelli A, Schrezenmeier H, Socie G, et al. A randomized controlled study in patients with newly diagnosed severe aplastic anemia receiving antithymocyte globulin (ATG), cyclosporine, with or without G-CSF: a study of the SAA Working Party of the European Group for Blood and Marrow Transplantation. Blood 2011;117(17):4434–41.

[24] Marsh J, Schrezenmeier H, Marin P, et al. Prospective randomized multicenter study comparing cyclosporin alone versus the combination of antithymocyte globulin and cyclosporin for treatment of patients with nonsevere aplastic anemia: a report from the European Blood and Marrow Transplant (EBMT) Severe Aplastic Anaemia Working Party. Blood 1999;93(7):2191–5.

[25] Kao SY, Xu W, Brandwein JM, et al. Outcomes of older patients (> or = 60 years) with acquired aplastic anaemia treated with immunosuppressive therapy. Br J Haematol 2008;143(5):738–43.

[26] Killick SB, Cavenagh JD, Davies JK, Marsh JC. Low dose antithymocyte globulin for the treatment of older patients with aplastic - anaemia. Leuk Res 2006;30(12):1517–20.

[27] Scheinberg P, Young NS. How I treat acquired aplastic anemia. Blood 2012;120(6):1185–96.

[28] Gupta V, Eapen M, Brazauskas R, et al. Impact of age on outcomes after bone marrow transplantation for acquired aplastic anemia using HLA-matched sibling donors. Haematologica 2010;95(12):2119–25.

[29] Gerull S, Stern M, Apperley J, et al. Syngeneic transplantation in aplastic anemia: pre-transplant conditioning and peripheral blood are associated with improved engraftment: an observational study on behalf of the Severe Aplastic Anemia and Pediatric Diseases Working Parties of the European Group for Blood and Marrow Transplantation. Haematologica 2013;98(11):1804–9.

[30] Worel N, Buser A, Greinix HT, et al. Suitability criteria for adult related donors: a consensus statement from the Worldwide Network for Blood and Marrow Transplantation Standing Committee on Donor Issues. Biol Blood Marrow Transplant 2015;21(12):2052–60.

[31] Lysák D, Kořístek Z, Gašová Z, Skoumalová I, Jindra P. Efficacy and safety of peripheral blood stem cell collection in elderly donors; does age interfere? J Clin Apheresis 2011;26(1):9–16.

[32] Killick SB, Bown N, Cavenagh J, et al. Guidelines for the diagnosis and management of adult aplastic anaemia. Br J Haematol 2016;172(2):187–207.

[33] Miano M, Dufour C. The diagnosis and treatment of aplastic anemia: a review. Int J Hematol 2015;101(6):527–35.

[34] Sangiolo D, Storb R, Deeg HJ, et al. Outcome of allogeneic hematopoietic cell transplantation from HLA-identical siblings for severe aplastic anemia in patients over 40 years of age. Biol Blood Marrow Transplant 2010;16(10):1411–8.

[35] Marsh JC, Gupta V, Lim Z, et al. Alemtuzumab with fludarabine and cyclophosphamide reduces chronic graft-versus-host disease after allogeneic stem cell transplantation for acquired aplastic anemia. Blood 2011;118(8):2351–7.

[36] Young ME, Potter V, Kulasekararaj AG, Mufti GJ, Marsh JC. Haematopoietic stem cell transplantation for acquired aplastic anaemia. Curr Opin Hematol 2013;20(6):515–20.

[37] Scheinberg P, Nunez O, Weinstein B, et al. Horse versus rabbit antithymocyte globulin in acquired aplastic anemia. N Engl J Med 2011;365(5):430–8.

[38] Marsh JC, Bacigalupo A, Schrezenmeier H, et al. Prospective study of rabbit antithymocyte globulin and cyclosporine for aplastic anemia from the EBMT Severe Aplastic Anaemia Working Party. Blood 2012;119(23):5391–6.

[39] Clé DV, Atta EH, Dias DSP, et al. Repeat course of rabbit antithymocyte globulin as salvage following initial therapy with rabbit antithymocyte globulin in acquired aplastic anemia. Haematologica 2015;100(9):e345–7.

[40] Olnes MJ, Scheinberg P, Calvo KR, et al. Eltrombopag and improved hematopoiesis in refractory aplastic anemia. New Engl J Med 2012;367(1):11–9.

[41] Desmond R, Townsley DM, Dunbar C, Young NS. Eltrombopag in aplastic anemia. Sem Hematol 2015;52(1):31–7.

[42] Platzbecker U, Wong RS, Verma A, et al. Safety and tolerability of eltrombopag versus placebo for treatment of thrombocytopenia in patients with advanced myelodysplastic syndromes or acute myeloid leukaemia: a multicentre, randomised, placebo-controlled, double-blind, phase 1/2 trial. Lancet Haematol 2015;2(10):e417–26.

[43] Uto Y, Fujiwara S, Arai N, et al. Age and bone marrow cellularity are associated with response to eltrombopag in Japanese adult immune thrombocytopenia patients: a retrospective single-center study. Rinsho byori 2015;63(5):548–56.

[44] Kroger N. Allogeneic stem cell transplantation for elderly patients with myelodysplastic syndrome. Blood 2012;119(24):5632–9.

13

Emerging New Therapies for Acquired Bone Marrow Failure Disorders

A.M. Risitano*, R. Peffault de Latour**

*Hematology, Department of Clinical Medicine and Surgery, Federico II University, Naples, Italy; **BMT Unit, French Reference Center for Aplastic Anemia and PNH, Saint-Louis Hospital, Paris, France

INTRODUCTION

Acquired idiopathic aplastic anemia (IAA) is the most typical form of immune-mediated bone marrow failure, which is characterized by peripheral blood pancytopenia associated with a deserted or a fatty bone marrow [1,2]. A number of experimental observations support the immune-mediated pathophysiology of IAA [3]; according to our current understanding, activation and expansion of oligoclonal pathogenic T cell may arise from unknown antigen-driven triggering [4], possibly in the context of a broader derangement of the immune system (i.e., reduced Tregs [5,6] and increased Th17 [6,7]). These T cells may damage hematopoietic progenitors and stem cells both directly and through cytokine production, eventually leading to the contraction of the hematopoietic stem

cell (HSC) pool and subsequently impairing the production of mature blood cells (see Chapter 2). As discussed elsewhere in this book the two key treatment options for IAA patients are allogeneic hematopoietic stem cell transplantation (HSCT) and immunosuppressive therapy (IST) (see Chapters 6–8). Both these strategies are based on the delivery of an appropriate immunosuppressive effect, which may eventually allow the expansion of residual surviving hematopoietic stem cells (in the case of IST) or of allogeneic hematopoietic stem cells transplanted from a healthy donor (in the case of HSCT). The choice between these two options relies on the patient's age and availability of an HLA-identical sibling donor. Indeed, HSCT is the preferred front-line treatment for young (<40 or even <50 years) patients with a sibling donor, while IST is the standard choice for all the others. Nowadays, the expected overall survival of IAA patients receiving either HSCT or IST is 70–80% in the long term; the outcome has improved over the last decades, but this is mostly due to better supportive care (i.e., improved management of the infectious risk, even in nonresponders) [8] rather than to the higher efficacy of IST regimens. Nevertheless, a number of different strategies have been investigated aiming to further improve the results of the standard treatment of IAA. Here we focus on alternative and novel therapies in the context of nontransplant treatment of IAA.

The standard nontransplant treatment of IAA exploits intensive IST based on the combination of a lymphocyte-depleting agent (antithymocyte globulin, ATG) with the calcineurine inhibitor, cyclosporine A (CsA), which prevents lymphocyte activation. The most effective IST regimen is based on the horse-ATG (h-ATG) preparation ATGAM (Pfizer; 40 mg/kg per day for 4 days) and CsA (see Chapter 6), which results in an overall hematologic response rate of 60–70% [9–11]. Nevertheless, about one-third of AA patients do not show any hematologic response after IST; furthermore, even among patients achieving a hematologic response, one-third may eventually relapse and one additional third may require continuous maintenance IST. Thus, especially in the context of non-HSCT treatment of IAA, unmet clinical needs remain and different emerging therapies have been investigated. These investigational treatments may be grouped in two main categories, according to their mechanism of actions: (1) novel or alternative strategies of immunosuppression and (2) nonimmunosuppressive strategies.

ALTERNATIVE STRATEGIES OF IMMUNOSUPPRESSION

Over the last two decades several alternative regimens of IST have been investigated, with the aim of improving the results achieved

by standard h-ATG and CsA [12,13]. Three different strategies to intensify IST have been exploited: (1) adding a third immunosuppressive agent on top of the platform h-ATG + CsA; (2) replacing the key lymphocyte-depleting agent (i.e., h-ATG) with alternative, more lymphotoxic, immunosuppressive agents; and (3) novel strategies without conventional lymphocyte depletion. Even if most of them (if not all) led to disappointing results, we describe the most relevant attempts in this chapter. While they remain scientifically remarkable for the information that they gave us, at the moment none of these strategies have resulted in a change of the standard care of IAA. Nevertheless, because of the obvious biologic activity shown by some of these regimens, at the moment these alternative IST strategies may be considered as worthy treatment options in patients failing standard IST (and lacking a low-risk transplant procedure).

Adding a Third Immunosuppressive Agent

The intensification of standard IST by adding a third immunosuppressive agent was investigated mostly at the National Institute of Health (NIH) exploiting an antimetabolite agent, mycophenolate mofetil (MMF) or an inhibitor of mammalian target (mTOR), rapamycin/sirolimus. Both these strategies were based on the hypothesis that an immunosuppressive agent with a different mechanism of action (as compared with h-ATG and CsA) may work synergistically with standard IST, eventually improving the response rate or decreasing the relapse rate.

Mycophenolate Mofetil

MMF is an inhibitor of the purine synthesis, which may affect T-cell proliferation; it was tested in a prospective single arm, open-label study in combination with standard h-ATG + CsA [14]. In a cohort of 104 untreated AA patients, this three-drug regimen resulted in a 6-month response rate of 62%; the relapse rate was 37% (despite of maintenance IST with MMF and CsA), and clonal evolution was 9%. When compared with historic controls, the addition of MMF to standard IST did not confer any advantage in terms of either hematologic response, relapse, or clonal evolution [14].

Rapamycin/Sirolimus

Sirolimus is an mTOR inhibitor, which has been developed in several autoimmune conditions, including graft versus host disease (GvHD). Sirolimus seems to exert its immunosuppressive effect through two different mechanism of actions. Sirolimus binds to immunophilins (FKBP12) blocking initial IL-2-dependent T-cell activation, similar to calcineurine inhibitors. Furthermore, once bound to its target sirolimus (the complex FKBP12/RAPA) inhibits mTOR, eventually modulating a number of intracellular pathways, which are needed for cell-cycle progression toward G1/S stage. Due to this mechanism of action, there is a strong rationale suggesting that sirolimus may work synergistically with CsA [15]. In the context of IAA, sirolimus was tested in a prospective randomized trial comparing a three-drug regimen h-ATG + CsA + sirolimus with the standard h-ATG + CsA IST [16]. Unfortunately, even this three-agent IST regimen did not show any benefit over the standard IST. The overall response rate at 1 year was 51% in the experimental arm (as compared with the 62% of the standard arm), with no difference in long-term relapse rate, clonal evolution, and overall survival [16].

Replacing h-ATG With Other Lymphocyte-Depleting Agents

The intensification of IST by replacing h-ATG with other lymphocyte-depleting agents was based on the assumption that there are alternative compounds that may lead to more profound and long-lasting lymphocyte depletion; theoretically, in an immune-mediated disease, such as IAA, this might lead to a better eradication of pathogenic T cells, eventually resulting in a better clinical outcome. Even if finally this dogma has been exploded by clinical observations because more "potent" lymphocyte-depleting agents did not result in any benefit, we further describe a list of agents that has been investigated: (1) other ATG preparations [rabbit ATG (r-ATG)], (2) cyclophosphamide (CTX), and (3) alemtuzumab.

Rabbit ATG

r-ATG was more or less systematically used in IAA since decades (see Chapter 6), because the difference among different ATG preparations was not entirely clear. It is now well established that r-ATG (Thymoglobuline, Sanofi) is an alternative ATG preparation which, in comparison to standard h-ATG, retains a more potent lymphocyte-depleting effect, as demonstrated in the transplant setting [17,18]. r-ATG results in a better clinical outcome in the setting of renal transplantation [17,18] and is the preferred ATG preparation in the setting of HSCT. Furthermore, it showed excellent clinical activity as salvage therapy for patients refractory to or relapsed after h-ATG [19,20]. With these premises, it was hypothesized that an IST regimen based on r-ATG rather than on h-ATG may result in a better clinical outcome. This was tested in a prospective, randomized, clinical trial comparing head-to-head of r-ATG + CsA + h-ATG (NCT00260689) [21]. Quite surprisingly, r-ATG led to a significantly lower response rate as compared with h-ATG (37% vs. 68% at 6 months, $p < 0.001$), with even a detrimental effect on overall survival at 3 years (70% vs. 94%, $p = 0.008$). These data were confirmed in a subsequent open label study from the European Group for Blood and Marrow Transplantation (EBMT; NCT00471848), that showed an overall response rate of 39% at 6 months (with only 7% complete responses), with an overall survival of 68% at 24 months (which in a retrospective matched-paired analysis resulted significantly lower in comparison to the 86% of h-ATG) [22]. As discussed in the Chapter 6 even if some other experiences reported that r-ATG may be equal to h-ATG [23–28], at the moment the standard lymphocyte-depleting agent in the context of IST for IAA remains h-ATG.

Cyclophosphamide

Due to its excellent antilymphocyte effect, CTX has been considered as a candidate lymphocyte-depleting agent for the treatment of IAA. The use CTX at high dose (50 mg/kg for 4 consecutive days, intravenously) was initially investigated at John Hopkins, where Brodsky et al. showed an overall response rate of 70% [29], with best results as front-line treatment. However, a subsequent randomized trial from the NIH comparing CTX (in combination with CsA) with the standard h-ATG + CsA did not confirm the initial promising data [30]. Indeed, at 6 months the overall response rate was 46% in comparison to the 75% of the standard arm; in addition, the investigational arm exhibited an excess morbidity and mortality secondary to infectious complications (mostly fungal infections), which eventually led to the early termination of the trial [30]. The long-term follow up of this study also demonstrated that CTX did not reduce the risk of relapse (25% in the CTX arm vs. 46% in the h-ATG arm), nor that of clonal evolution [31]. Despite of these data coming from a prospective, randomized study, the open label experience from John Hopkins continues to show good results; the latest report including 67 AA patients (44 untreated) exhibited an overall response rate of 70% (with 43% of complete responses) as front-line treatment, with an overall survival of 88%. Results were slightly inferior when CTX was used in patients refractory to IST; here the overall response rate was 47% (with 21% of complete responses) and overall survival was 61% [32]. Irrespective of a previous IST course, the risk of clonal evolution was negligible [32]. However, the main problem of this treatment remains the longer time to response (median 20 months), which eventually impacts the risk of life-threatening infectious complications (fungal infections were 18% in untreated patients and even 45% in refractory ones) [32]. To reduce the possible toxicity of high-dose CTX on committed hematopoietic progenitors, this agent was also tested at lower doses (30 mg/kg on 4 consecutive days). Unfortunately, initial promising data coming from China [33] were not confirmed in a subsequent study from the NIH, in which 22 untreated IAA patients showed an overall response rate of only 41%, again with unacceptable long-lasting neutropenias that led to 6 episodes of severe fungal infections [34].

Alemtuzumab

Another alternative to the polyclonal ATG is the use of monoclonal antibodies (mAb) targeting antigens highly expressed on lymphocytes or even lymphocyte subsets. As IAA is considered a T-cell mediated disease, the most appealing mAb was the anti-CD52 alemtuzumab, which has been investigated for the treatment of IAA in different trials and settings. The most systematic experience was performed at the NIH, within three distinct studies, which exploited alemtuzumab (as single agent) at the dose of 10 mg/day, intravenously for 10 days, in refractory, relapsed, and untreated IAA [35]. Alemtuzumab resulted biologically effective in all these conditions: the overall response rate was 37% in refractory patients ($n = 54$) and even higher in relapsed patients (56%, $n = 25$). In refractory patients the response rate was similar to that seen with r-ATG (33%), which was the control arm of this randomized study. Nevertheless, as for r-ATG, this excellent activity as second-line treatment did not translate into similar efficacy as front-line treatment; indeed, in a large prospective randomized trial for untreated IAA [21] the alemtuzumab arm was prematurely closed because of lack of efficacy [35]. Of the 16 patients who received alentuzumab as first-line treatment for their IAA at the NIH, only 3 (19%)

achieved a hematologic response, eventually failing the primary endpoint of the study (which aimed to demonstrate the superiority of the investigational arm) [35]. The biologic activity of alemtuzumab in IAA was demonstrated in other studies also. A small prospective trial from the EBMT investigated alemtuzumab administrated subcutaneously at the total dose of 103 mg (3–10–30–30–30 in 5 consecutive days) in 11 IAA patients (6 untreated), in combination with low-dose CsA as maintenance IST [36]. The cumulative incidence of overall response was as high as 84%, with 27% complete responses [36]; the follow up of this study showed a 3-year overall survival of 74%, even if failure-free survival was only 40%, mostly due to subsequent relapses [37]. Notably, in contrast to CTX, these studies demonstrated an excellent safety profile, with infectious complications not emerging as a major risk irrespective of the long-lasting lymphocytopenia (especially for the CD4 compartment) [37,38]. These excellent results were confirmed in two other studies from Korea (response rate of 35%, 2-year overall survival of 81%, $n = 17$) [39] and from Mexico [40]. In this latter study alemtuzumab was used at a lower dose (50 mg in total) in combination with CsA (at 2 mg/kg twice a day) in 14 IAA patients; the overall response rate was 57% (complete responses were 14%), with no subsequent relapse or clonal evolution, and an overall survival at 38 months of 71% [40].

The Lesson From Alternative Lymphocyte-Depleting Agents

The use of lymphocyte killers other than h-ATG clearly demonstrates that they have a biologic activity in IAA because of their lympholytic effect on T cells. However, in contrast with the easiest anticipation, even if they resulted in a more profound lymphocyte depletion this was not associated with any clinical benefit. Indeed, in comparison to standard h-ATG, the achievement of a more profound and durable lymphocyte depletion did not lead to a higher response

rate nor prevented relapses (for all r-ATG, CTX, and alemtuzumab), and rather resulted in unacceptable infectious-related morbidity (at least for CTX). The actual explanations for this dichotomy between lymphocyte depletion and clinical efficacy remain elusive, even if several hypotheses have been raised to interpret the better results seen with h-ATG: (1) selective activity on distinct lymphocyte subsets (e.g., Tregs or Th17) [21], (2) other immunomodulatory effects delivered independently from lymphocyte depletion (e.g., cytokine polarization or inflammatory milieu) [41]; and (3) direct effect on hematopoietic stem cells or progenitors (e.g., unknown nonimmunosuppressive off-target effect) [42–44]. Nevertheless, the disappointing and unexpected results seen with these more potent immunosuppressive agents [21,31,35] clearly enable us to conclude that they (and any other experimental IST) must be offered only within well-designed, prospective studies (even more because of the rarity of the disease).

Novel Strategies Without Conventional Lymphocyte Depletion

Several novel strategies of immunosuppression are currently under investigation in different autoimmune disease, even if they have not been tested yet in IAA, it is possible that this may be the case in the near future. The list of these novel agents would be long, however, a few examples are presented here.

As the role of cytokines in the pathophysiology of IAA has been extensively proven, the use of mAbs, targeting specific cytokines has been postulated. Anti-TNF agents are already in the clinic, with three different compounds: etanercept (a TNF-receptor/Ig fusion protein), infliximab (a chimeric anti-TNF-α mAb), and adalimumab (a fully humanized anti-TNF-α mAb). Even if systematic investigation in IAA is lacking, some anecdotic data already have been reported [45]. Cytokines also play a key role in T-cell activation, maturation, and polarization;

mAbs interfering with them (e.g., ABT-874, a fully humanized anti-IL12/IL23) might be of interest even in IAA.

In addition to mAbs leading to cytolytic effect (e.g., alemtuzumab), there are other mAbs that target lymphocyte antigens required for lymphocyte activation, proliferation, or even trafficking. They may include, among the others, alefacept, efalizumab, natalizumab, and begedina. The first three agents have been already approved for different autoimmune disorders, whereas begedina is currently under investigation in steroid-refractory acute GvHD [46]. Alefacept is a fully human lymphocyte function-associated antigen-3 (LFA-3)/immunoglobulin G1 fusion protein that binds to CD2 eventually impairing the costimulatory signaling. This results in inhibition of T-cell activation and proliferation, with subsequent apoptosis [47]. Efalizumab is a fully humanized anti-CD11a mAb that inhibits lymphocyte function associated antigen-1 (LFA-1)/ICAM-1 interaction, thereby impairing lymphocyte adhesion and possibly T-cell activation and migration [48]. Natalizumab is a humanized mAb targeting the cell adhesion molecule α4-integrin, which plays a major role in T-cell trafficking, thereby preventing the extravasation of T cells to target organs [49]. Begedina is a fully humanized anti-CD26 mAb, which disables the CD26-mediated signaling required for proper activation, proliferation, and tissue migration of functional T cells. All these agents are potentially of interest in several autoimmune diseases, including IAA, even if for some of them a clinical application in IAA would be questionable for the risk of life-threatening complications (i.e., progressive multifocal leukoencephalopathy associated with natalizumab and efalizumab).

Finally, even cell therapy has been investigated for the treatment of immune-mediated disorders. Mesenchymal stem cells (MSCs) have been used with excellent result as salvage treatment of GvHD [50,51]. The use of allogeneic MSC for the treatment of refractory IAA has been already postulated [52], and it may be even more appropriate considering previous in vitro data that suggest the impairment of the immuno-modulatory capability of autologous MSC may play a role in the immune derangement embedded with the pathophysiology of IAA [53].

NONIMMUNOSUPPRESSIVE STRATEGIES

Androgens

Historically, androgens have been used for the treatment of IAA since the initial studies investigating the effect of IST [54], even if their actual role in promoting hematopoietic recovery was unclear [55]. The systematic use of androgens in combination with ATG as first-line treatment of IAA was discouraged because of the results from one relatively large randomized trial comparing ATG alone versus ATG + oxymetholone [56]. The overall response rate was 44% in the ATG arm ($n = 27$) and 42% in the ATG + oxymetholone arm ($n = 26$), without differences in terms of quality of response and long-term survival [56]. Even if these data seem to suggest that androgens are not required to respond to ATG and they do not add any hematologic benefit over ATG treatment, other studies continued to support a possible direct effect of androgens in the treatment of IAA. For instance, a large EBMT retrospective study showed even a survival benefit in patients receiving androgens in addition to ATG and steroids (77%, $n = 91$) as compared with ATG and steroids alone (44%, $n = 71$). However, the conclusions from this study are weakened by the lack of treatment homogeneity [57]. These positive results were somehow confirmed in a small prospective study from Germany [58], as well as, in a subsequent randomized trial from the EBMT [59]. This latter study randomized 134 patients with IAA to receive h-ATG (Lymphoglobulin) and methylprednisolone either alone or in combination with oxymetholone. The addition of oxymetholone resulted in a significantly higher response rate (56% vs. 40% at 4 months), with a more evident effect in female patients (78% vs. 27%) [59].

More recently, an additional argument suggesting the potential effect of androgens on hematopoiesis came from in vitro data showing that androgens are able to stimulate telomerase activity [60]. As telomere attrition has been implicated in HSC exhaustion in inherited but possibly also in acquired forms of bone marrow failure (see Chapter 2) [61], androgens may work to rescue hematopoiesis in some IAA patients who may fail hematologic recovery after standard IST because of a severe impairment of the residual HSC compartment [62]. In a recent retrospective study from the Genoa groups, the effects of testosterone ondecanoato (andriol 40 mg/day, orally) were investigated in 44 IAA patients who had failed a previous IST course [63]. The overall response rate was as high as 57%, with 31% complete responses and 36% partial responses; responses were more frequent in female patients (63%) as compared to male ones (44%). Androgens were well tolerated, even in combination with CsA, and could be tapered over time without losing hematologic responses. This therapeutic effect of androgens resulted in an excellent failure-free survival of 57% at 5 years; failure-free survival was even better in patients showing hematologic response to androgens (92% at 5 years). Thus, even if their systematic use is not recommended and their mechanistic effects are not fully understood, androgens represent a potential treatment option for some IAA patients who fail standard IST.

Eltrombopag

Eltrombopag is a synthetic, orally available, nonpeptide thrombopoietin (TPO) mimetic, which was initially developed for autoimmune thrombocytopenia [64]. Eltrombopag binds to the TPO receptor c-mpl at the level of its transmembrane and juxtamembrane domains,

without competing with endogenous TPO, which rather binds to a distinct, extracellular, binding site [65]. TPO is a glycoprotein produced by the liver (and at a lower extent by stromal marrow cells), which works as a hematopoietic cytokine, mostly stimulating megakaryopoiesis [65]. However, the TPO receptor c-mpl is also expressed on primitive hematopoietic stem and progenitor cells [66], where it remains fully functional, eventually promoting cell proliferation in presence of additional cytokines [67,68]. Loss of function mutations of c-MPL are associated with the human disease congenital amegakaryocytic thrombocytopenia, which is characterized by a bone marrow failure initially starting as isolated thrombocytopenia, eventually leading to progressive pancytopenia due to HSC exhaustion [69]. TPO is considered as a key cytokine for proper functioning of the bone marrow, but it does not seem to play a major role in the pathophysiology of IAA, as even serum levels of TPO are increased in IAA patients [70,71]. Nevertheless, because of its possible effect on HSCs and early hematopoietic progenitors, TPO was investigated as salvage treatment in IAA patients refractory to IST; indeed, in this condition even a mild increase just of platelet counts may be clinically beneficial.

The first pilot experience with eltrombopag for IAA was a small phase II trial in patients refractory to at least one course of IST (administered at least 6 months in advance), with platelet count $< 30 \times 10^3$ μL^{-1} (NCT00922883) [72]; the majority of patients required both red blood cell and platelet transfusions. Eltrombopag was used as a single agent at the starting dose of 50 mg daily, with dose escalation of 25 mg every 2 weeks up to the maximum of 150 mg; in this cohort of heavily pretreated patients the overall response rate at 12 weeks was 44% [73]. Eleven out of the 25 enrolled patients showed some hematologic benefit, with 9 patients no longer requiring platelet transfusions, 6 patients showing improvement in their anemia (3 becoming transfusion-independent), and 9 patients exhibiting increased neutrophil counts [73]. Thus, the hematologic response sometimes was not limited to the platelet count and rather included more hematopoietic lineages, eventually providing evidence that eltrombopag may act at the level of early multipotent hematopoietic progenitors (or even HSCs). This nonrandomized, open label, phase II study eventually continued to enroll patients; according to the latest update, 43 patients were included [74]. The overall response rate was 40%, with several multilineage responses; responses were seen at 3–4 months from treatment and for patients remaining on eltrombopag, blood counts continued to show progressive improvement. The restoration of hematopoiesis was confirmed by serial bone marrow biopsies, which showed improved cellularity, in absence of any increase in reticulin staining [74]. In some patients ($n = 6$) achieving trilineage responses eltrombopag was even discontinued, without loss of response. Due to a possible effect of eltrombopag on HSCs carrying detrimental somatic mutations, clonal evolution was systematically investigated in these patients. Indeed, 18% of these advanced patients showed a clonal evolution, mostly in case of a lack of response (all but two cases of clonal evolution were seen in nonresponder patients), possibly suggesting that clonal evolution may be due to the underlying bone marrow disorders (with clonal hematopoiesis eventually acquiring somatic mutations (see Chapter 6) rather than to a mechanistic effect of eltrombopag. In the setting of IAA, eltrombopag showed a favorable safety profile, with no dose-limiting toxicity, and excellent tolerability; drug-related transaminitis was observed in a few patients, but it remained transient and self-limiting.

All together, these data provide evidence that eltrombopag may lead to a hematologic response in AA patients; its broad effect on all blood lineages supports the concept that eltrombopag acts directly on early hematopoietic progenitors and HSCs spared by the immune-mediated damage (Fig. 13.1) [75]. The observation

that hematologic responses may be retained even after treatment discontinuation suggests that eltrombopag is able to amplify the number of functional, cycling HSCs, which then can sustain hematopoiesis irrespective of further exposure to the drug. Based on these observations, eltrombopag has received marketing authorization both in the United States and in Europe for the treatment of IAA patients refractory to IST. However, once the biologic efficacy of eltrombopag has been proven, the next step for the scientific community is to incorporate eltrombopag in the treatment algorithm of IAA in the early disease phase. Theoretically, in untreated IAA patients, eltrombopag may be seen as an alternative to IST or as a complementary treatment. Given the excellent results achieved with IST, the scientific community has started to investigate the feasibility of a combination of IST with eltrombopag [75].

COMBINATION STRATEGIES

After a long list of failures, eltrombopag seems to be the first agent that, quite unexpectedly, resulted in a substantial hematologic benefit in IAA patients who have failed IST. As discussed earlier, eltrombopag does not work as an immunosuppressive agent, and rather it seems to exert its effect by rescuing residual hematopoiesis. In other words, eltrombopag acts directly at the level of the hematopoietic stem cells through their TPO receptors, independently from any ongoing immune-mediated damage over hematopoiesis. It is not yet demonstrated whether eltrombopag protects hematopoietic stem cells from immune-mediated lethal damage (or functional impairment), or if it promotes the recruitment and the proliferation of the few surviving hematopoietic stem cells. Moreover, it is not clear how repopulating hematopoietic clones are selected by eltrombopag, as in different contexts, eltrombopag may also have antiproliferative effects, possibly

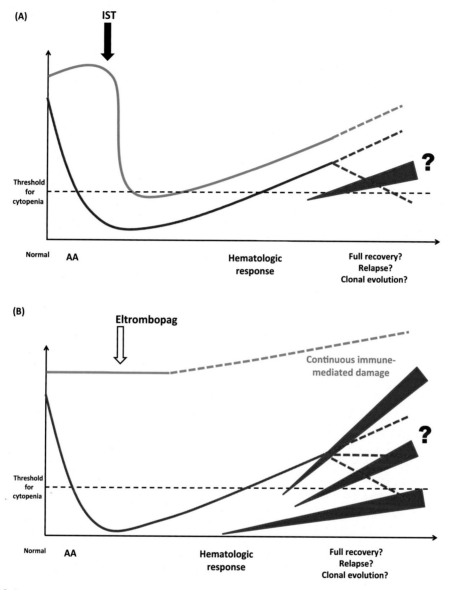

FIGURE 13.1 Diagram of hematopoiesis and (auto-)immunity in response to different treatment strategies. The *red line* represents functional hematopoiesis, the *green line* depicts the immune response eventually causing the disease, and the *blue arrows* represent possible clones acquiring somatic mutations eventually leading to malignant transformation. Once an (auto-)immune attack damages the hematopoietic compartment, cytopenia arises when the hematopoietic stem cell (HSC) mass reaches a critical threshold, then etiologic treatments may restore hematopoiesis. (A) Immunosuppression alone: by destroying the immune cells damaging the HSC compartment *IST* allows a slow recovery of the HSC compartment, possibly leading to improvement and even normalization of blood counts. (B) Eltrombopag alone: by stimulating the HSCs and early hematopoietic progenitors, eltrombopag may expand the HSC mass, possibly resulting in improved blood counts irrespective of an overt, ongoing, immune-mediated damage. (C) Immunosuppression and eltrombopag: while IST disables the undergoing autoimmune attack on hematopoiesis, eltrombopag may work synergistically by expanding the HSC mass. This eventually leads to faster and more robust hematologic responses. Clonal evolution may arise at any time during the disease course. In the context of oligoclonal hematopoiesis, a further proliferative stress (as induced by eltrombopag) and/or a continuous inflammatory milieu (as induced by an overt immune-mediated damage) may theoretically increase such risk. *AA*, Aplastic anemia; *IST*, immunosuppressive therapy.

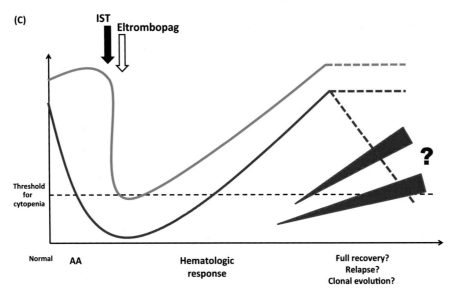

FIGURE 13.1   (cont.)

independently from c-mpl [76]. Nevertheless, it is conceivable that eltrombopag overcomes the therapeutic limitations of IST by reducing the functional impairment of hematopoiesis irrespective of persistent immune damage. Thus, eltrombopag may work synergistically with IST, eventually improving its efficacy even by combining two different mechanisms of action. The rationale for combining IST and eltrombopag is very strong and different clinical trials are currently ongoing to investigate whether this strategy can improve hematologic response rates and survival in untreated IAA patients. A single arm phase II clinical trial combining h-ATG, CsA, and eltrombopag is currently ongoing at the NIH (NCT01623167) [77]. Preliminary data of this study have been presented at the 58th American Society of Hematology Meeting in Orlando [78]; three different patient cohorts have been reported, which differ for the timing and length of eltrombopag treatment. In the first two cohorts, eltrombopag was administered from day 15 until either 6 months ($n = 30$) or 3 months ($n = 31$) and in the third cohort it was started at day 1 and continued until

6 months ($n = 27$). The overall response rates at 6 months in the three cohorts were 77, 77, and 90% (globally 80%), with complete response rates of 33, 26, and 54% (globally 34%). These results were largely superior to historic data, where the 6-month overall response is around 63%, with a complete response rate of 12%. With the limited follow up available, clonal evolution did not emerge as a possible concern, eventually proving the concept that combination of IST with eltrombopag is safe and potentially effective as an initial treatment of IAA patients. The same combination strategy is currently under investigation in Europe; within a large phase III randomized studied performed by the EBMT (NCT02099747) [79]. In this study, the standard IST regimen of h-ATG (ATGAM) and CsA is compared in a one-to-one randomization with the same regimen plus eltrombopag, administered from day 14. The aim of the study is to demonstrate an improvement of the quality and the timing of hematologic response in the experimental arm, possibly providing definite evidence for a change in the standard of care of IAA patients lacking a sibling donor.

CONCLUSIONS

Several alternative therapies have been tested in the last decades in IAA because irrespective of the excellent and improved results some patients continue to die due to the disease. These efforts focused both on patients with refractory/relapsed disease, as well as, on front-line treatment, to decrease the number of failures after standard IST. The most relevant improvement was seen in HSCT because nowadays transplant strategies result in excellent long-term outcome irrespective of donor type (sibling vs. unrelated), and even in older patients. More recently, the possibility to use even HLA-haploidentical donors [80], or even cord-blood units [81], extended the access to HSC for all patients within the appropriate age. Nevertheless, even non-transplant strategies have been implemented over time. Several attempts to improve the initial response and the long-term outcome by intensifying the IST regimens have failed, even if the spectrum of biologically active agents has grown and may be used as salvage treatment (especially when a HSCT option is not feasible). More recently, the field of IAA has been livened up by the introduction of the TPO-mimetic agent, eltrombopag. Available data are exciting because beyond the unexpected activity in refractory disease [74], this compound, for the first time in the last 30 years, seems to add a significant benefit even in the front-line treatment of IAA, once added to the standard IST [78]. Of course the final conclusion cannot be drawn at this stage and it will only be possible through the careful analysis of the data produced by the ongoing clinical trials that we will address the pending open questions. Indeed, the molecular mechanism of action is not fully understood and thereby even the possible impact on the most feared complication of IAA, namely clonal evolution to malignant disorders, needs to be further investigated. Then, the role of eltrombopag in the treatment algorithm of IAA will have to be defined. At the moment, it is the best salvage

treatment for patients lacking a transplant procedure. However, ongoing studies will define its role upstream in the disease course, even as front-line treatment. Even if eltrombopag is the most promising emerging agent, the three-agent regimen h-ATG + CsA + eltrombopag represents the most intriguing treatment strategy that promises to become the future standard of care of IAA in patients lacking a sibling donor.

Nevertheless, the addition of eltrombopag to h-ATG and CsA is not the end of the story. Current insights suggest that we may increase the response rate (as compared to IST alone), but we cannot predict the long-term outcome of these patients, as well as, possible biologic changes of their disease (i.e., clonal evolution; Fig. 13.1). The clonal nature of hematopoiesis in IAA, together with the potential acquisition of somatic mutations eventually driving disease evolution, will have to be investigated even more in the context of the exposure to eltrombopag [82]. At the moment, it is not known whether the expanded HSC mass after exposure to eltrombopag retains (or reacquires) a polyclonal pattern or whether it is derived from the expansion of a limited number of clones. In addition, irrespective of initial responses, we are going to experience possible relapses; biologically, relapses after eltrombopag + IST may be different from those seen so far after IST alone. Hopefully the number of treatment failures will be lower, but in this setting modulation of exposure to maintenance IST or to continuous HSC stimulation (i.e., retreatment with eltrombopag) will have to be investigated. The current hope is that the synergistic effect of eltrombopag may be confirmed in the ongoing studies, with these excellent initial responses confirmed in the long-term outcomes. If this is the case, one may also anticipate that this novel standard regimen may be considered even in patients with a sibling donor, while on the other hand a failure after eltrombopag + IST may represent an indication to immediate IST irrespective of the type of donor (including haplos). Novel therapies are emerging quickly. The scientific community has the duty to design the most appropriate future clinical investigations to optimize their use and define their role in the treatment of IAA.

References

[1] Young NS. Acquired aplastic anemia. In: Young NS, editor. Bone marrow failure syndromes. Philadelphia: WB Saunders; 2000. p. 1–46.

[2] Young NS, Maciejewski J. The pathophysiology of acquired aplastic anemia. New Engl J Med 1997;336: 1365–72.

[3] Young NS. Current concepts in the pathophysiology and treatment of aplastic anemia. Am Soc Hematol Educ Program Book 2013;2013:76–81.

[4] Risitano AM, Maciejewski JP, Green S, Plasilova M, Zeng W, Young NS. In-vivo dominant immune responses in aplastic anaemia: molecular tracking of putatively pathogenetic T-cell clones by TCR beta-CDR3 sequencing. Lancet 2004;364:355–64.

[5] Solomou EE, Rezvani K, Mielke S, Malide D, Keyvanfar K, Visconte V, Kajigaya S, Barrett AJ, Young NS. Deficient CD4+ CD25+ FOXP3+ T regulatory cells in acquired aplastic anemia. Blood 2007;110(5):1603–6.

[6] Kordasti S, Marsh J, Al-Khan S, Jiang J, Smith A, Mohamedali A, Abellan PP, Veen C, Costantini B, Kulasekararaj AG, Benson-Quarm N, Seidl T, Mian SA, Farzaneh F, Mufti GJ. Functional characterization of CD4+ T cells in aplastic anemia. Blood 2012;119(9):2033–43.

[7] de Latour RP, Visconte V, Takaku T, Wu C, Erie AJ, Sarcon AK, Desierto MJ, Scheinberg P, Keyvanfar K, Nunez O, Chen J, Young NS. Th17 immune responses contribute to the pathophysiology of aplastic anemia. Blood 2010;116(20):4175–84.

[8] Valdez JM, Scheinberg P, Nunez O, Wu CO, Young NS, Walsh TJ. Decreased infection-related mortality and improved survival in severe aplastic anemia in the past two decades. Clin Infect Dis 2011;52(6):726–35.

[9] Scheinberg P, Wu CO, Nunez O, Young NS. Predicting response to immunosuppressive therapy and survival in severe aplastic anaemia. Br J Haematol 2009;144: 206–16.

[10] Scheinberg P, Wu CO, Nunez O, Scheinberg P, Boss C, Sloand EM, Young NS. Treatment of severe aplastic anemia with a combination of horse antithymocyte globulin and cyclosporine, with or without sirolimus: a prospective randomized study. Haematologica 2009;94:348–54.

[11] Tichelli A, Schrezenmeier H, Socie G, et al. A randomized controlled study in newly-diagnosed severe aplastic anemia patients receiving antithymocyte globulin, ciclosporin, with or without G-CSF. Blood 2011;117:4434–41.

[12] Risitano AM. Immunosuppressive therapies in the management of immune-mediated marrow failures in adults: where we stand and where we are going. Br J Haematol 2011;152(2):127–40.

[13] Risitano AM. Immunosuppressive therapies in the management of acquired immune-mediated marrow failures. Curr Opin Hematol 2012;19(1):3–13.

[14] Scheinberg P, Wu CO, Nunez O, Scheinberg P, Boss C, Sloand EM, Young NS. Treatment of severe aplastic anemia with a combination of horse antithymocyte globulin and cyclosporine, with or without sirolimus: a prospective randomized study. Haematologica 2009;94:348–54.

[15] Sehgal SN. Rapamune® (RAPA, rapamycin, sirolimus): mechanism of action immunosuppressive effect results from blockade of signal transduction and inhibition of cell cycle progression. Clin Biochem 1998;31:335–40.

[16] Scheinberg P, Nunez O, Wu C, Young NS. Treatment of severe aplastic anaemia with combined immunosuppression: antithymocyte globulin, ciclosporin and mycophenolate mofetil. Br J Haematol 2006;133:606–11.

[17] Brennan DC, Flavin K, Lowell JA, et al. A randomized, double-blinded comparison of Thymoglobulin versus Atgam for induction immunosuppressive therapy in adult renal transplant recipients. Transplantation 1999;67:1011–8. [Erratum, Transplantation 1999;67:1386.].

[18] Gaber AO, First MR, Tesi RJ, et al. Results of the double-blind, randomized, multicenter, phase III clinical trial of Thymoglobulin versus Atgam in the treatment of acute graft rejection episodes after renal transplantation. Transplantation 1998;66:29–37.

[19] Di Bona E, Rodeghiero F, Bruno B, Gabbas A, Foa P, Locasciulli A, Rosanelli C, Camba L, Saracco P, Lippi A, Iori AP, Porta F, De Rossi G, Comotti B, Iacopino P, Dufour C, Bacigalupo A. Rabbit antithymocyte globulin (r-ATG) plus cyclosporine and granulocyte colony stimulating factor is an effective treatment for aplastic anaemia patients unresponsive to a first course of intensive immunosuppressive therapy. Gruppo Italiano Trapianto di Midollo Osseo (GITMO). Br J Haematol 1999;107:330–4.

[20] Scheinberg P, Nunez O, Young NS. Re-treatment with rabbit antithymocyte globulin and ciclosporin for patients with relapsed or refractory severe aplastic anaemia. Br J Haematol 2006;133:622–7.

[21] Scheinberg P, Nunez O, Weinstein BP, et al. Horse versus rabbit antithymocyte globulin in acquired aplastic anemia. New Engl J Med 1997;365:430–8.

[22] Marsh JC, Bacigalupo A, Schrezenmeier H, Tichelli A, Risitano AM, Passweg JR, Killick SB, Warren AJ, Foukaneli T, Aljurf M, Al-Zahrani HA, Schafhausen P, Roth A, Franzke A, Brummendorf TH, Dufour C, Oneto R, Sedgwick P, Barrois A, Kordasti S, Elebute MO, Mufti GJ, Socie G. European Blood and Marrow

Transplant Group Severe Aplastic Anaemia Working Party. Prospective study of rabbit antithymocyte globulin and cyclosporine for aplastic anemia from the EBMT Severe Aplastic Anaemia Working Party. Blood 2012;119(23):5291–6.

[23] Kadia T, Ravandi F, Garcia-Manero G, et al. Updated results of combination cytokine immunotherpay in the treatment of aplastic anaemia and myelodysplastic syndroems (MDS). Blood 2010;116:2920.

[24] Afable MG, Shaik M, Sugimoto Y, et al. Efficacy of rabbit antithymocyte globulin in severe aplastic anemia. Haematologica 2011;96:1269–75.

[25] Atta EA, D Dias SP, Marra VLN, de Azevedo AM. Comparison between horse and rabbit antithymocyte globulin as first line treatment for patients with severe aplastic anemia: a single centre retrospective study. Ann Hematol 2010;89:851–9.

[26] Halkes CJM, Brand A, von dem Borne PA, et al. Increasing the dose of rabbit-ATG does not lead to a higher response rate in the first-line treatment of severe aplastic anaemia. Bone Marrow Transplant 2011;46: S90–S389.

[27] Saracco P, Lorenzati A, Oneto R, AIEOP Bone Marrow Failure Study Group. et al. Italian registry of pediatric acquired aplastic anaemia: a retrospective study. Bone Marrow Transplant 2011;46:S90–S389.

[28] Vallejo C, Montesinos P, Rosell A, et al. Comparison between lymphoglobuline- and thymoglobuline-based immunosuppressive therapy as first-line treatment for patients with aplastic anemia. Blood 2009;114(22):3194.

[29] Brodsky RA, Sensenbrenner LL, Jones RJ. Complete remission in severe aplastic anemia after high-dose cyclophosphamide without bone marrow transplantation. Blood 1996;87:491–4.

[30] Tisdale JF, Dunn DE, Geller N, Plante M, Nunez O, Dunbar CE, Barrett AJ, Walsh TJ, Rosenfeld SJ, Young NS. High-dose cyclophosphamide in severe aplastic anaemia: a randomised trial. Lancet 2000;356:1554–9.

[31] Tisdale JF, Maciejewski JP, Nunez O, Rosenfeld SJ, Young NS. Late complications following treatment for severe aplastic anemia (SAA) with high-dose cyclophosphamide (Cy): follow-up of a randomized trial. Blood 2002;100:4668–70.

[32] Brodsky RA, Chen AR, Dorr D, Fuchs EJ, Huff CA, Luznik L, Smith BD, Matsui WH, Goodman SN, Ambinder RF, Jones RJ. High-dose cyclophosphamide for severe aplastic anemia: long-term follow-up. Blood 2010;115:2136–41.

[33] Zhang F, Zhang L, Jing L, et al. High-dose cyclophosphamide compared with antithymocyte globulin for treatment of acquired severe aplastic anemia. Exp Hematol 2013;41(4):328–34.

[34] Scheinberg P, Townsley D, Dumitriu B, Scheinberg P, Weinstein B, Daphtary M, Rios O, Wu CO, Young NS. Moderate-dose cyclophosphamide for severe aplastic anemia has significant toxicity and does not prevent relapse and clonal evolution. Blood 2014;124(18): 2820–3.

[35] Scheinberg P, Nunez O, Weinstein B, Scheinberg P, Wu CO, Young NS. Activity of alemtuzumab monotherapy in treatment-naive, relapsed, and refractory severe acquired aplastic anemia. Blood 2012;119(2):345–54.

[36] Risitano AM, Selleri C, Serio B, Torelli GF, Kulagin A, Maury S, Halter J, Gupta V, Bacigalupo A, Sociè G, Tichelli A, Schrezenmeier H, Marsh J, Passweg J, Rotoli B. Working Party Severe Aplastic Anaemia (WPSAA) of the European Group for Blood and Marrow Transplantation (EBMT). Alemtuzumab is safe and effective as immunosuppressive treatment for aplastic anaemia and single-lineage marrow failure: a pilot study and a survey from the EBMT WPSAA. Br J Haematol 2010;148:791–6.

[37] Risitano AM, Selleri C, Serio B, et al. Alemtuzumab for aplastic anemia and related immune-mediated bone marrow failures: long-term follow up of a pilot study. Haematologica 2011;96(2):105.

[38] Scheinberg P, Fischer SH, Li L, Nunez O, Wu CO, Sloand EM, Cohen JI, Young NS, John Barrett A. Distinct EBV and CMV reactivation patterns following antibody-based immunosuppressive regimens in patients with severe aplastic anemia. Blood 2007;109(8):3219–24.

[39] Kim H, Min YJ, Baek JH, Shin SJ, Lee EH, Noh EK, Kim MY, Park JH. A pilot dose-escalating study of alemtuzumab plus cyclosporine for patients with bone marrow failure syndrome. Leuk Res 2009;33:222–31.

[40] Gómez-Almaguer D, Jaime-Pérez JC, Garza-Rodríguez V, Chapa-Rodríguez A, Tarín-Arzaga L, Herrera-Garza JL, Ruiz-Argüelles GJ, López-Otero A, González-Llano O, Rodríguez-Romo L. Subcutaneous alemtuzumab plus cyclosporine for the treatment of aplastic anemia. Ann Hematol 2010;89:299–303.

[41] LaCorcia G, Swistak M, Lawendowski C, et al. Polyclonal rabbit antithymocyte globulin exhibits consistent immunosoppressive capabilities beyond cell depletion. Transplantation 2009;87:966–74.

[42] Kawano Y, Nissen C, Gratwohl A, Würsch A, Speck B. Cytotoxic and stimulatory effects of antilymphocyte globulin (ALG) on hematopoiesis. Blut 1990;60: 297–300.

[43] Barbano GC, Schenone A, Roncella S, et al. Anti-lymphocyte globulin stimulates normal human T cells to proliferate and to release lymphokines in vitro: a study at the clonal level. Blood 1988;72:956–63.

[44] Flynn J, Cox CV, Rizzo S, Foukaneli T, Rice K, Murphy M, Welsh J, Rutherford TR, Gordon-Smith EC, Gibson FM. Direct binding of antithymoctye globulin to haemopoietic progenitor cells in aplastic anaemia. Brit J Haematol 2003;122:289–97.

[45] Dufour C, Giacchino R, Ghezzi P, Tonelli R, Ferretti E, Pitto A, Pistoia V, Lanza T, Svahn J. Etanercept as

a salvage treatment for refractory aplastic anemia. Pediatr Blood Cancer 2009;52:522–5.

[46] Available from: https://clinicaltrials.gov/ct2/show/NCT02411084?term=begedina&rank=1

[47] Krueger GG. Selective targeting of T cell subsets: focus on alefacept—a remittive therapy for psoriasis. Expert Opin Biol Ther 2002;2:431–41.

[48] Cather JC, Cather JC, Menter A. Modulating T cell responses for the treatment of psoriasis: a focus on efalizumab. Expert Opin Biol Ther 2003;3:361–70.

[49] Rice GP, Hartung HP, Calabresi PA. Anti-alpha4 integrin therapy for multiple sclerosis: mechanisms and rationale. Neurology 2005;64(8):1336–42.

[50] Le Blanc K, Frassoni F, Ball L, Locatelli F, Roelofs H, Lewis I, Lanino E, Sundberg B, Bernardo ME, Remberger M, Dini G, Egeler RM, Bacigalupo A, Fibbe W, Ringdén O. Developmental Committee of the European Group for Blood and Marrow Transplantation. Mesenchymal stem cells for treatment of steroid-resistant, severe, acute graft-versus-host disease: a phase II study. Lancet 2008;371(9624):1579–86.

[51] Introna M, Lucchini G, Dander E, Galimberti S, Rovelli A, Balduzzi A, Longoni D, Pavan F, Masciocchi F, Algarotti A, Micò C, Grassi A, Deola S, Cavattoni I, Gaipa G, Belotti D, Perseghin P, Parma M, Pogliani E, Golay J, Pedrini O, Capelli C, Cortelazzo S, D'Amico G, Biondi A, Rambaldi A, Biagi E. Treatment of graft versus host disease with mesenchymal stromal cells: a phase I study on 40 adult and pediatric patients. Biol Blood Marrow Transplant 2014;20(3):375–81.

[52] Fouillard L, Bensidhoum M, Bories D, Bonte H, Lopez M, Moseley AM, Smith A, Lesage S, Beaujean F, Thierry D, Gourmelon P, Najman A, Gorin NC. Engraftment of allogeneic mesenchymal stem cells in the bone marrow of a patient with severe idiopathic aplastic anemia improves stroma. Leukemia 2003;17:474–6.

[53] Bacigalupo A, Valle M, Podestà M, Pitto A, Zocchi E, De Flora A, Pozzi S, Luchetti S, Frassoni F, Van Lint MT, Piaggio G. T-cell suppression mediated by mesenchymal stem cells is deficient in patients with severe aplastic anemia. Exp Hematol 2005;33:819–27.

[54] Gluckman E, Devergie A, Faille A, Barrett AJ, Bonneau M, Boiron M, Bernard J. Treatment of severe aplastic anemia with antilymphocyte globulin and androgens. Exp Hematol 1978;6(8):679–87.

[55] Gluckman E, Devergie A, Poros A, Degoulet P. Results of immunosuppression in 170 cases of severe aplastic anaemia. Report of the European Group of Bone Marrow Transplant (EGBMT). Br J Haematol 1982;51(4):541–50.

[56] Champlin RE, Ho WG, Feig SA, Winston DJ, Lenarsky C, Gale RP. Do androgens enhance the response to antithymocyte globulin in patients with aplastic anemia? A prospective randomized trial. Blood 1985;66(1):184–8.

[57] Bacigalupo A, Van Lint MT, Congiu M, Pittaluga PA, Occhini D, Marmont AM. Treatment of SAA in Europe 1970-1985: a report from the SAA Working Party. Bone Marrow Transplant 1986;1(S1):19–21.

[58] Kaltwasser JP, Dix U, Schalk KP, Vogt H. Effect of androgens on the response to antithymocyte globulin in patients with aplastic anaemia. Eur J Haematol 1988;40(2):111–8.

[59] Bacigalupo A, Chaple M, Hows J, Van Lint MT, McCann S, Milligan D, Chessells J, Goldstone AH, Ottolander J, van't Veer ET, et al. Treatment of aplastic anaemia (AA) with antilymphocyte globulin (ALG) and methylprednisolone (MPred) with or without androgens: a randomized trial from the EBMT SAA working party. Br J Haematol 1993;83(1):145–51.

[60] Calado RT, Yewdell WT, Wilkerson KL, Regal JA, Kajigaya S, Stratakis CA, Young NS. Sex hormones, acting on the TERT gene, increase telomerase activity in human primary hematopoietic cells. Blood 2009;114(11):2236–43.

[61] Calado RT, Young NS. Telomere maintenance and human bone marrow failure. Blood 2008;111:4446–55.

[62] Maciejewski JP, Risitano AM. Aplastic anemia: management of adult patients. Am Soc Hematol Educ Program Book 2005;110–7.

[63] Annunziata S, Van Lint MT, Lamparelli T, Gualandi F, Di Grazia C, Dominietto A, Bregante S, Bacigalupo. Androgens may boost response to anti-thymocyte globulin in acquired aplastic anemia. Bone Marrow Transplant 2010;45(S2):S168.

[64] Bussel JB, Cheng G, Saleh MN, Psaila B, Kovaleva L, Meddeb B, Kloczko J, Hassani H, Mayer B, Stone NL, Arning M, Provan D, Jenkins JM. Eltrombopag for the treatment of chronic idiopathic thrombocytopenic purpura. N Engl J Med 2007;357(22):2237–47.

[65] Garnock-Jones KP, Keam SJ. Eltrombopag. Drugs 2009;69(5):567–76.

[66] Zeigler FC, de Sauvage F, Widmer HR, Keller GA, Donahue C, Schreiber RD, Malloy B, Hass P, Eaton D, Matthews W. In-vitro megakaryocytopoietic and thrombopoietic activity of c-mpl ligand (TPO) on purified murine hematopoietic stem-cells. Blood 1994;84(12):4045–52.

[67] Ku H, Yonemura Y, Kaushansky K, Ogawa M. Thrombopoietin, the ligand for the Mpl receptor, synergizes with steel factor and other early acting cytokines in supporting proliferation of primitive hematopoietic progenitors of mice. Blood 1996;87(11):4544–51.

[68] Sitnicka E, Lin N, Priestley GV, Fox N, Broudy VC, Wolf NS, Kaushansky K. The effect of thrombopoietin on the proliferation and differentiation of murine hematopoietic stem cells. Blood 1996;87(12):4998–5005.

[69] Ihara K, Ishii E, Eguchi M, Takada H, Suminoe A, Good RA, Hara T. Identification of mutations in the c-mpl

gene in congenital amegakaryocytic thrombocytopenia. Proc Natl Acad Sci USA 1999;96(6):3132–6.

[70] Emmons RV, Reid DM, Cohen RL, Meng G, Young NS, Dunbar CE, Shulman NR. Human thrombopoietin levels are high when thrombocytopenia is due to megakaryocyte deficiency and low when due to increased platelet destruction. Blood 1996;87(10):4068–71.

[71] Feng X, Scheinberg P, Wu CO, Samsel L, Nunez O, Prince C, Ganetzky RD, McCoy JP Jr, Maciejewski JP, Young NS. Cytokine signature profiles in acquired aplastic anemia and myelodysplastic syndromes. Haematologica 2011;96(4):602–26.

[72] ClinicalTrials.gov. A service of the U.S. National Institutes of Health. Available from: http://clinicaltrials.gov/ct2/show/NCT00922883

[73] Olnes MJ, Scheinberg P, Calvo KR, Desmond R, Tang Y, Dumitriu B, Parikh AR, Soto S, Biancotto A, Feng X, Lozier J, Wu CO, Young NS, Dunbar CE. Eltrombopag and improved hematopoiesis in refractory aplastic anemia. N Engl J Med. 2012;367(1):11–9.

[74] Desmond R, Townsley DM, Dumitriu B, Olnes MJ, Scheinberg P, Bevans M, Parikh AR, Broder K, Calvo KR, Wu CO, Young NS, Dunbar CE. Eltrombopag restores trilineage hematopoiesis in refractory severe aplastic anemia that can be sustained on discontinuation of drug. Blood 2014;123(12):1818–25.

[75] Desmond R, Townsley DM, Dunbar C, Young NS. Eltrombopag in aplastic anemia. Semin Hematol 2015;52(1):31–7.

[76] Roth M, Will B, Simkin G, Narayanagari S, Barreyro L, Bartholdy B, Tamari R, Mitsiades CS, Verma A, Steidl U. Eltrombopag inhibits the proliferation of leukemia cells via reduction of intracellular iron and induction of differentiation. Blood 2012;120(2):386–94.

[77] ClinicalTrials.gov. A service of the U.S. National Institutes of Health. Available from: http://clinicaltrials.gov/ct2/show/NCT01623167

[78] Townsley DM, Dumitriu B, Scheinberg P, Desmond R, Feng X, Rios O, Weinstein B, Valdez J, Winkler T, Desierto M, Leuva H, Wu C, Calvo KR, Larochelle A, Dunbar CE, Young NS. Eltrombopag added to standard immunosuppression for aplastic anemia accelerates count recovery and increases response rates. Blood 2015;126. 23(LBA-2).

[79] ClinicalTrials.gov. A service of the U.S. National Institutes of Health. Available from: http://clinicaltrials.gov/ct2/show/NCT02099747

[80] Marotta S, Pagliuca S, Risitano AM. Hematopoietic stem cell transplantation for aplastic anemia and paroxysmal nocturnal hemoglobinuria: current evidence and recommendations. Expert Rev Hematol 2014;7(6):775–89.

[81] Peffault de Latour R, Purtill D, Ruggeri A, Sanz G, Michel G, Gandemer V, Maury S, Kurtzberg J, Bonfim C, Aljurf M, Gluckman E, Socié G, Passweg J, Rocha V. Influence of nucleated cell dose on overall survival of unrelated cord blood transplantation for patients with severe acquired aplastic anemia: a study by eurocord and the aplastic anemia working party of the European group for blood and marrow transplantation. Biol Blood Marrow Transplant 2011;17(1):78–85.

[82] Yoshizato T, Dumitriu B, Hosokawa K, Makishima H, Yoshida K, Townsley D, Sato-Otsubo A, Sato Y, Liu D, Suzuki H, Wu CO, Shiraishi Y, Clemente MJ, Kataoka K, Shiozawa Y, Okuno Y, Chiba K, Tanaka H, Nagata Y, Katagiri T, Kon A, Sanada M, Scheinberg P, Miyano S, Maciejewski JP, Nakao S, Young NS, Ogawa S. Somatic mutations and clonal hematopoiesis in aplastic anemia. N Engl J Med 2015;373(1):35–47.

Bone Marrow Failure in Paroxysmal Nocturnal Hemoglobinuria

R. Peffault de Latour*, A.M. Risitano**

*BMT Unit, French Reference Center for Aplastic Anemia and PNH, Saint-Louis
Hospital, Paris, France; **Hematology, Department of Clinical Medicine and Surgery,
Federico II University, Naples, Italy

INTRODUCTION

The diagnosis of aplastic anemia (AA) is defined as the association of pancytopenia with persistent and unexplained reduced marrow hematopoietic cellularity in the absence of dysplastic features. There are thus no specific biologic markers and the diagnosis is reached by exclusion of other reasonable entities. The diagnosis of idiopathic aplastic anemia (IAA) can be difficult due to the overlapping morphologic characteristics with other bone marrow failure (BMF) disorders, especially Fanconi anemia [1–3]. In this context, the detection of a paroxysmal nocturnal hemoglobinuria (PNH) clone is helpful as it is generally the signature of an autoimmune-mediated process, eliminating per se a constitutional disorder. However,

the presence of a PNH clone in the context of BMF is most of the time confusing for hematologists, especially when the question of a specific treatment of PNH arises. Indeed, the diagnosis of classic PNH is based on additional clinical manifestations (i.e., overt hemolysis), which often does not occur in the context of BMF, where usually the diagnosis of an AA/PNH syndrome or of subclinical PNH is more appropriate [4]. The pathophysiology of this particular disease is also not very clear and is the subject of intense research worldwide. This paper will successively address the pathophysiology of such disease, the significance of a PNH clone in the context of BMF and the treatment approach consisting of hematopoietic stem cell transplantation (SCT) or immunosuppressive therapy/eculizumab.

Congenital and Acquired Bone Marrow Failure
http://dx.doi.org/10.1016/B978-0-12-804152-9.00014-2

PATHOPHYSIOLOGY OF BMF IN PNH

As discussed further, BMF is one of the three typical manifestations of PNH, the other two being complement-mediated hemolysis and thrombophilia. PNH patients are anemic, as they often exhibit cytopenias involving other blood lineages, which result from an impaired production from the bone marrow rather than from increased turnover (as for erythrocytes). The impairment of the bone marrow in PNH is confirmed by in vitro studies showing reduced numbers of lineage-committed hematopoietic progenitors (CFU-E, BFU-E, CFU-GM, and CFU-GEMM) [5,6] and also a contraction of the best in vitro surrogate for hematopoietic stem cells (HSCs), the so-called long-term culture initiating cells (LTC-IC) [6]. Thus, in PNH the underlying bone marrow disorder is not only qualitative (i.e., the production of mature blood cells lacking all GPI-linked proteins on their surface), but also quantitative. This BMF represents a bridge between PNH and the most typical acquired BMF, IAA. Clinically, many PNH patients develop mild to severe cytopenia (up to frank IAA) during the disease course [7,8], as well as, many IAA patients harbor PNH cell populations (and may even develop clinical PNH) [9–12]. The association between these diseases is well known since half a century, with the initial description by Dr. Lewis and Sir Dacie [13] and even the pathogenic similarities observed directed the investigators to consider AA and PNH as two faces of the same coin rather than two distinct diseases [14]. Several evidence suggest that an immune-mediated pathophysiology may play a role in BMF in PNH and it eventually may also be involved in the expansion of the PIG-A-mutated hematopoiesis, which is important for the development of the clinical disease. All these aspects are discussed in this chapter.

Derangement of the Immune System

The immune-mediated pathophysiology of BMF is well established in IAA. A plethora of in vitro and ex vivo data demonstrate that an immune-mediated attack over normal hematopoiesis may damage HSCs and hematopoietic progenitors [15,16]. While it is still unknown what the target antigens are and why such a response eludes physiologic control of self-reactive immunity, the key role of preferentially expanded clonal T cells has been documented [17]. Due to the clinical overlap between IAA and PNH, several investigators have searched for evidence for antigen-driven immune responses even in PNH. Indeed, oligoclonality of the T-cell pool has been initially reported by Karadimitris et al. [18], and subsequently confirmed by other groups [17,19]. As for IAA, highly homologous TCRβ sequences were found in different PNH patients, consistent with an antigen-driven public immune response [20]. Interestingly, these homologous TCRs were shared among patients irrespective of their HLA background, possibly suggesting that the driving antigen may be nonpeptidic (i.e., non-HLA restricted), such as glycolipids [20]. As in IAA, these clonal populations often exhibit an effector cytotoxic phenotype, characterized by expression of CD8 and CD57, with possible imbalance of the activating- and inhibitory surface receptors [21]. In a few cases, these T-cell clonal expansions may functionally resemble those seen in large granular lymphocyte [22,23].

The Dual Pathophysiology Theory

The immunologic abnormalities found in PNH patients are crucial not only because they support an immune-mediated BMF, such in IAA, but also because they may shed light on the key pathogenic events in PNH, namely the expansion of the aberrant PIG-A-mutated HSCs. Indeed, while it is well accepted that a somatic mutation in HSCs is necessary to develop PNH, several observations support the concept that the mutation itself is not sufficient to cause the disease [24]. Indeed, it is true that a PIG-A mutation (or more recently very rare inactivating mutations of other genes involved in the same pathway) is found in all PNH patients, yet the same mutation can be detected even in subject without PNH. A few blood cells with the PNH phenotype (namely, a complete or partial deficiency in all GPI-linked proteins) and harboring inactivating PIG-A mutations may be found even in normal individuals, even at a very low frequency (ranging between 10 and 50 cells per million) [25]. Thus, at least in these subjects a PIG-A mutation was not sufficient to cause PNH, possibly because it arose in blood cells without self-renewal capability (i.e., not in a HSC). This observation was also confirmed in the animal model. In all the sophisticated attempts aiming to generate a mouse model of PNH, the pig-a mutated hematopoiesis usually disappear over time, eventually suggesting that other mechanisms are required to sustain the expansion of PNH-like hematopoiesis [26–28]. The possibility that such additional factors exist arises from clinical observations in PNH patients. It is well known that the same PNH patient may harbor distinct PNH clones with different PIG-A mutations [29,30]. Thus, it is conceivable that a PIG-A mutation simply confers a biologic phenotype (the GPI deficiency) that eventually requires additional PIG-A-independent factors/events for further clonal expansion necessary for the disease development of PNH. In fact, the expansion of distinct clones carrying the same, albeit molecularly heterogeneous, functional defect seems to not fit with a random process, as also supported by the observation that in case of relapse of PNH clones harboring PIG-A mutations different from those identified at diagnosis can be found [31].

To reconcile all these observations, in 1989 Rotoli and Luzzatto drafted the theory of the "dual pathophysiology" of PNH [24], which is also known as the "relative advantage" [32] or "escape" theory" [16]. According to this theory, a PIG-A mutation is required but not sufficient to promote the expansion of GPI-deficient hematopoiesis because it does not confer any intrinsic proliferation advantage on normal hematopoiesis [25]. Then, additional events

may occur, which eventually promote the expansion of PNH HSCs over normal ones; this second event is likely an immune-mediated attack over hematopoiesis, similar to that causing IAA. However, such an immune-attack for some reasons may spare PNH HSCs, eventually resulting in their relative expansion over normal, damaged hematopoiesis. Thus, PNH hematopoiesis may finally limit the development of clinical BMF but also result in PNH with its typical the clinical phenotype [16,24,32]. This scenario is supported by the immune derangement discussed earlier and even more by gene-expression profile data on PNH and non-PNH CD34+ cells isolated from PNH patients. In this study, while PNH (GPI-deficient) CD34+ showed a gene-expression profile almost indistinguishable from healthy individuals, phenotypically normal (GPI-positive) CD34+ cells harbored diffuse abnormalities, with overexpression of genes involved in apoptosis and immune activity, paralleling the findings seen in CD34+ cells of AA patients [33]. These data strongly support the presence of an immune-mediated extrinsic, sublethal damage, which selectively pertain to normal HSCs in PNH patients and apparently spares PNH HSCs [33]. Thus, PNH hematopoiesis eventually expands as a result of this persistent selective pressure; the reasons accounting for the escape of PNH HSC are still unknown. It has been hypothesized that the antigen targeted by the immune attack may be absent in PNH cells. Quite recently, GPI-specific clonal CD8+ T cells have been found in several PNH patients, possibly suggesting that the GPI anchor itself (thus a glycolipid antigen presented via C1d rather than HLA molecules) may be the target of the autoimmune process causing BMF in PNH patients (and even the expansion of PNH HSCs, eventually causing the clinical phenotype of PNH) [34].

The Role of Somatic Mutations

The presence of an immune-mediated attack sparing PNH hematopoiesis reconciles with most of the observations described earlier. However, a definitive proof of this pathogenic mechanism is still lacking, eventually raising alternative hypotheses. Indeed, the selective advantage of the PNH HSCs may result from secondary genetic events that may confer an absolute growth advantage. Even if the demonstration that the PIG-A mutation does not confer any intrinsic genetic susceptibility [25] argued against this hypothesis, recent data may now change this view. A few cases of PNH patients harboring well-defined mutations in the 3′ end of the HMGA2 gene have been reported [35]. However, even if HMGA2 overexpression may lead to a proliferative advantage, this mutation has not been found in a larger series of patients. More recently, the availability of next generation–sequencing techniques allowed the identification of additional somatic mutations in several PNH patients [36]. Notably, even if mutations were not recurrent, they may affect genes also involved in the pathophysiology of myeloid neoplasm, such as, TET2, SUZ12, U2AF1, and JAK2 [36]. The definition of the hierarchical clonal architecture of hematopoiesis in PNH patients defined a complex scenario where accessory genetic events may occur before or after the PIG-A mutation, similar to the stepwise clonal evolution seen in hematologic malignancies [36].

However, these data have to be interpreted in the context of our novel understanding of clonal hematopoiesis, which have clearly demonstrated that some of these somatic mutations do not necessarily imply an obvious malignant transformation [37,38]. Indeed, these mutations have been found also in IAA and their pathogenic meaning remains elusive [39]; even if in the context of PNH pathophysiology they might account for selective clonal expansion, eventually arguing against the need of an immune-mediated selection, further investigations are needed to define their possible pathogenic role.

PNH CLONE IN PATIENTS WITH BMF

PNH Disease Subcategories

PNH is characterized by a unique triad of clinical features: intravascular hemolysis, thromboembolic events, and cytopenia [4,7,8,40]. However, not all three manifestations are presents in all the patients and the individual presentation of each patient may greatly vary according to the most dominant signs and symptoms. Thus, many investigators have tried to classify PNH according to the most typical clinical presentations. However, distinct categories are hard to define for a disease with such an unpredictable presentation and evolution.

The most-adopted classification of PNH was proposed by the International PNH Interest Group in 2005 [4], whereby PNH patients were grouped according to the presence of hemolysis and of an underlying bone marrow disorder. Accordingly, three distinct subtypes are identified: (1) classic PNH, characterized by hemolysis without other marrow disorder (i.e., hemolytic PNH patients without relevant cytopenia); (2) PNH in the setting of another bone marrow disorder, characterized by hemolysis associated with an underlying marrow disorder, usually AA or MDS (i.e., hemolytic PNH patients with cytopenia; AA or MDS may be concomitant or have preceded PNH); and (3) subclinical PNH, characterized by the presence of PNH cells in the absence of any clinical or laboratory sign of hemolysis, in the setting of other hematologic disorders (i.e., AA or MDS patients with GPI-AP deficient cells, but not clinical PNH). This classification does not completely take into account that, by definition, PNH carries an underlying bone marrow disorder and, as a result, most PNH patients have cytopenia or some signs of marrow failure. In fact, a recent registry study [8] made the point that many PNH patients do not fit in either of the two major categories and a fourth subgroup has been included (intermediate PNH, characterized by hemolysis and mild cytopenia not qualifying for the diagnosis of AA or typical classic PNH). However, even this classification seems to fail the goal of identifying patient subgroups with distinct clinical outcome mainly because of the overlapping form of the disease (evolution from an aplastic form to an hemolytic form and vice versa during evolution) [8]. Some other groups have used a different classification in the past [41], where the category of AA/PNH patients is restricted to those with concomitant severe AA and clinically meaningful hemolysis, who require more intensive care and are supposed to have a worse prognosis. According to this classification, classic PNH patients are further grouped into hyperplastic and hypoplastic (based on peripheral counts and bone marrow analysis), AA/PNH patients are only those with severe marrow failure and concomitant clinically relevant hemolysis, whereas subclinical PNH patients are characterized by small PNH clone(s) (even with minimal signs of hemolysis) associated with either AA or thromboembolic disease. This latter condition is very rare, and includes a few patients where thromboembolic disease can be found in absence of hemolysis. This rare condition may also be described as "myeloid PNH" or "white PNH," and seems to imply the possibility that the expansion of a GPI-deficient hematopoiesis may be restricted to specific blood lineages (i.e., mutations occurring in committed progenitors or PIG-A mutated HSCs, which preferentially contribute to myelopoiesis or megakaryocytopoiesis), eventually driving the subsequent clinical phenotype (thrombophilia would be explained directly by GPI-deficient platelets that are unable to control complement activation on their surface).

For clinical purposes, it seems reasonable to consider each disease (AA with subclinical PNH, AA/PNH syndrome, and hemolytic PNH) separately that helps discussing treatment indications and management.

The Clinical Relevance of a PNH Clone in the Context of BMF

Pathophysiology

As described before, the persistence of a PNH clone is underlying an on going autoimmune process in the bone marrow. The diagnosis of AA is defined by the association of pancytopenia with persistent and unexplained reduced marrow hematopoietic cellularity, with no major dysplastic features. There are thus no specific biologic markers and the diagnosis is reached by exclusion of other reasonable entities. The diagnosis is thus difficult especially to eliminate a constitutional disorder especially in young patients. The presence of a significant PNH clone (1% or more) is in favor of an underlying autoimmune process and thus an IAA. A PNH clone is diagnosed in 30–40% of patients with IAA at diagnosis or during evolution [9–12]. This justifies testing of all PNH patients with AA at least once a year or in the presence of clinical sign of hemolysis (abdominal pain, nausea, and red urine).

Risk of Thrombosis

Thrombosis is the third typical manifestation of PNH, which develops in about 40% of all patients; accordingly, PNH is the medical condition with the higher risk of thrombosis. While the pathophysiology of thrombosis occurrence in PNH is still not clear, presentation is quite unique because it mostly occurs at venous sites, which are unusual for other non-PNH-related thrombosis. Intraabdominal veins are the most frequent sites, followed by cerebral and limb veins; other possible sites include dermal veins, the lungs—with pulmonary embolus—and the arteries, leading to arterial thrombosis. Thrombotic disease may be life threatening and is the main cause of death for PNH patients [7,8]. Typical severe presentations of thrombotic PNH include hepatic venous (Budd–Chiari syndrome) [42], portal, mesenteric, and renal vein thrombosis. Usually patients are asymptomatic, until clinical manifestations appear, especially pain. Other signs and symptoms are specific according to the vessel involved (e.g., ascites, varices, and splenomegaly in hepatic/portal thrombosis or stroke in cerebral vein thrombosis). According to most series, they generally develop in patients with large clones and massive hemolysis [43,44] but may also complicate a significant number of patients with BMF and a PNH clone (i.e., the 10-year cumulative incidence of thrombosis is 30% in those patients [8]). This is the reason why patients with BMF and a PNH clone are exposed to a higher risk of thrombosis and should be explored quickly in case they have symptoms that might be related to this complication (headache, abdominal pain, etc.).

Evolution to Classic PNH

Hemolysis is the most typical manifestation of PNH, which by definition, affects all patients with clinical PNH. However, the extent of hemolysis varies among patients, according to the size of the PNH clone(s), the type of PNH erythrocytes (type II vs. type III), and possibly the level of complement activation (which may vary according to specific medical conditions or patient-specific features). Thus, patients with BMF and a PNH clone have generally either no level or a low level of hemolysis and are not classically suffering from clinical signs of intravascular hemolysis. However, 20% of patients diagnosed with PNH in the context of BMF will develop further on a typical classic highly hemolytic PNH [8]. This is the reason why such patients should be monitored carefully at the clinical level (hemoglobinuria, abdominal pain, and fatigue). At the biologic level, the typical biochemical marker of hemolysis is the increase in lactate dehydrogenase (LDH), which may be as high as 10-fold the upper normal value; additional intraerythrocytary components may also increase, such as aminotransferases (especially the alanine aminotransferase). As in other hemolytic disorders, unconjugated bilirubin levels may increase, even up to frank jaundice; compensatory erythropoiesis is usually demonstrated by very high reticulocyte counts, even if the latter value is rarely present in patients with BMF. Clinical and biologic follow-up of patients with BMF and a PNH clone should thus include this particular follow-up.

TREATMENT

The management of marrow failure in PNH patients is the same as for AA patients and represents the main challenge for physicians dealing with the treatment of this condition [45]. In addition to supportive strategies, such as, anti-infectious, antithrombotic, and antihemorrhagic prophylaxis and/or treatments, etiologic therapies can also be attempted. According to their conditions, patients with BMF and PNH may benefit from SCT, immunosuppresive therapy (IST), or eculizumab [46]. In the absence of intravascular hemolysis (normal LDH), there is no reason to consider another treatment algorithm than the one applied to patients with IAA with a PNH clone (please see specific sections of this handbook). Several reports have suggested that the presence of a PNH population may even increase the chance of response to standard IST [11,47]. Conversely, patients with BMF and a PNH clone with a significant level of intravascular hemolysis (LDH more than 2 times higher than normal) may be more challenging to manage as eculizumab might be useful too on top of the treatment for BMF.

Immunosuppressive Therapy

According to the pathogenic mechanisms and the dual hypothesis described earlier, an immune-mediated inhibition of hematopoiesis is postulated in PNH, similar to that demonstrated in AA. Thus, immunosuppressive strategies have been reasonably utilized in PNH patients, even if large prospective studies are lacking. Cyclosporine A has led to some improvement in a few series [48,49]. More intensive regimens (as those recommended in severe AA) using the antithymocyte globulin, associated with high dose prednisone and cyclosporin A, have also been exploited; however, the available results are quite heterogeneous [50,51]. Alternative immunosuppressive agents, such as, the anti-CD52 monoclonal antibody alemtuzumab [52] may be an alternate option (as salvage treatment). In the setting of alemtuzumab-based treatment, there is no concern over the potential selection risk for PNH hematopoiesis, given that GPI-linked CD52 is not expressed on HSCs [52]. Cyclophosphamide has been suggested by some [53] but never reproduced and is potentially toxic (for a review, see [54]). Regardless of the specific immunosuppressive regimen, when this etiologic treatment leads to an improvement of the underlying bone marrow impairment, usually normal (non-PNH) hematopoiesis may be restored, possibly resulting in a progressive dilution (or even extinction) of the PNH clone.

Hematopoietic Stem Cell Transplantation

The only curative strategy for PNH is allogeneic SCT. SCT has been exploited since the late 1980s and has proven effective in eradicating the abnormal PNH clone possibly leading to definitive cure of PNH, even if morbidity and mortality remain a major limitation. All young PNH patients with BMF should be considered for transplantation if they have an HLA-matched donor [1] or even if they have an unrelated donor (later in their disease course). Marrow failure of PNH patients has to be treated as AA by either immunosuppression or allogeneic stem cell transplantation, regardless of the presence of the PNH clone(s) [45]. Most reports in the literature refer to single cases or small series from single institutions [55–57]. The International Bone Marrow Transplant Registry reported 57 consecutive SCT performed for

PNH (16 AA/PNH) between 1978 and 1995 [58], showing a 2-year survival of 56% in 48 HLA-identical sibling transplants (the median follow-up was 44 months). The incidence of grade II or severe acute GvHD was 34% and that of chronic GvHD of 33%; graft failure ($n = 7$) and infections ($n = 3$) were the most common causes of treatment failure. Another retrospective study from the Italian Transplant Group (GITMO) on 26 PNH patients (4 AA/PNH) transplanted between 1998 and 2006 showed a 57% survival rate at 10 years. Acute and chronic GvHD were 42% (grade III–IV 12%) and 50% (extensive 16%), respectively [59]. The largest study published so far was the report of the European group for Blood and Marrow Transplantation (EBMT) experience in 2012 [60]. The characteristics and overall survival of 211 patients transplanted for PNH in 83 EBMT centers from 1978 to 2007 were analyzed. The three main indications for SCT were AA ($n = 100$, 48%), severe recurrent hemolytic crisis ($n = 64$, 30%) and thrombosis ($n = 47$, 22%). Engraftment failed in 14 (7%) of the 202 transplanted patients for whom there was documentation on this aspect. Eighty-five patients developed acute GvHD leading to a CIF of grade II–IV acute GvHD of 40% (95% CI, 34–47%). Fifty-seven patients developed chronic GvHD (extensive, $n = 24$) leading to a CIF of 29% (95% CI, 23–36%) at 5 years. After a median (±standard error) follow-up time of 61 ± 6 months, 64 patients had died and the 5-year OS probability was $68 \pm 3\%$. Infections and GvHD were the main causes of death. None of the variables investigated for an association with transplant outcome was a statistically significant predictor of survival, except for the indication for SCT with outcome being worse if the indication for SCT was thromboembolism ($p = 0.03$). The conclusions of this large study were: (1) HSCT could no longer be considered as a standard of care of PNH patients with thromboembolism when eculizumab is available; (2) regarding the good results of SCT in the case of recurrent hemolytic crises, SCT can be a valuable option for patients living in countries who cannot afford eculizumab regardless of the type of donor; and (3) PNH patients with BMF are appropriate candidates for SCT if they have indication for treatment (i.e., severe AA or moderate AA but transfused).

At the moment, the main indication for SCT in PNH patients is thus an underlying BMF. As for AA patients, SCT may be performed as first-line therapy in the presence of an HLA-identical sibling donor, or in case of treatment failure in patients with an HLA-matched unrelated donor [45]. The patient's age largely drives the choice of treatment, given that transplant-related mortality and morbidity increase with age. Refractoriness to transfusions and life-threatening thrombosis were also indications to SCT in the past, but nowadays they rather represent indications to anticomplement treatment, with the exception of countries where eculizumab is not available (yet). Poor response to eculizumab, albeit rare, can be another indication to SCT, even if alternative [61] in this setting [62].

Due to the retrospective nature of available studies, several aspects about the best way to perform a SCT in PNH remain open. However, as the main indication to SCT is BMF, several information can be drawn from SCT for AA (please see specific sections of this handbook). The two main open questions concern the HSC source and the conditioning regimens. As for AA, GvHD is the most feared complication after SCT for PNH and a plethora of data demonstrate that bone marrow should be preferred over peripheral blood as a stem cell source [63,64]. Some investigators have pointed out that a "graft versus PNH" effect may be needed to eradicate the PNH clone, especially in nonhypoplastic PNH [65]. However, as PNH is not a cancer, this does not seem enough to justify the increased risk of GvHD expected with the use peripheral blood stem cells. Strategies of GvHD prophylaxis should parallel those used in the context of AA, eventually guided by donor type and his HLA compatibility (please see specific sections of this handbook). Even for the conditioning regimen

(i.e., myeloablative vs. reduced-intensity regimens), objective data are lacking and information can only be drawn from the AA experience (please see specific sections of this handbook). PNH patients transplanted for concomitant AA should follow the same conditioning regimens used for AA (i.e., cyclophosphamide/antithymocyte globulin for sibling transplants, fludarabine-based RIC for unrelated transplants (please see specific sections of this handbook). Conversely, there are no specific guidelines for patients transplanted for a nonhypoplastic PNH; based on available data [60], myeloablative conditioning should be recommended, even busulphan-based [56]. However, fludarabine-based reduced intensity regimens can be considered in older patients or in presence of relevant comorbidities, as they have been proven effective as well [65].

Eculizumab Treatment

However, some patients with AA and a PNH clone may experience hemolysis concomitantly to severe cytopenia. Most of the time, the level of hemolysis is low (less than 1.5 times the normal) and no specific treatment is required. Conversely, in rare cases, patients with AA and PNH present high level of hemolysis with in some cases, clinical signs (abdominal pain, hemoglobinuria) and even thrombosis. There are very few reports in the literature of the association [66] and no guidelines can be drawn. The main recommendation is to treat first the most severe presentation (i.e., BMF or intravascular hemolysis). Sequential treatment should be preferred over concomitant delivery of IST and eculizumab. Patients who are transfused in platelets or with low neutrophils count should be primary treated with SCT or immunosuppression irrespective of hemolysis (please see specific sections of this handbook). Eculizumab is a C5 inhibitor and has no role in the treatment of BMF; nevertheless, its efficacy in the treatment of intravascular hemolysis of PNH is largely established, being the standard treatment for classic PNH [67,68]. Indeed, by controlling intravascular hemolysis, eculizumab results in Hb stabilization and transfusion-independence in at least half of PNH patients with severe hemolytic anemia. Furthermore, eculizumab significantly reduced the risk of thromboembolic events [69], eventually impacting the long-term survival of PNH patients [70]. Thus, eculzumab treatment should be considered in case of symptomatic hemolysis, aiming to control the clinical symptomatology, and possibly to improve Hb levels eventually reducing the need of red blood cell transfusions. However, it has to be remarked that in the context of BMF the hematologic benefit of eculizumab may be limited because anemia of AA/PNH patients may result mostly from impaired erythropoiesis rather than from intravascular hemolysis. Reticulocyte count is always very informative to predict the possible benefit from anticomplement treatment in these patients. Nevertheless, the use of anticomplement treatment may be useful to handle the risk of thromboembolic complications, which may occur even in AA/PNH patients [8]; again the presence of specific risk factors for thrombophilia (e.g., previous thrombotic events, extent of hemolysis, size of PNH population [44], and additional genetic or acquired factors) may guide the therapeutic decision. In all these conditions, the use of eculizumab concomitantly to IST represents a medical challenge. Even if the clinical outcome may be excellent [66], this treatment should be handled only in centers with the highest expertise in IST and anticomplement treatment.

CONCLUSIONS

The presence of a PNH population in the context of a BMF syndrome is a common finding, carrying specific pathophysiologic and therapeutic implications. Indeed, typical immune-mediated AA should be suspected, ruling out constitutional forms of BMF. Nevertheless, the management of a BMF associated with a PNH clone may be challenging because specific clinical presentations may appear anytime during the disease course; usually intravascular hemolysis, possibly thromboembolic events. The specific clinical presentation guides treatment choice in PNH, according to the main (and prognostically more severe) manifestation. Indeed, in presence of severe BMF patients should be treated as those affected by AA (either IST or SCT, based on age and donor availability). In contrast, when BMF is clinically less meaningful, other manifestations of PNH may be treated (i.e., eculizumab for hemolytic and/or thromboembolic PNH). Sometimes the clinical picture may be more complex because more manifestations may be present at the same time. In these conditions these different treatment approaches may be considered sequentially or even concomitantly. Nevertheless, these cases remain very challenging and should be handled with the active involvement of reference centers for AA and PNH.

References

[1] Young NS, Calado RT, Scheinberg P. Current concepts in the pathophysiology and treatment of aplastic anemia. Blood 2006;108:2509–19.

[2] Shimamura A. Clinical approach to marrow failure. Hematology Am Soc Hematol Educ Program 2009;329–37.

[3] Young NS. Acquired aplastic anemia. Ann Intern Med 2002;136:534–46.

[4] Parker C, Omine M, Richards S, et al. Diagnosis and management of paroxysmal nocturnal hemoglobinuria. Blood 2005;106:3699–709.

[5] Rotoli B, Robledo R, Scarpato N, Luzzatto L. Two populations of erythroid cell progenitors in paroxysmal nocturnal hemoglobinuria. Blood 1984;64:847–51.

[6] Maciejewski JP, Sloand EM, Sato T, Anderson S, Young NS. Impaired hematopoiesis in paroxysmal nocturnal hemoglobinuria/aplastic anemia is not associated with a selective proliferative defect in the glycosylphosphatidylinositol-anchored protein-deficient clone. Blood 1997;89:1173–81.

[7] Hillmen P, Lewis SM, Bessler M, Luzzatto L, Dacie JV. Natural history of paroxysmal nocturnal hemoglobinuria. N Engl J Med 1995;333:1253–8.

[8] de Latour RP, Mary JY, Salanoubat C, et al. Paroxysmal nocturnal hemoglobinuria: natural history of disease subcategories. Blood 2008;112:3099–106.

[9] Nissen C, Tichelli A, Gratwohl A, et al. High incidence of transiently appearing complement-sensitive bone marrow precursor cells in patients with severe aplastic anemia—A possible role of high endogenous IL-2 in their suppression. Acta Haematol 1999;101:165–72.

[10] Mukhina GL, Buckley JT, Barber JP, Jones RJ, Brodsky RA. Multilineage glycosylphosphatidylinositol anchor-deficient haematopoiesis in untreated aplastic anaemia. Br J Haematol 2001;115:476–82.

[11] Sugimori C, Chuhjo T, Feng X, et al. Minor population of CD55-CD59- blood cells predicts response to immunosuppressive therapy and prognosis in patients with aplastic anemia. Blood 2006;107:1308–14.

[12] Scheinberg P, Marte M, Nunez O, Young NS. Paroxysmal nocturnal hemoglobinuria clones in severe aplastic anemia patients treated with horse anti-thymocyte globulin plus cyclosporine. Haematologica 2010;95:1075–80.

[13] Lewis SM, Dacie JV. The aplastic anaemia—paroxysmal nocturnal haemoglobinuria syndrome. Br J Haematol 1967;13:236–51.

[14] Dameshek W. Riddle: what do aplastic anemia, paroxysmal nocturnal hemoglobinuria (PNH) and "hypoplastic" leukemia have in common? Blood 1967;30:251–4.

[15] Young NS, Maciejewski J. The pathophysiology of acquired aplastic anemia. N Engl J Med 1997;336:1365–72.

[16] Young NS, Maciejewski JP. Genetic and environmental effects in paroxysmal nocturnal hemoglobinuria: this little PIG-A goes "Why? Why? Why?". J Clin Invest 2000;106:637–41.

[17] Risitano AM, Maciejewski JP, Green S, Plasilova M, Zeng W, Young NS. In-vivo dominant immune responses in aplastic anaemia: molecular tracking of putatively pathogenetic T-cell clones by TCR beta-CDR3 sequencing. Lancet 2004;364:355–64.

[18] Karadimitris A, Manavalan JS, Thaler HT, et al. Abnormal T-cell repertoire is consistent with immune process underlying the pathogenesis of paroxysmal nocturnal hemoglobinuria. Blood 2000;96:2613–20.

[19] Plasilova M, Risitano AM, O'Keefe CL, et al. Shared and individual specificities of immunodominant cytotoxic T-cell clones in paroxysmal nocturnal hemoglobinuria as determined by molecular analysis. Exp Hematol 2004;32:261–9.

[20] Gargiulo L, Lastraioli S, Cerruti G, et al. Highly homologous T-cell receptor beta sequences support a common target for autoreactive T cells in most patients with paroxysmal nocturnal hemoglobinuria. Blood 2007;109:5036–42.

[21] Poggi A, Negrini S, Zocchi MR, et al. Patients with paroxysmal nocturnal hemoglobinuria have a high frequency of peripheral-blood T cells expressing activating isoforms of inhibiting superfamily receptors. Blood 2005;106:2399–408.

[22] Karadimitris A, Li K, Notaro R, et al. Association of clonal T-cell large granular lymphocyte disease and paroxysmal nocturnal haemoglobinuria (PNH): further evidence for a pathogenetic link between T cells, aplastic anaemia and PNH. Br J Haematol 2001;115:1010–4.

[23] Risitano AM, Maciejewski JP, Muranski P, et al. Large granular lymphocyte (LGL)-like clonal expansions in paroxysmal nocturnal hemoglobinuria (PNH) patients. Leukemia 2005;19:217–22.

[24] Rotoli B, Luzzatto L. Paroxysmal nocturnal haemoglobinuria. Baillieres Clin Haematol 1989;2:113–38.

[25] Araten DJ, Nafa K, Pakdeesuwan K, Luzzatto L. Clonal populations of hematopoietic cells with paroxysmal nocturnal hemoglobinuria genotype and phenotype are present in normal individuals. Proc Natl Acad Sci USA 1999;96:5209–14.

[26] Tremml G, Dominguez C, Rosti V, et al. Increased sensitivity to complement and a decreased red blood cell life span in mice mosaic for a nonfunctional Piga gene. Blood 1999;94:2945–54.

[27] Keller P, Payne JL, Tremml G, et al. FES-Cre targets phosphatidylinositol glycan class A (PIGA) inactivation to hematopoietic stem cells in the bone marrow. J Exp Med 2001;194:581–9.

[28] Jasinski M, Keller P, Fujiwara Y, Orkin SH, Bessler M. GATA1-Cre mediates Piga gene inactivation in the erythroid/megakaryocytic lineage and leads to circulating red cells with a partial deficiency in glycosyl phosphatidylinositol-linked proteins (paroxysmal nocturnal hemoglobinuria type II cells). Blood 2001;98:2248–55.

[29] Endo M, Ware RE, Vreeke TM, et al. Molecular basis of the heterogeneity of expression of glycosyl phosphatidylinositol anchored proteins in paroxysmal nocturnal hemoglobinuria. Blood 1996;87:2546–57.

[30] Nishimura J, Inoue N, Wada H, et al. A patient with paroxysmal nocturnal hemoglobinuria bearing four independent PIG-A mutant clones. Blood 1997;89:3470–6.

[31] Nafa K, Bessler M, Castro-Malaspina H, Jhanwar S, Luzzatto L. The spectrum of somatic mutations in the PIG-A gene in paroxysmal nocturnal hemoglobinuria includes large deletions and small duplications. Blood Cells Mol Dis 1998;24:370–84.

[32] Luzzatto L, Bessler M, Rotoli B. Somatic mutations in paroxysmal nocturnal hemoglobinuria: a blessing in disguise? Cell 1997;88:1–4.

[33] Chen G, Zeng W, Maciejewski JP, Kcyvanfar K, Billings EM, Young NS. Differential gene expression in hematopoietic progenitors from paroxysmal nocturnal hemoglobinuria patients reveals an apoptosis/immune response in 'normal' phenotype cells. Leukemia 2005;19:862–8.

[34] Gargiulo L, Papaioannou M, Sica M, et al. Glyco-sylphosphatidylinositol-specific, CD1d-restricted T cells in paroxysmal nocturnal hemoglobinuria. Blood 2013;121:2753–61.

[35] Inoue N, Izui-Sarumaru T, Murakami Y, et al. Molecular basis of clonal expansion of hematopoiesis in 2 patients with paroxysmal nocturnal hemoglobinuria (PNH). Blood 2006;108:4232–6.

[36] Shen W, Clemente MJ, Hosono N, et al. Deep sequencing reveals stepwise mutation acquisition in paroxysmal nocturnal hemoglobinuria. J Clin Invest 2014;124:4529–38.

[37] Steensma DP, Bejar R, Jaiswal S, et al. Clonal hematopoiesis of indeterminate potential and its distinction from myelodysplastic syndromes. Blood 2015;126:9–16.

[38] Malcovati L, Cazzola M. The shadowlands of MDS: idiopathic cytopenias of undetermined significance (ICUS) and clonal hematopoiesis of indeterminate potential (CHIP). Hematology Am Soc Hematol Educ Program 2015;2015:299–307.

[39] Yoshizato T, Dumitriu B, Hosokawa K, et al. Somatic mutations and clonal hematopoiesis in aplastic anemia. N Engl J Med 2015;373:35–47.

[40] Dunn DE, Tanawattanacharoen P, Boccuni P, et al. Paroxysmal nocturnal hemoglobinuria cells in patients with bone marrow failure syndromes. Ann Intern Med 1999;131:401–8.

[41] Luzzatto L, Gianfaldoni G, Notaro R. Management of paroxysmal nocturnal haemoglobinuria: a personal view. Br J Haematol 2011;153:709–20.

[42] Hoekstra J, Leebeek FW, Plessier A, et al. Paroxysmal nocturnal hemoglobinuria in Budd-Chiari syndrome: findings from a cohort study. J Hepatol 2009;51:696–706.

[43] Hall C, Richards S, Hillmen P. Primary prophylaxis with warfarin prevents thrombosis in paroxysmal nocturnal hemoglobinuria (PNH). Blood 2003;102:3587–91.

[44] Moyo VM, Mukhina GL, Garrett ES, Brodsky RA. Natural history of paroxysmal nocturnal haemoglobinuria using modern diagnostic assays. Br J Haematol 2004;126:133–8.

[45] Risitano AM. Immunosuppressive therapies in the management of immune-mediated marrow failures in adults: where we stand and where we are going. Br J Haematol 2010;152:127–40.

[46] Brodsky RA. How I treat paroxysmal nocturnal hemoglobinuria. Blood 2009;113:6522–7.

[47] Scheinberg P, Wu CO, Nunez O, Young NS. Predicting response to immunosuppressive therapy and survival in severe aplastic anaemia. Br J Haematol 2009;144:206–16.

[48] Stoppa AM, Vey N, Sainty D, et al. Correction of aplastic anaemia complicating paroxysmal nocturnal haemoglobinuria: absence of eradication of the PNH clone and dependence of response on cyclosporin A administration. Br J Haematol 1996;93:42–4.

[49] van Kamp H, van Imhoff GW, de Wolf JT, Smit JW, Halie MR, Vellenga E. The effect of cyclosporine on haematological parameters in patients with paroxysmal nocturnal haemoglobinuria. Br J Haematol 1995;89:79–82.

[50] Tichelli A, Gratwohl A, Nissen C, Signer E, Stebler Gysi C, Speck B. Morphology in patients with severe aplastic anemia treated with antilymphocyte globulin. Blood 1992;80:337–45.

[51] Sanchez-Valle E, Morales-Polanco MR, Gomez-Morales E, Gutierrez-Alamillo LI, Gutierrez-Espindola G, Pizzuto-Chavez J. Treatment of paroxysmal nocturnal hemoglobinuria with antilymphocyte globulin. Rev Invest Clin 1993;45:457–61.

[52] Risitano AM, Selleri C, Serio B, et al. Alemtuzumab is safe and effective as immunosuppressive treatment for aplastic anaemia and single-lineage marrow failure: a pilot study and a survey from the EBMT WPSAA. Br J Haematol 2009;148:791–6.

[53] Brodsky RA, Chen AR, Dorr D, et al. High-dose cyclophosphamide for severe aplastic anemia: long-term follow-up. Blood 2010;115:2136–41.

[54] Peffault de Latour R. Cyclophosphamide in severe aplastic anemia? Blood 2014;124:2758–60.

[55] Bemba M, Guardiola P, Garderet L, et al. Bone marrow transplantation for paroxysmal nocturnal haemoglobinuria. Br J Haematol 1999;105:366–8.

[56] Raiola AM, Van Lint MT, Lamparelli T, et al. Bone marrow transplantation for paroxysmal nocturnal hemoglobinuria. Haematologica 2000;85:59–62.

[57] Saso R, Marsh J, Cevreska L, et al. Bone marrow transplants for paroxysmal nocturnal haemoglobinuria. Br J Haematol 1999;104:392–6.

[58] Matos-Fernandez NA, Abou Mourad YR, Caceres W, Kharfan-Dabaja MA. Current status of allogeneic hematopoietic stem cell transplantation for paroxysmal nocturnal hemoglobinuria. Biol Blood Marrow Transplant 2009;15:656–61.

[59] Santarone S, Bacigalupo A, Risitano AM, et al. Hematopoietic stem cell transplantation for paroxysmal nocturnal hemoglobinuria: long-term results of a retrospective study on behalf of the Gruppo Italiano Trapianto Midollo Osseo (GITMO). Haematologica 2010;95:983–8.

[60] Peffault de Latour R, Schrezenmeier H, Bacigalupo A, et al. Allogeneic stem cell transplantation in paroxysmal nocturnal hemoglobinuria. Haematologica 2012;97:1666–73.

[61] Schrezenmeier H, Passweg JR, Marsh JC, et al. Worse outcome and more chronic GVHD with peripheral blood progenitor cells than bone marrow in HLA-matched sibling donor transplants for young patients with severe acquired aplastic anemia. Blood 2007;110:1397–400.

[62] Risitano AM. Dissecting complement blockade for clinic use. Blood 2015;125:742–4.

[63] Bacigalupo A, Socie G, Schrezenmeier H, et al. Bone marrow versus peripheral blood as the stem cell source for sibling transplants in acquired aplastic anemia: survival advantage for bone marrow in all age groups. Haematologica 2012;97:1142–8.

[64] Eapen M, Le Rademacher J, Antin JH, et al. Effect of stem cell source on outcomes after unrelated donor transplantation in severe aplastic anemia. Blood 2011;118:2618–21.

[65] Takahashi Y, McCoy JP Jr, Carvallo C, et al. In vitro and in vivo evidence of PNH cell sensitivity to immune attack after nonmyeloablative allogeneic hematopoietic cell transplantation. Blood 2004;103:1383–90.

[66] Marotta S, Pagliuca S, Risitano AM. Hematopoietic stem cell transplantation for aplastic anemia and paroxysmal nocturnal hemoglobinuria: current evidence and recommendations. Expert Rev Hematol 2014;7:775–89.

[67] Hillmen P, Young NS, Schubert J, et al. The complement inhibitor eculizumab in paroxysmal nocturnal hemoglobinuria. N Engl J Med 2006;355:1233–43.

[68] Brodsky RA, Young NS, Antonioli E, et al. Multicenter phase 3 study of the complement inhibitor eculizumab for the treatment of patients with paroxysmal nocturnal hemoglobinuria. Blood 2008;111:1840–7.

[69] Hillmen P, Muus P, Duhrsen U, et al. Effect of the complement inhibitor eculizumab on thromboembolism in patients with paroxysmal nocturnal hemoglobinuria. Blood 2007;110:4123–8.

[70] Kelly RJ, Hill A, Arnold LM, et al. Long-term treatment with eculizumab in paroxysmal nocturnal hemoglobinuria: sustained efficacy and improved survival. Blood 2011;117:6786–92.

Telomere Biology and Disease

J.N. Cooper, N.S. Young

National Institutes of Health, National Heart, Lung, and Blood Institute,
Bethesda, MD, United States

INTRODUCTION

Telomere biology and disorders of telomere maintenance are important in both adult and pediatric bone marrow failure syndromes. Defects in multiples genes responsible for telomere maintenance underlie many of the inherited bone marrow failure syndromes seen in childhood. Additionally, a subset of adult patients presenting with bone marrow failure have inherited defects in genes responsible for telomere maintenance. This chapter will review the basic biology of telomere regulation, as well as, the clinical implications of dysregulation of telomeres in bone marrow failure syndromes.

MOLECULAR BIOLOGY OF TELOMERES AND TELOMERASE

The telomere complex consists of a repetitive six-base DNA sequence at the ends of chromosomes (TTAGGG in vertebrates) called telomeres and multiple associated proteins [1]. The telomeric region of chromosomes serves multiple roles critical to all eukaryotic cells. First, the telomere-associated proteins cap the free ends of the chromosomes. Capping prevents the ends from being recognized by DNA repair machinery as double stranded breaks that could otherwise lead to chromosomal instability via end joining or recombination of the free chromosome ends [2,3].

Second, the telomere complex solves the "end-replication problem." During DNA replication, DNA polymerase cannot replicate the final 50–200 base pairs at the 3' ends of the chromosome. DNA polymerases require a 3'-hydroxyl group to initiate replication that is provided by a temporary RNA template. When this 3'-RNA template is absent from newly synthesized DNA, it is not possible to fill in the resulting gap, which leads to a shorter daughter strand of DNA [4]. Thus telomeres serve as a buffer of noncoding DNA that can be lost with cell division without compromising cellular function. Telomeres also serve as a check on uncontrolled cell proliferation. Somatic cell division eventually results in a critical shortening of telomeres, activation of the DNA damage repair machinery, cell senescence, and/or apotoposis [5,6]. This limit on cell division is so important that cancer

Congenital and Acquired Bone Marrow Failure
http://dx.doi.org/10.1016/B978-0-12-804152-9.00015-4

cells must acquire a mechanism to elongate critically shortened telomeres [7,8].

For stem cells and other cells that must avoid critical telomeric shortening, the telomerase complex enables 3' elongation of telomeric DNA. The ribonucleoprotein enzyme complex telomerase is composed of a reverse transcriptase (TERT) and its RNA template (TERC) (Fig. 15.1). Telomerase specifically binds to the

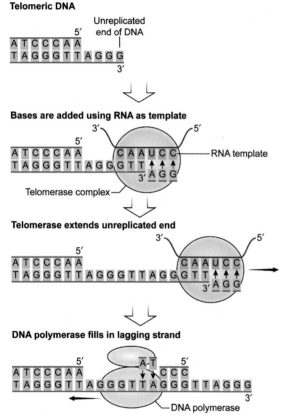

FIGURE 15.1 Telomerase complex mediated telomere elongation. The 3' ends of mammalian chromosomes cannot be fully replicated by DNA polymerase. The *telomerase complex* consists of a protein component (TERT), which can elongate DNA in a 3' to 5' fashion and a *RNA template* (TERC). The *RNA template* binds the complementary free 3'-DNA overhang to serve as a primer for TERT-mediated 3' addition of DNA bases. Once telomerase has extended the 3' overhang, *DNA polymerase* can fill in the remaining gap left after the TERC primer is removed.

free 3'-TTAGGG nucleotide overhangs generated during DNA replication. Using the RNA template (TERC) the reverse transcriptase, TERT processively extends the 3'-end of chromosomes and reduces replication-associated telomere shortening (Fig. 15.2) [9,10]. Proper synthesis of telomerase requires associated proteins that form the dyskerin complex (dyskerin, GAR1, NOP10, and NHP2). This small nucleolar ribonucleoprotein complex is required for RNA modifications of TERC and is essential for the formation of a functional telomerase complex [11,12]. Telomerase expression is highly regulated. Almost all adult somatic cells lack detectable telomerase expression [13]. Somatic cells experience age-associated telomere loss [14]. The loss is due in part to age-associated cell divisions but additional mechanisms, such as oxidative stress, also contribute to telomere attrition [15]. Even germ cells, activated lymphocytes, and somatic stems cells, in which telomerase is active, have age-associated telomere loss [16–18]. Only during embryogenesis does telomerase significantly lengthen telomeres [19]. Not surprisingly, loss-of-function mutations in the components of the telomerase or dyskerin complex dramatically accelerate aging-related telomeric loss [20–22].

The other regulator of telomeres is the shelterin complex. Components of the shelterin complex (TRF1, TRF2, and POT1) directly recognize and bind to both the double and single stranded portions of telomeres, allowing formation of a telomeric loop (T-loop) that "hides" the free 3'-DNA end within the double stranded telomeric region [23,24]. The shelterin complex thus prevents the 3'-end from recognition by the DNA repair machinery. Loss of function in this complex leads to end-to-end fusions of chromosomes and activation of the DNA repair machinery [25]. Shelterin also regulates access of telomerase to telomeres and provides a mechanism to control telomere length by a negative feedback loop [11]. When telomeres are of adequate length, the shelterin complex binds telomeres and blocks access of telomerase. As

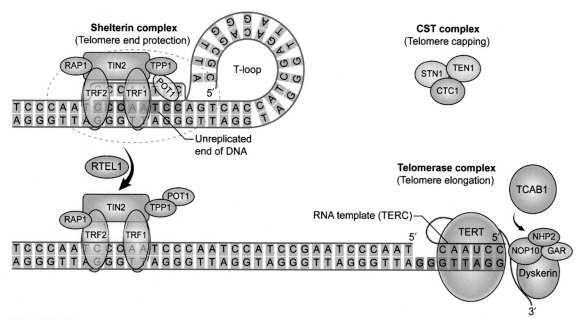

FIGURE 15.2 **Essential component for telomere maintenance.** The *shelterin complex* binds to telomeres to cap the ends of chromosomes. *TRF1* and *TRF2* recognize and directly bind to the double stranded portion of telomeres. *POT1* binds to the 3′-single stranded telomeric overhang in order to "hide" the single stranded DNA within the double stranded telomeric region. RTEL1 is a DNA helicase that can unwind the T-loop formed by shelterin to allow telomerase access to the free 3′ end of the telomere. *TCAB1* is necessary for trafficking of telomerase to telomeres. The *CST complex* has multiple roles in telomere maintenance including trafficking of telomerase, binding single stranded DNA, and interacting with DNA polα-primase. Gene mutations in the proteins colored in *purple* have been implicated in human telomeropathies.

telomeres shorten, shelterin is less likely to bind telomeres and thus allows telomerase to access and elongate ends of chromosomes [26].

GENOTYPE AND PHENOTYPE IN TELOMERE DISEASE

The finding that germline mutations in *DKC1*, the gene encoding dyskerin, are responsible for many cases of dyskeratosis congenita first established a direct link between telomere biology and bone marrow failure syndromes [22]. Since this discovery, a total of 11 genes related to telomere maintenance have been identified in patients with inherited forms of bone marrow failure (Table 15.1). The correlation between these mutations and clinical phenotype is complex. Most individuals, but

TABLE 15.1 Genes and Inheritance Patterns of Telomere-Related Diseases

Inheritance mode	Gene mutated
X-linked	*DKC1*
Autosomal dominant	*TERC, TERT, TINF2, RTEL1*
Autosomal recessive	*DKC1, NOP10, NHP2, WRAP53, PARN, CTC1, RTEL1*
De novo mutations	*TINF2, DKC1*

not all, have shortened telomeres significantly below the first percentile, adjusted for age [27]. Yet telomere length alone is insufficient to predict a patient's clinical phenotype or the age of disease onset [28]. Asymptomatic family members of affected patients with the same inherited mutation can have very short telomeres but without evidence of disease. Furthermore, the same gene mutations do not necessarily produce the same phenotypes. Variable penetrance is well documented within pedigrees bearing telomere mutations. Environmental and epigenetic factors are hypothesized to have a significant role in disease development.

There is a wide spectrum of clinical features observed in telomere-associated diseases ranging from single organ dysfunction to severe multisystem organ involvement as seen in dyskeratosis congenita. Terms, such as, "telomere disorders," "telomeropathies," or "telomere syndromes" have all been used to encompass this wide range of clinical phenotypes resulting from inherited mutations in telomere maintenance genes. Dyskeratosis congenita is a term reserved to describe patients presenting with the classic triad of dystrophic nails with clubbing, skin hyperpigmentation, and oral leukoplakia. The most frequent mutations in classic dyskeratosis congenita occur in *DKC1*, but mutations in *TIN2*, *TERT*, and *TERC* are also common [29]. Dyskeratosis congenita usually, but not exclusively, presents in childhood. Mutations in *TERT* or *TERC* can also result in isolated presentations of bone marrow failure, idiopathic pulmonary fibrosis (IPF), or cryptogenic cirrhosis in adults. These patients often have a family history of unexplained hematologic, pulmonary, or hepatic diseases.

Dyskeratosis congenita variably impacts different organ systems. Cytopenias and progressive bone marrow failure are the hematologic hallmarks of the disease and frequently lead to early mortality directly related to cytopenias or due to the development of myelodysplastic syndrome (MDS) or acute myeloid leukemia (AML). Severe pulmonary disease and nonhematologic malignancies are additional common causes of premature death [30,31]. A detailed family history frequently identifies multiple family members with features of an underlying telomere disorder without frank presentation of dyskeratosis congenita. Even without overt evidence of a telomeropathy, carriers of telomere-related gene mutations have significantly short telomeres. All

their offspring will inherit shortened telomeres. Children who also bear a mutation in telomere maintenance genes are more likely to have clinical manifestations of a telomere disease at younger ages [32]. Disease anticipation is frequently observed within kindreds. The inheritance pattern of dyskeratosis congenita depends on the underlying gene mutation and can be X-linked, autosomal dominant, or autosomal recessive (Table 15.1).

There are distinct and severe forms of dyskeratosis congenita in childhood. In the Hoyeraal–Hreidarsson syndrome, children have cerebellar hypoplasia, immunodeficiency, progressive bone marrow failure, and intrauterine growth retardation in addition to the typical cutaneous findings of dyskeratosis congenita [33]. Although nearly all patients have very short telomeres, mutations within known telomere genes are present in only 60% of cases. No single mutation is uniquely linked to the syndrome. In Revesz syndrome, bilateral exudative retinopathy occurs with short telomeres and marrow failure; mutations within *TINF2* are present in this variant syndrome [34]. Coats plus syndrome includes the findings of bilateral retinopathy and retinal telangiectasis of Coats disease and additional features of intracranial calcifications and cysts, bone abnormalities, and gastrointestinal vascular ectasias. Shortened telomeres, as well as, mutations within *CTC1* are present in some of these patients [35]. Although telomeric dysfunction is clearly central to many of the clinical manifestations of these variant syndromes, the mutated proteins all have additional cellular functions unrelated to telomere maintenance, which may account for the defining features of these unique phenotypes.

Telomerase Complex Mutations

Disruptions of essential elements of the telomerase complex, TERT or TERC, lead to autosomal dominant or recessive telomeropathies. In classic cases of dyskeratosis congenita, mutations in either *TERT* or *TERC* are identified in about 10% of cases [20,21,36]. Mutations within *TERT* and *TERC* have been identified in adults with aplastic anemia, as well as, IPF and liver cirrhosis [37–39]. Mutations in *TERT* occur across multiple regions of the protein, including the reverse transcriptase (RT) domain, as well as, the N-terminus and C-terminus portions of the protein [11,40,41]. Some mutations directly impair the ability of telomerase to add telomeric repeats to chromosomes. However, many *TERT* mutations do not completely abolish in vitro telomerase activity [42]. The precise mechanisms by which *TERT* variants lead to telomere shortening are still unclear. Variants within the RT domain directly impair catalytic activity; in some cases, the catalytic mutants may exert a dominant negative effect on telomerase activity by impairing the function of wildtype copy of TERT [43,44]. Alternatively, the availability of TERC may be the rate-limiting step in the production of functional telomerase. Thus expression of a variant copy of TERT would reduce formation of functional telomerase complex by competing with the wildtype copy of TERT for available TERC RNA templates [45]. Variants outside of the RT domain may indirectly disrupt telomerase activity without affecting the enzymatic activity. The C-terminal domain of TERT is necessary for proper trafficking and recruitment of telomerase to telomeres [46]. Additionally, the C-terminal domain is required for telomerase to processively add multiple TTAGGG repeats to telomeres [47]. Loss-of-function mutations within the N-terminus of TERT impair interactions between the RNA template and telomeric DNA [41,48,49]. These non-RT domain mutations may result in the inability to form a functional telomerase complex or disrupt the normal recruitment of telomerase to telomeres.

As with *TERT*, mutations within the RNA template gene *TERC* result in bone marrow failure, as well as, isolated pulmonary fibrosis. Mutations often abolish the formation of a critical psuedoknot structure that is necessary for the formation of the RNA template within telomerase [50].

Additional mutations in alternative domains, as well as, in the promoter of the *TERC* gene have also been described [51–53]. Experimental evidence suggests that even a 50% percent reduction in *TERC* expression is sufficient to disrupt proper telomere maintenance in dividing cells [54].

Proper assembly of the telomerase complex requires processing and trafficking of the RNA component TERC by the dyskerin complex. Mutations in multiple components of this complex occur in telomere syndromes. X-linked dyskeratosis congenita is due to *DKC1* missense mutations, which result in a loss-of-function in dyskerin. *NOLA3* and *NOLA2*, which encode for the NOP10 and NHP2 proteins, respectively, are mutated in autosomal recessive dyskeratosis congenita [55,56]. Together with dyskerin, these proteins form a H/ACA small nucleolar ribonucleoportein complex necessary to process TERC during biogenesis of the telomerase complex [57]. Telomerase Cajal body protein 1 (TCAB1), encoded by *WRAP53*, is not part of the dyskerin complex but also is involved in telomerase biogenesis [58]. TCAB1 is necessary for proper trafficking of TERC within the cell. Autosomal recessive mutations occur within the *WRAP53* gene in motifs necessary for binding of TCAB1 to TERC [59]. Loss of any of these proteins, which are required for proper TERC trafficking and processing, disrupts formation of functional telomerase within a cell.

Shelterin Complex Mutations

Of the six subunits present in the shelterin complex, only mutations in *TINF2*, encoding for TRF1-interacting nuclear factor (TIN2) and *ACD*, encoding for telomere-binding protein TPP1, have been described [60,61]. *TINF2* mutations were originally found in a family with autosomal dominant dyskeratosis congenita. However, most cases result from de novo mutations in *TINF2*, likely due to the early and severe presentation of marrow failure, so that few affected individuals reach reproductive age [62,63]. Only a third of *TINF2*-mutated patients have the classic signs and symptoms of dyskeratosis congenita. A few cases of both Hoyeraal–Hreidarsson and Revesz syndromes have *TINF2* mutations.

TPP1, encoded by the *ACD* gene, is another protein critical to the shelterin complex and telomere maintenance. TPP1 is necessary for recruitment of telomerase to telomeres, as well as, maintenance of processivity [64]. TPP1 contains a TEL patch domain that is critical for telomerase to interact with TPP1 [65,66]. In two separately reported cases of patients with either the Hoyeraal–Hreidarsson or isolated aplastic anemia bearing *ACD* mutations, both shared the same amino acid deletion within the TEL patch domain, resulting in the inability of TPP1 to recruit telomerase to telomeres [60,61].

Additional Mutations

Mutations outside of telomerase or shelterin genes occur in inherited telomeropathies. *RTEL1* encodes a DNA helicase required for the unwinding of the T-loop formed by the shelterin complex [67]. Mutations within this gene have recently been reported in cases of both dyskeratosis congenita (Hoyeraal–Hreidarsson variant) and IPF [68–71]. Mutations within *CTC1* are responsible for the Coats plus syndrome [42,72]. Mutations in the *PARN* gene were recently described in three families with autosomal recessive dyskeratosis congenita [68,73]. These mutations impact a poly(A)-specific ribonuclease leading to decreased mRNA expression of multiple telomere-associated genes (*TERC*, *DKC1*, *RTEL1*, and *TERF1*), as well as, very short telomeres.

BONE MARROW, ORGAN FAILURE, AND MALIGNANCY IN TELOMEROPATHIES

With the exception of the rare severe variants of dyskeratosis congenita, the major organs affected in telomeropathies are the bone marrow,

lungs, and liver. Patient presentation may range from mild involvement of a single organ involvement to severe disease in all three organs. Within kindreds, carriers of the same gene mutation may manifest distinct clinical features and different severities. With pulmonary and hepatic disease, additional environmental exposures, such as, alcohol or tobacco likely explain some clinical variance. Recognition that telomeropathies are multiorgan syndromes is important, as individuals initially exhibiting single organ dysfunction may develop signs of additional organ failure over time.

Bone Marrow Failure

Bone marrow failure is a common but not inevitable consequence of telomere disease. For children with dyskeratosis congenita, cytopenias typically develop within the first decade of life. By the second or third decade, progressive pancytopenia meeting the criteria for severe aplastic anemia may occur. Severe marrow failure can be the presenting symptom of dyskeratosis congenita if mucocutenaous features were not previously recognized or in the variant syndromes, such as Hoyeraal–Hreidarsson syndrome, in which severe aplastic anemia can be the initial symptom [33]. The occurrence of marrow failure and the severity of disease presentation cannot be predicted by an individual's telomere length, the gene affected, or the site of gene mutation [28]. Unaffected family members with the same mutation may have very mild nonprogressive cytopenias, macrocytosis, and a hypocellular bone marrow but never develop severe aplastic anemia.

A subset of adults presenting with aplastic anemia will have an underlying telomeropathy. A careful personal and family history and physical examination often reveals suggestive features, such as, early hair graying or unexplained anemia, cirrhosis, or pulmonary disease. Telomere length less than the first percentile for age is specific for telomeropathies but lacks sensitivity. Patients with telomeropathies frequently have a history of macrocytosis, mild to moderate thrombocytopenia misdiagnosed as immune-mediated thrombocytopenia, or moderate aplastic anemia. Leukopenia alone is infrequent. Blood counts may be stable for years or progress to severe cytopenias. Acute presentations of marrow failure are also possible with telomeropathies.

Treatment for telomeropathy-associated bone marrow failure includes androgen therapy or hematopoietic stem cell transplantation. Sex hormone therapy has been used for decades in acquired aplastic anemia, as well as, for patients with dyskeratosis congenita, without a known mechanism of action [74–77]. Estrogen-responsive elements are present within the promoter of *TERT*, which enable upregulation of telomerase expression by estrogen or progesterone [78,79]. In vitro exposure of lymphocytes from patients with heterozygous *TERT* mutations to androgens results in a normalization of telomerase activity. In a mouse model of bone marrow failure resulting from a shelterin gene mutation, testosterone therapy elongated telomeres and improved survival [80]. In patients with aplastic anemia and telomeres less than the first percentile for age and/or a known telomere-associated gene mutation, danazol therapy produced hematologic responses in nearly all patients. Furthermore, patients' peripheral blood leukocyte telomere lengths were elongated. Future studies will be needed to address the minimum dose necessary to mitigate the accelerated telomere attrition in telomeropathies and potentially to prevent the development of severe bone marrow failure. The use of immunosuppression with horse antithymocyteglobulin and cyclosporine to treat children with dyskeratosis congenita has not been systematically addressed, but the limited data available does not support this approach [81]. In our marrow failure clinic, adults with aplastic anemia and *TERT* mutations have responded to immunosuppressive therapy (unpublished).

Reduced intensity allogeneic stem cell transplantation (SCT) remains the only definitive therapy for bone marrow failure due to telomeropathies but remains an imperfect treatment opition [82]. Mortality rates from SCT in patients with dyskeratosis congenita approach 60% due to high incidences of graft failure and pulmonary toxicities [83–85]. Primary or secondary graft failure occurs in nearly a third of transplant recipients. Patients are at a risk for both acute respiratory failure and the late development of severe pulmonary fibrosis even in the absence of pretransplant pulmonary dysfunction. Matched sibling donors are preferred but should not carry a telomere-associated gene mutation. Use of an unrelated donor is associated with a significantly higher risk of graft failure and death. Systematic data of SCT of adults with telomeropathies without features of dyskeratosis congenita are lacking. Similar concern for pulmonary toxicities and late organ failures is warranted when transplanting these patients. SCT is an imperfect solution to telomeropathies as many patients maybe excluded due to multisystem disease and it does not address the pulmonary or hepatic consequences of telomere disease.

Pulmonary Disease

The initial link between bone marrow failure and pulmonary disease was observed in children with classic dyskeratosis congenita, in whom, nearly 20% have loss of pulmonary function with a decrease in diffusion capacity or evidence of restrictive lung disease [86]. This association raised the possibility that some inherited forms of pulmonary fibrosis might be related to telomere disease. Multiple groups have identified mutations within *TERT, TERC, TINF2, RTEL1*, and *DKC1* in familial cases of isolated IPF [37,69,87–91]. While these patients lacked overt features of dyskeratosis congenita and did not have evidence of bone marrow failure, family members without IPF had cryptogenic liver disease or aplastic anemia, again demonstrating the variable penetrance typical of telomere disease. Even asymptomatic carriers of *TERT* mutations may have subtle subclinical evidence of pulmonary disease, such as, decreased diffusion capacity and radiographic signs of lung fibrosis [92].

There is no clear mechanism to explain how mutations within telomere-associated genes lead to pulmonary fibrosis [93]. While knockout mouse models of telomere genes can recapitulate bone marrow failure, pulmonary disease only occurs with addition of environmental toxic exposures, such as tobacco smoke [91]. Even then, these mice develop emphysema instead of pulmonary fibrosis. *TERT*-deficient mice exposed to bleomycin, paradoxically, are less likely to develop lung fibrosis compared to wildtype mice [94].

For patients with telomeropathies, monitoring of pulmonary function and awareness of potential pulmonary complications is important. Screening of patients lacking symptomatic pulmonary disease may reveal abnormalities in pulmonary function or evidence of subclinical pulmonary fibrosis on high-resolution computed tomography scanning. All patients with telomere disease should avoid tobacco use; smoking appears to increase the rates of pulmonary fibrosis among *TERT*-mutation carriers [95]. New or worsening dyspnea in telomeropathy patients may indicate progression of pulmonary fibrosis but alternative diagnoses must be considered. Cardiac echocardiography or pulmonary computed tomography scans may reveal the presence of pulmonary arteriovenous malformation [96]. Dyspnea in the absence of pulmonary disease or out of proportion to a patient's anemia may suggest the presence of hepatopulmonary syndrome due to liver cirrohosis, a finding that may precede or coexist with pulmonary fibrosis [97].

Individuals undergoing allogeneic stem cell transplantation for a telomeropathy are at increased risk of pulmonary complications, and a significant percentage of posttransplant mortality is due to respiratory failure [83]. The absence of pretransplant pulmonary dysfunction does not preclude the development of either early or late pulmonary complications after transplant. Reduced intensity preparative regimens have gained favor due to their lower risk of early pulmonary complications. Long-term survivors of transplant remain at risk for the development of fatal pulmonary fibrosis years following transplant.

Liver Disease

Similar to pulmonary fibrosis, cryptogenic cirrhosis does present as the sole manifestation of telomere disease or accompany pulmonary disease and marrow failure. Original descriptions of dyskeratosis congenita noted 7% of patients who developed liver disease, initially attributed to transfusion-associated iron overload [31,86]. Familial cases of liver cirrhosis and aplastic anemia were also reported before the relationship between marrow failure and telomere disorders was known [98]. Later it was demonstrated that in some kindreds with a history of isolated cryptogenic liver cirrhosis, mutations in *TERT* or *TERC* were present [38]. As with IPF, purposeful screening of cohorts of patients with liver cirrhosis uncovered mutations within *TERT* and *TERC* [99,100]. Many of these patients had concomitant viral hepatitis exposure or alcohol abuse suggesting that, as in pulmonary disease, environmental exposures play some role in the development of fibrosis. The presentation of liver disease in telomeropathies can range from mild elevation in transaminases to liver failure requiring transplantation. Liver histology varies from case-to-case without a single unifying pattern seen in liver biopsy. Heterogeneous features of necrosis, inflammation, as well as, fibrosis may be observed.

Management of hepatic complications of telomeropathies is similar to pulmonary disease. Avoidance of additional hepatic insults, such as alcohol, is advised. If not already immune, vaccination against hepatitis A and B is warranted. Involvement of a hepatologist is warranted to manage portal hypertension and esophageal varices. Ongoing studies will clarify the role of male androgens as potential therapy to attenuate progressive liver disease.

Telomeres and Malignancy Risk

Patients with classic dyskeratosis congenita have a dramatically increased risk of developing cancer with nearly 10% of patients in registries developing malignancies including AML, head and neck squamous-cell carcinomas, tongue cancer, and MDS [30,31]. Germline loss-of-function mutations within *TERT* seen in telomeropathies are a risk factor for the development of adult but not pediatric AML [101,102]. Multiple genome-wide association studies have found a small increased risk of multiple types of malignancies in individuals bearing common polymorphisms in the *TERT* gene [103].

Telomere dynamics also play a role in the development of myeloid cancers in patients with bone marrow failure syndromes not associated with a telomeropathy. In adult patients with severe aplastic anemia undergoing immunosuppressive therapy without a known telomeropathy, pretreatment-telomere length in the bottom quartile for age was a significant risk factor for evolution to MDS [104]. In this cohort, patients with the shortest telomeres had significantly more uncapped telomere-free chromosome ends as compared to the patients with the longest telomeres [105]. Analysis of patients without telomeropathies and normal telomere length at the time of diagnosis who developed to monosomy 7 found that these patients had dramatically accelerated telomere attrition before developing MDS [106]. This rapid telomere loss led to an accumulation of individual chromosomes bearing extremely short telomeres prior to the development of monosomy 7. All patients with rapid telomere loss had evidence of oligoclonal hematopoiesis. Dependence on a limited stem cell pool to support hematopoiesis would require an increased rate of cell division

and accelerate telomere attrition. Perhaps therapies that augment telomerase activity, such as sex hormones, might prevent this replication-associated telomere loss and reduce the risk of clonal evolution to MDS.

DIAGNOSIS OF TELOMERE DISEASE

A detailed personal and family history and physical examination is often sufficient to strongly suspect the diagnosis of a telomeropathy. De novo telomeropathy mutations are uncommon and most patients have some family members with histories of unexplained cytopenias, pulmonary disease, or liver disease often attributed to tobacco or alcohol use. In adults, hair graying in the twenties, which may be masked with hair dye, is common. Mucocutaneous features of nail dystrophy, clubbing, and leukoplakia are typical of dyskeratosis congenita in children but are occasionally seen in adults.

Measurement of peripheral blood telomere length is helpful to confirm or rule in a telomeropathy associated with bone marrow failure, cryptogenic cirrhosis, or IPF. Telomere length below the first percentile for age with a clinical phenotype consistent with telomere dysfunction (early hair graying, pulmonary disease, and strong family history) is sufficient to diagnose a telomeropathy even in the absence of a known telomeropathy-associated gene mutation [27]. Knowledge of all genes responsible for these syndromes is incomplete and the consideration of noncoding mutations or epigenetic causes of a telomeropathy is possible [51,107]. Measurement of telomere length is also appropriate in the workup of unexplained thrombocytopenia, moderate aplastic anemia, unexplained marrow hypocellularity, or macrocytosis.

Multiple assays exist to measure telomere length. Southern blot was the original method to measure average chromosome telomere length and remains the standard [108]. This method has the advantage of being a direct measure of telomere length but is time consuming and difficult to scale for high-throughput studies. The other two common methods of telomere measurement, quantitate polymerase chain reaction (qPCR), and flow-FISH, are measures of average telomere content within cells rather than a direct measure of telomere length. Both methods only measure telomere content (TTAGGG repeats) as compared to Southern blot, in which variable amounts of subtelomeric DNA are also measured. With both methods, the average telomeric content is converted into an estimated telomere length based on laboratory-specific internal controls, which complicates comparisons of measures from different laboratories [109]. Flow-FISH allows for simultaneous measurement of telomere content in both lymphocyte- and granulocyte fractions of peripheral blood. Flow-FISH also has a higher correlation with Southern blot measurements and a higher sensitivity and specificity to detect short telomeres as compared to qPCR [110]. Allele-specific single-telomere length analysis (STELA) or quantitative fluorescent in situ hybridization (Q-FISH) can detect and quantify the amount of critically shortened telomeres in a cell population [111–114].

Interpretation of an individual patient's telomere length has some important caveats. Although in dyskeratosis congenita, lymphocyte telomere length below the first percentile for age is highly sensitive and specific, adults presenting with telomeropathies may have telomere lengths between the 1st and 10th percentile [37,39]. Additionally, marrow failure alone without clinical or molecular evidence of a telomeropathy results in shortened telomeres (but rarely below the first percentile) [115–119]. Measurement of telomere content in both the lymphocyte- and granulocyte fractions by flow-FISH in these cases may identify a discrepancy between the two cell populations [120]. Lymphocyte-telomere content has a higher specificity for telomeropathy-associated gene mutations. Finally, both qPCR and flow-FISH only represent the mean telomeric content

[121]. Individual chromosomes may have critically short telomere lengths despite an overall normal mean telomere length of all chromosomes [111,122]. Either STELA or Q-FISH can measure the burden of critically shortened telomeres but these methods are only available in the research setting.

CONCLUSIONS

Individuals with telomeropathies represent an important subset of both adult and pediatric patients presenting with bone marrow failure. Diagnosis of a telomeropathy significantly impacts the prognosis and the management of patients with bone marrow failure. Special attention to the multisystem nature of these disorders is required when monitoring patients and planning treatment. Current therapies to treat telomere-related disorders are not optimal but ongoing research will focus on how to prevent telomere attrition and the development of multisystem organ failure.

References

[1] Moyzis RK, Buckingham JM, Cram LS, et al. A highly conserved repetitive DNA sequence, (TTAGGG)$_n$, present at the telomeres of human chromosomes. Proc Natl Acad Sci USA 1988;85(18):6622–6.

[2] Artandi SE, Chang S, Lee SL, et al. Telomere dysfunction promotes non-reciprocal translocations and epithelial cancers in mice. Nature 2000;406(6796):641–5.

[3] Maciejowski J, Li Y, Bosco N, Campbell PJ, de Lange T. Chromothripsis and kataegis induced by telomere crisis. Cell 2015;163(7):1641–54.

[4] de Lange T. How telomeres solve the end-protection problem. Science 2009;326(5955):948–52.

[5] Harley CB, Futcher AB, Greider CW. Telomeres shorten during ageing of human fibroblasts. Nature 1990;345(6274):458–60.

[6] d'Adda di Fagagna F, Reaper PM, Clay-Farrace L, et al. A DNA damage checkpoint response in telomere-initiated senescence. Nature 2003;426(6963):194–8.

[7] Pickett HA, Reddel RR. Molecular mechanisms of activity and derepression of alternative lengthening of telomeres. Nat Struct Mol Biol 2015;22(11):875–80.

[8] Sharpless NE, DePinho RA. Telomeres, stem cells, senescence, and cancer. J Clin Invest 2004;113(2):160–8.

[9] Greider CW, Blackburn EH. Identification of a specific telomere terminal transferase activity in *Tetrahymena* extracts. Cell 1985;43(2 Pt. 1):405–13.

[10] Shippen-Lentz D, Blackburn EH. Functional evidence for an RNA template in telomerase. Science 1990;247(4942):546–52.

[11] Hockemeyer D, Collins K. Control of telomerase action at human telomeres. Nat Struct Mol Biol 2015;22(11):848–52.

[12] Angrisani A, Vicidomini R, Turano M, Furia M. Human dyskerin: beyond telomeres. Biol Chem 2014;395(6):593–610.

[13] Kim NW, Piatyszek MA, Prowse KR, et al. Specific association of human telomerase activity with immortal cells and cancer. Science 1994;266(5193):2011–5.

[14] Aubert G, Lansdorp PM. Telomeres and aging. Physiol Rev 2008;88(2):557–79.

[15] Blackburn EH, Epel ES, Lin J. Human telomere biology: A contributory and interactive factor in aging, disease risks, and protection. Science 2015;350(6265):1193–8.

[16] Wang JC, Warner JK, Erdmann N, Lansdorp PM, Harrington L, Dick JE. Dissociation of telomerase activity and telomere length maintenance in primitive human hematopoietic cells. Proc Natl Acad Sci USA 2005;102(40):14398–403.

[17] Kalmbach KH, Fontes Antunes DM, Dracxler RC, et al. Telomeres and human reproduction. Fertil Steril 2013;99(1):23–9.

[18] Wright WE, Piatyszek MA, Rainey WE, Byrd W, Shay JW. Telomerase activity in human germline and embryonic tissues and cells. Dev Genet 1996;18(2):173–9.

[19] Liu L, Bailey SM, Okuka M, et al. Telomere lengthening early in development. Nat Cell Biol 2007;9(12):1436–41.

[20] Vulliamy T, Marrone A, Goldman F, et al. The RNA component of telomerase is mutated in autosomal dominant dyskeratosis congenita. Nature 2001;413(6854):432–5.

[21] Armanios M, Chen JL, Chang YP, et al. Haploinsufficiency of telomerase reverse transcriptase leads to anticipation in autosomal dominant dyskeratosis congenita. Proc Natl Acad Sci USA 2005;102(44):15960–4.

[22] Mitchell JR, Wood E, Collins K. A telomerase component is defective in the human disease dyskeratosis congenita. Nature 1999;402(6761):551–5.

[23] Griffith JD, Comeau L, Rosenfield S, et al. Mammalian telomeres end in a large duplex loop. Cell 1999;97(4):503–14.

[24] de Lange T. T-loops and the origin of telomeres. Nat Rev Mol Cell Biol 2004;5(4):323–9.

[25] Sfeir A, de Lange T. Removal of shelterin reveals the telomere end-protection problem. Science 2012;336(6081):593–7.

[26] Smogorzewska A, van Steensel B, Bianchi A, et al. Control of human telomere length by TRF1 and TRF2. Mol Cell Biol 2000;20(5):1659–68.

[27] Alter BP, Baerlocher GM, Savage SA, et al. Very short telomere length by flow fluorescence in situ hybridization identifies patients with dyskeratosis congenita. Blood 2007;110(5):1439–47.

[28] Vulliamy TJ, Kirwan MJ, Beswick R, et al. Differences in disease severity but similar telomere lengths in genetic subgroups of patients with telomerase and shelterin mutations. PLoS One 2011;6(9):e24383.

[29] Dokal I. Dyskeratosis congenita. Hematology Am Soc Hematol Educ Program 2011;2011:480–6.

[30] Alter BP, Giri N, Savage SA, Rosenberg PS. Cancer in dyskeratosis congenita. Blood 2009;113(26):6549–57.

[31] Dokal I. Dyskeratosis congenita in all its forms. Br J Haematol 2000;110(4):768–79.

[32] Vulliamy T, Marrone A, Szydlo R, Walne A, Mason PJ, Dokal I. Disease anticipation is associated with progressive telomere shortening in families with dyskeratosis congenita due to mutations in TERC. Nat Genet 2004;36(5):447–9.

[33] Glousker G, Touzot F, Revy P, Tzfati Y, Savage SA. Unraveling the pathogenesis of Hoyeraal-Hreidarsson syndrome, a complex telomere biology disorder. Br J Haematol 2015;170(4):457–71.

[34] Berthet F, Caduff R, Schaad UB, et al. A syndrome of primary combined immunodeficiency with microcephaly, cerebellar hypoplasia, growth failure and progressive pancytopenia. Eur J Pediatr 1994;153(5):333–8.

[35] Anderson BH, Kasher PR, Mayer J, et al. Mutations in CTC1, encoding conserved telomere maintenance component 1, cause Coats plus. Nat Genet 2012;44(3):338–42.

[36] Vulliamy TJ, Walne A, Baskaradas A, Mason PJ, Marrone A, Dokal I. Mutations in the reverse transcriptase component of telomerase (TERT) in patients with bone marrow failure. Blood Cells Mol Dis 2005;34(3):257–63.

[37] Armanios MY, Chen JJ, Cogan JD, et al. Telomerase mutations in families with idiopathic pulmonary fibrosis. N Engl J Med 2007;356(13):1317–26.

[38] Calado RT, Regal JA, Kleiner DE, et al. A spectrum of severe familial liver disorders associate with telomerase mutations. PLoS One 2009;4(11):e7926.

[39] Yamaguchi H, Calado RT, Ly H, et al. Mutations in TERT, the gene for telomerase reverse transcriptase, in aplastic anemia. N Engl J Med 2005;352(14):1413–24.

[40] Banik SS, Guo C, Smith AC, et al. C-terminal regions of the human telomerase catalytic subunit essential for in vivo enzyme activity. Mol Cell Biol 2002;22(17):6234–46.

[41] Moriarty TJ, Huard S, Dupuis S, Autexier C. Functional multimerization of human telomerase requires an RNA interaction domain in the N terminus of the catalytic subunit. Mol Cell Biol 2002;22(4):1253–65.

[42] Zaug AJ, Crary SM, Jesse Fioravanti M, Campbell K, Cech TR. Many disease-associated variants of hTERT retain high telomerase enzymatic activity. Nucleic Acids Res 2013;41(19):8969–78.

[43] Sauerwald A, Sandin S, Cristofari G, Scheres SH, Lingner J, Rhodes D. Structure of active dimeric human telomerase. Nat Struct Mol Biol 2013;20(4):454–60.

[44] Lingner J, Hughes TR, Shevchenko A, Mann M, Lundblad V, Cech TR. Reverse transcriptase motifs in the catalytic subunit of telomerase. Science 1997;276(5312):561–7.

[45] Errington TM, Fu D, Wong JM, Collins K. Disease-associated human telomerase RNA variants show loss of function for telomere synthesis without dominant-negative interference. Mol Cell Biol 2008;28(20):6510–20.

[46] Sarek G, Marzec P, Margalef P, Boulton SJ. Molecular basis of telomere dysfunction in human genetic diseases. Nat Struct Mol Biol 2015;22(11):867–74.

[47] Huard S, Moriarty TJ, Autexier C, The C. terminus of the human telomerase reverse transcriptase is a determinant of enzyme processivity. Nucleic Acids Res 2003;31(14):4059–70.

[48] Akiyama BM, Parks JW, Stone MD. The telomerase essential N-terminal domain promotes DNA synthesis by stabilizing short RNA-DNA hybrids. Nucleic Acids Res 2015;43(11):5537–49.

[49] Robart AR, Collins K. Human telomerase domain interactions capture DNA for TEN domain-dependent processive elongation. Mol Cell 2011;42(3):308–18.

[50] Zhang Q, Kim NK, Feigon J. Architecture of human telomerase RNA. Proc Natl Acad Sci USA 2011;108(51):20325–32.

[51] Aalbers AM, Kajigaya S, van den Heuvel-Eibrink MM, van der Velden VH, Calado RT, Young NS. Human telomere disease due to disruption of the CCAAT box of the TERC promoter. Blood 2012;119(13):3060–3.

[52] Takeuchi J, Ly H, Yamaguchi H, et al. Identification and functional characterization of novel telomerase variant alleles in Japanese patients with bone-marrow failure syndromes. Blood Cell Mol Dis 2008;40(2):185–91.

[53] Theimer CA, Finger LD, Trantirek L, Feigon J. Mutations linked to dyskeratosis congenita cause changes in the structural equilibrium in telomerase RNA. Proc Natl Acad Sci USA 2003;100(2):449–54.

[54] Greider CW. Telomerase RNA levels limit the telomere length equilibrium. Cold Spring Harb Symp Quant Biol 2006;71:225–9.

[55] Trahan C, Martel C, Dragon F. Effects of dyskeratosis congenita mutations in dyskerin, NHP2 and NOP10 on assembly of H/ACA pre-RNPs. Hum Mol Genet 2010;19(5):825–36.

[56] Vulliamy T, Beswick R, Kirwan M, et al. Mutations in the telomerase component NHP2 cause the premature ageing syndrome dyskeratosis congenita. Proc Natl Acad Sci USA 2008;105(23):8073–8.

[57] Schmidt JC, Cech TR. Human telomerase: biogenesis, trafficking, recruitment, and activation. Genes Dev 2015;29(11):1095–105.

[58] Venteicher AS, Abreu EB, Meng Z, et al. A human telomerase holoenzyme protein required for Cajal body localization and telomere synthesis. Science 2009;323(5914):644–8.

[59] Zhong F, Savage SA, Shkreli M, et al. Disruption of telomerase trafficking by TCAB1 mutation causes dyskeratosis congenita. Genes Dev 2011;25(1):11–6.

[60] Guo Y, Kartawinata M, Li J, et al. Inherited bone marrow failure associated with germline mutation of ACD, the gene encoding telomere protein TPP1. Blood 2014;124(18):2767–74.

[61] Kocak H, Ballew BJ, Bisht K, et al. Hoyeraal-Hreidarsson syndrome caused by a germline mutation in the TEL patch of the telomere protein TPP1. Genes Dev 2014;28(19):2090–102.

[62] Walne AJ, Vulliamy T, Beswick R, Kirwan M, Dokal I. TINF2 mutations result in very short telomeres: analysis of a large cohort of patients with dyskeratosis congenita and related bone marrow failure syndromes. Blood 2008;112(9):3594–600.

[63] Savage SA, Giri N, Baerlocher GM, Orr N, Lansdorp PM, Alter BP. TINF2, a component of the shelterin telomere protection complex, is mutated in dyskeratosis congenita. Am J Hum Genet 2008;82(2):501–9.

[64] Wang F, Podell ER, Zaug AJ, et al. The POT1-TPP1 telomere complex is a telomerase processivity factor. Nature 2007;445(7127):506–10.

[65] Zhong FL, Batista LF, Freund A, Pech MF, Venteicher AS, Artandi SE. TPP1 OB-fold domain controls telomere maintenance by recruiting telomerase to chromosome ends. Cell 2012;150(3):481–94.

[66] Nandakumar J, Bell CF, Weidenfeld I, Zaug AJ, Leinwand LA, Cech TR. The TEL patch of telomere protein TPP1 mediates telomerase recruitment and processivity. Nature 2012;492(7428):285–9.

[67] Vannier JB, Sandhu S, Petalcorin MI, et al. RTEL1 is a replisome-associated helicase that promotes telomere and genome-wide replication. Science 2013;342(6155):239–42.

[68] Stuart BD, Choi J, Zaidi S, et al. Exome sequencing links mutations in PARN and RTEL1 with familial pulmonary fibrosis and telomere shortening. Nat Genet 2015;47(5):512–7.

[69] Kannengiesser C, Borie R, Menard C, et al. Heterozygous RTEL1 mutations are associated with familial pulmonary fibrosis. Eur Respir J 2015;46(2):474–85.

[70] Walne AJ, Vulliamy T, Kirwan M, Plagnol V, Dokal I. Constitutional mutations in RTEL1 cause severe dyskeratosis congenita. Am J Hum Genet 2013;92(3):448–53.

[71] Ballew BJ, Yeager M, Jacobs K, et al. Germline mutations of regulator of telomere elongation helicase 1, RTEL1, in Dyskeratosis congenita. Hum Genet 2013;132(4):473–80.

[72] Chen LY, Majerska J, Lingner J. Molecular basis of telomere syndrome caused by CTC1 mutations. Genes Dev 2013;27(19):2099–108.

[73] Tummala H, Walne A, Collopy L, et al. Poly(A)-specific ribonuclease deficiency impacts telomere biology and causes dyskeratosis congenita. J Clin Invest 2015;125(5):2151–60.

[74] Islam A, Rafiq S, Kirwan M, et al. Haematological recovery in dyskeratosis congenita patients treated with danazol. Br J Haematol 2013;162(6):854–6.

[75] Najean Y. Long-term follow-up in patients with aplastic anemia. A study of 137 androgen-treated patients surviving more than two years. Joint Group for the Study of Aplastic and Refractory Anemias. Am J Med 1981;71(4):543–51.

[76] Androgen therapy in aplastic anaemia: a comparative study of high and low-doses and of 4 different androgens. French Cooperative Group for the Study of Aplastic and Refractory Anemias. Scand J Haematol 1986;36(4): 346–52.

[77] Camitta BM, Thomas ED. Severe aplastic anaemia: a prospective study of the effect of androgens or transplantation on haematological recovery and survival. Clin Haematol 1978;7(3):587–95.

[78] Kyo S, Takakura M, Kanaya T, et al. Estrogen activates telomerase. Cancer Res 1999;59(23):5917–21.

[79] Wang Z, Kyo S, Takakura M, et al. Progesterone regulates human telomerase reverse transcriptase gene expression via activation of mitogen-activated protein kinase signaling pathway. Cancer Res 2000;60(19):5376–81.

[80] Bar C, Huber N, Beier F, Blasco MA. Therapeutic effect of androgen therapy in a mouse model of aplastic anemia produced by short telomeres. Haematologica 2015;100(10):1267–74.

[81] Al-Rahawan MM, Giri N, Alter BP. Intensive immunosuppression therapy for aplastic anemia associated with dyskeratosis congenita. Int J Hematol 2006;83(3):275–6.

[82] Peffault de Latour R, Peters C, Gibson B, et al. Recommendations on hematopoietic stem cell transplantation for inherited bone marrow failure syndromes. Bone Marrow Transplant 2015;50(9):1168–72.

[83] Gadalla SM, Sales-Bonfim C, Carreras J, et al. Outcomes of allogeneic hematopoietic cell transplantation in patients with dyskeratosis congenita. Biol Blood Marrow Transplant 2013;19(8):1238–43.

[84] Dietz AC, Orchard PJ, Baker KS, et al. Disease-specific hematopoietic cell transplantation: nonmyeloablative conditioning regimen for dyskeratosis congenita. Bone Marrow Transplant 2011;46(1):98–104.

[85] Rocha V, Devergie A, Socie G, et al. Unusual complications after bone marrow transplantation for dyskeratosis congenita. Br J Haematol 1998;103(1):243–8.

[86] Knight S, Vulliamy T, Copplestone A, Gluckman E, Mason P, Dokal I. Dyskeratosis Congenita (DC) Registry: identification of new features of DC. Br J Haematol 1998;103(4):990–6.

[87] Tsakiri KD, Cronkhite JT, Kuan PJ, et al. Adult-onset pulmonary fibrosis caused by mutations in telomerase. Proc Natl Acad Sci USA 2007;104(18):7552–7.

[88] Kropski JA, Mitchell DB, Markin C, et al. A novel dyskerin (DKC1) mutation is associated with familial interstitial pneumonia. Chest 2014;146(1):e1–7.

[89] Alder JK, Stanley SE, Wagner CL, Hamilton M, Hanumanthu VS, Armanios M. Exome sequencing identifies mutant TINF2 in a family with pulmonary fibrosis. Chest 2015;147(5):1361–8.

[90] Alder JK, Parry EM, Yegnasubramanian S, et al. Telomere phenotypes in females with heterozygous mutations in the dyskeratosis congenita 1 (DKC1) gene. Human Mutat 2013;34(11):1481–5.

[91] Alder JK, Guo N, Kembou F, et al. Telomere length is a determinant of emphysema susceptibility. Am J Respir Crit Care Med 2011;184(8):904–12.

[92] Diaz de Leon A, Cronkhite JT, Yilmaz C, et al. Subclinical lung disease, macrocytosis, and premature graying in kindreds with telomerase (TERT) mutations. Chest 2011;140(3):753–63.

[93] Calado RT. Telomeres in lung diseases. Prog Mol Biol Transl Sci 2014;125:173–83.

[94] Liu T, Chung MJ, Ullenbruch M, et al. Telomerase activity is required for bleomycin-induced pulmonary fibrosis in mice. J Clin Invest 2007;117(12):3800–9.

[95] Diaz de Leon A, Cronkhite JT, Katzenstein AL, et al. Telomere lengths, pulmonary fibrosis and telomerase (TERT) mutations. PLoS One 2010;5(5):e10680.

[96] Samuel BP, Duffner UA, Abdel-Mageed AS, Vettukattil JJ. Pulmonary arteriovenous malformations in dyskeratosis congenita. Pediatr Dermatol 2015;32(4):e165–6.

[97] Gorgy AI, Jonassaint NL, Stanley SE, et al. Hepatopulmonary syndrome is a frequent cause of dyspnea in the short telomere disorders. Chest 2015;148(4):1019–26.

[98] Qazilbash MH, Liu JM, Vlachos A, et al. A new syndrome of familial aplastic anemia and chronic liver disease. Acta Haematologica 1997;97(3):164–7.

[99] Hartmann D, Srivastava U, Thaler M, et al. Telomerase gene mutations are associated with cirrhosis formation. Hepatology 2011;53(5):1608–17.

[100] Calado RT, Brudno J, Mehta P, et al. Constitutional telomerase mutations are genetic risk factors for cirrhosis. Hepatology 2011;53(5):1600–7.

[101] Aalbers AM, Calado RT, Young NS, et al. Telomere length and telomerase complex mutations in pediatric acute myeloid leukemia. Leukemia 2013;27(8):1786–9.

[102] Calado RT, Regal JA, Hills M, et al. Constitutional hypomorphic telomerase mutations in patients with acute myeloid leukemia. Proc Natl Acad Sci USA 2009;106(4):1187–92.

[103] Mocellin S, Verdi D, Pooley KA, et al. Telomerase reverse transcriptase locus polymorphisms and cancer risk: a field synopsis and meta-analysis. J Natl Cancer Inst 2012;104(11):840–54.

[104] Scheinberg P, Cooper JN, Sloand EM, Wu CO, Calado RT, Young NS. Association of telomere length of peripheral blood leukocytes with hematopoietic relapse, malignant transformation, and survival in severe aplastic anemia. JAMA 2010;304(12):1358–64.

[105] Calado RT, Cooper JN, Padilla-Nash HM, et al. Short telomeres result in chromosomal instability in hematopoietic cells and precede malignant evolution in human aplastic anemia. Leukemia 2012;26(4):700–7.

[106] Dumitriu B, Feng X, Townsley DM, et al. Telomere attrition and candidate gene mutations preceding monosomy 7 in aplastic anemia. Blood 2015;125(4):706–9.

[107] Collopy LC, Walne AJ, Cardoso S, et al. Triallelic and epigenetic-like inheritance in human disorders of telomerase. Blood 2015;126(2):176–84.

[108] Kimura M, Stone RC, Hunt SC, et al. Measurement of telomere length by the Southern blot analysis of terminal restriction fragment lengths. Nat Protocols 2010;5(9):1596–607.

[109] Aubert G, Hills M, Lansdorp PM. Telomere length measurement-caveats and a critical assessment of the available technologies and tools. Mutat Res 2012;730(1-2):59–67.

[110] Gutierrez-Rodrigues F, Santana-Lemos BA, Scheucher PS, Alves-Paiva RM, Calado RT. Direct comparison of flow-FISH and qPCR as diagnostic tests for telomere length measurement in humans. PLoS One 2014;9(11):e113747.

[111] Baird DM, Rowson J, Wynford-Thomas D, Kipling D. Extensive allelic variation and ultrashort telomeres in senescent human cells. Nat Genet 2003;33(2):203–7.

[112] Bendix L, Horn PB, Jensen UB, Rubelj I, Kolvraa S. The load of short telomeres, estimated by a new method, Universal STELA, correlates with number of senescent cells. Aging Cell 2010;9(3):383–97.

[113] Lansdorp PM, Verwoerd NP, van de Rijke FM, et al. Heterogeneity in telomere length of human chromosomes. Hum Mol Genet 1996;5(5):685–91.

[114] Canela A, Vera E, Klatt P, Blasco MA. High-throughput telomere length quantification by FISH and its application to human population studies. Proc Natl Acad Sci USA 2007;104(13):5300–5.

[115] Alter BP, Giri N, Savage SA, Rosenberg PS. Telomere length in inherited bone marrow failure syndromes. Haematologica 2015;100(1):49–54.

[116] Calado RT, Graf SA, Wilkerson KL, et al. Mutations in the SBDS gene in acquired aplastic anemia. Blood 2007;110(4):1141–6.

[117] Pavlaki KI, Kastrinaki MC, Klontzas M, Velegraki M, Mavroudi I, Papadaki HA. Abnormal telomere shortening of peripheral blood mononuclear cells and granulocytes in patients with chronic idiopathic neutropenia. Haematologica 2012;97(5):743–50.

[118] Du HY, Pumbo E, Ivanovich J, et al. TERC and TERT gene mutations in patients with bone marrow failure and the significance of telomere length measurements. Blood 2009;113(2):309–16.

[119] Ball SE, Gibson FM, Rizzo S, Tooze JA, Marsh JC, Gordon-Smith EC. Progressive telomere shortening in aplastic anemia. Blood 1998;91(10):3582–92.

[120] Alter BP, Rosenberg PS, Giri N, Baerlocher GM, Lansdorp PM, Savage SA. Telomere length is associated with disease severity and declines with age in dyskeratosis congenita. Haematologica 2012;97(3):353–9.

[121] Vera E, Blasco MA. Beyond average: potential for measurement of short telomeres. Aging 2012;4(6):379–92.

[122] Hemann MT, Strong MA, Hao LY, Greider CW. The shortest telomere, not average telomere length, is critical for cell viability and chromosome stability. Cell 2001;107(1):67–77.

Fanconi Anemia

R. Peffault de Latour*, A.M. Risitano**, E. Gluckman†

*BMT Unit, French Reference Center for Aplastic Anemia and PNH, Saint-Louis
Hospital, Paris, France; **Hematology, Department of Clinical Medicine and Surgery,
Federico II University, Naples, Italy; †Eurocord, Hospital Saint-Louis, Paris, France

INTRODUCTION

Aplastic anemia (AA) is a rare disease in children that is most commonly idiopathic and less frequently a hereditary disorder. Hereditary bone marrow (BM) failure (BMF) syndromes, however, should be considered both in children and adults before the institution of any therapeutic treatment plan. They represent a very heterogeneous group of diseases with different mutations and pathophysiology but the exact diagnosis is essential because it will interfere with clinical decisions [1,2]. Genomic instability in the presence of clastogenic agents is the hallmark of Fanconi anemia (FA), which represents today as the most classic inherited BMF syndrome.

FA is characterized by congenital abnormalities, progressive BMF, chromosome breakage, and susceptibility to cancer. FA is an inherited disease resulting from mutations of 1 of the 18 *FANC* genes, the most frequently seen being FANCA, FANCC, FANCG, and FANCD2 [3–5]. The products of those genes interact in the unique FA/BRCA pathway, which is involved in the response to cellular stress and DNA damage and maintains genome integrity. FA cells have chromosome fragility both spontaneous and induced by interstrand crosslinking agents, this feature being central for patient diagnosis [6,7]. FA patients display progressive BMF during childhood, the pathophysiologic mechanisms of which have been elusive to date. This paper summarizes the diagnosis- and treatment approaches used for FA patients in 2016.

DIAGNOSIS AND STAGING

Clinical Presentation

FA patients often have skeletal thumb or limb abnormalities, and abnormal skin pigmentation (*café au lait* spots). Other organ systems commonly involved include cardiac, renal, and auditory systems. Low birth weight and growth retardation are frequent but some patients do not exhibit any congenital defect. The hematologic consequences of FA often develop in the first decade of life but absence of malformations and reversion

Congenital and Acquired Bone Marrow Failure
http://dx.doi.org/10.1016/B978-0-12-804152-9.00016-6

due to somatic mosaicism can result in delayed or failed diagnosis in a small proportion of patients. Death, however, often results from the complications of BMF or occurrence of malignancy. The most frequent is AML with cytogenetic BM clonal abnormalities; older patients are at a high risk of squamous cell carcinomas of the esophagus, head and neck, and urogenital tract [8].

An underlying FA diagnosis should thus also be suspected in any case in young patients with de novo BMF, myelodysplastic syndrome (MDS), or acute myeloid leukemia (AML), and even more if suggestive features are present [physical abnormalities; family or personal history; high-mean corpuscular volume or thrombocytopenia in previous CBCs; spontaneous chromatid breaks; or radials or unbalanced 1q, 3q, or 7q translocations on BM karyotype at an MDS or AML workup]. The unexpected occurrence of malignancy in young patients, as well as, excessive toxicity of usual chemotherapy during an AML induction is also in favor of an underlying FA disease.

Evaluating New Onset Cytopenia in Children [9]

The following items should be addressed carefully:

1. Family history	6. BM aspiration with cytogenetics and BM biopsy if necessary
2. Malformations	
3. Date of onset	7. Chromosome breaks with DEB or MMC
4. Liver function tests	
5. Blood counts	8. α-Fetoprotein
	9. HLA typing
	10. FANCD2 test[a]

BM, Bone marrow; DEB, diepoxy butane; MMC, mitomycin C.
[a] FANCD2 test: FANCD2 monoubiquitination by western blot on peripheral blood lymphocytes in order to evaluate the ability of the FA-core complex to monoubiquitinate FANCD2 and the level of expression of the FANCD2 protein.

Diagnosis and Staging

The clinical diagnosis of FA, being a heterogeneous disease, is not always sufficient to assess the correct diagnosis in children or young adults with AA. Other constitutional AA may have similar congenital abnormalities and FA patients may have no abnormalities [7].

Diagnosis is suspected with:	• Blood counts: pancytopenia with macrocytic anemia • Raised α-fetoprotein and hemoglobin F
Diagnosis is confirmed with:	• Peripheral blood lymphocyte cytogenetics with clastogenic agents: DEB or MMC showing increased chromosome breaks with tri- and quadriradial figures • Study of the cell cycle showing a G2/M arrest increased by incubation with clastogenic agents
Other tests:	• BM cytogenetic abnormalities for diagnosis of leukemia or MDS, with abnormalities in chromosomes 1, 3, 7, 5, 8, and 11 being the most common
New tests not for routine use:	• Ubiquitination of FANCD2: this test is specific and sensitive, if negative skin fibroblasts may be positive for FA and this confirms the existence of mosaicism • Identification of the complementation group with retroviral or lentiviral vectors • Sequencing and identification of the mutation—this test is useful for preimplantation diagnosis and possibly for assessing prognosis

MDS, Myelodysplastic syndrome.

HEMATOPOIETIC STEM CELL TRANSPLANTATION (HSCT)

HSCT is the only curative therapy for the hematologic manifestations of FA, including AA, MDS, and AML [10]. Donor stem cells may be obtained from BM, peripheral blood (following stimulation of donor hematopoiesis with G-CSF), or cord blood. Better results have been shown using BM as source of stem cells [11]. Ideally the HSCT should be performed prior to onset of MDS/leukemia and before multiple transfusions have been given for hematopoietic support. On the other hand because HSCT also carries the risk of fatal complications, FA patients without meaningful hematologic manifestations should not be referred for HSCT since they may not benefit from it. HSCT should be performed by centers with specific expertise in HSCT for FA.

FA anemia cells are hypersensitive to DNA crosslinking agents. Cellular exposure to genotoxic agents, including, cyclophosphamide (CY), busulfan, or irradiation increases chromosome breaks and tissue damage [12]. Graft-versus-host disease (GvHD) induces severe tissue damage and absence of repair [13]. For this reason, it has been recognized very early that reduced dose conditioning should be used in these patients, justifying the need for an accurate diagnosis [12]. Graft failure, historically a major impediment to FA transplantation, has been largely ameliorated by the use of fludarabine (FLU). Additionally, improvements in donor selection, especially HLA typing and supportive care, have led to greater use of alternative donors and decreased differences in the outcomes between sibling transplants and unrelated donor transplants [11,14,15].

Individuals whose hematologic manifestations have been successfully treated with HSCT appear to be at an increased risk for solid tumors, particularly squamous cell carcinoma of the tongue. In one study the risk was increased fourfold and the median age of onset was 16 years younger than in persons with FA who were not transplanted [16].

HLA Identical Sibling Transplants

First of all, it is important that all potential related donors are tested for the causative gene of the disorder, identified in the index patient, to avoid the use of a donor destined to develop BMF. The first large series of HSCT in patients with FA used a protocol of conditioning with low-dose CY 40 mg/kg with 4 Gy thoracoabdominal irradiation [17]; 5-year survival was 85%. In general, most series have reported that the following factors are associated with better survival after transplantation: younger patient age, higher pretransplant platelet counts,

absence of previous treatment with androgens, normal pretransplant liver function tests, and limited malformations.

In an attempt to reduce the potential impact of irradiation and GvHD on the risk of late effects, including cancer, newer regimens have replaced thoracoabdominal irradiation with FLU in combination with low-dose CY, together with antithymocyte globulin (ATG) or T-cell depletion to reduce the risk of GvHD. Pasquini et al. on behalf of CIBMTR compared the early outcome of HSCT using nonirradiation-containing regimens ($n = 71$) to the outcome of regimens with irradiation ($n = 77$) for FA patients transplanted with HLA identical sibling donors. Hematopoietic recovery, acute- and chronic GvHD, and mortality were similar after the two regimens [18]; the 5-year probability of overall survival was 78% after irradiation and 81% after the nonirradiation regimen. We recently reported the Saint Louis experience in 20 patients who received the combination of FLU and low-dose CY (40 mg/kg total dose) without in vivo or ex vivo T-cell depletion, which led to excellent results with 95% overall survival with a median follow-up of 2 years [19] (Fig. 16.1).

While promising, longer follow-up is needed to determine whether nonirradiation-based regimens will reduce the risk of late effects, such as, infertility, cataracts, endocrinopathies, and secondary malignancies. Of note, some pregnancies

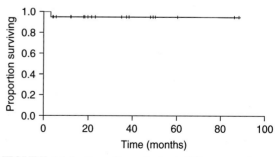

FIGURE 16.1 Overall survival of sibling transplantation in Fanconi anemia (FA) patients with bone marrow failure (BMF).

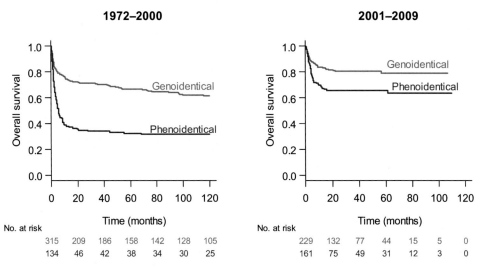

FIGURE 16.2   Overall survival according to donor type and transplantation period.

have been observed in women after HSCT even with irradiation-containing regimens [20]. In the light of these data, the most frequently used conditioning regimen for HLA identical sibling HSCT is low-dose CY and FLU. The source of stem cells should be BM, to decrease the risk of GvHD. Whether T-cell depletion should be used or not is still not clear regarding the balance of benefit/risk. T-cell depletion is indeed associated with profound immunosuppression during the early months post-HSCT with the potential for higher transplantation-related morbidity and mortality due to opportunistic infections and graft rejection, justifying its use today mainly in unrelated setting.

Alternative Donor Transplants

Matched Unrelated Donors

Results of matched unrelated donor transplant for FA have been improving with time [11,14,15]. Since the first results published in 2007, survival has increased from 33% to 80% in selected centers. Drastic improvements have been reported since 2000 in patients transplanted from unrelated donors due to better HLA

typing and the use of FLU-based reduced intensity-conditioning regimens, with or without T-cell depletion [11,14,15] (Fig. 16.2). Moreover, MacMillan et al. recently reported 130 FA patients who underwent alternative donor HSCTs between 1995 and 2012. All patients received CY, single-fraction total body irradiation (TBI), and ATG with or without FLU, followed by T-cell-depleted BM or unmanipulated umbilical cord blood transplantation [21]. The incidences of grades II–IV acute and chronic GvHD were 20 and 10%, respectively. Severe toxicity was highest in patients over 10 years of age or those with a history of opportunistic infections or transfusions before HSCT. Patients without a history of opportunistic infections or transfusions before HSCT and who received conditioning with TBI 300 cGy, CY, FLU, and ATG have a 5-year survival probability of 94%. Of note, 37 patients (38%) were transplanted due to clonal abnormality before HSCT and 10 patients (8%) only because of late MDS/AML.

Based on these results, significant changes in practice were suggested: use of a FLU-containing conditioning regimen in the context of in vivo or in vitro T-cell depletion and earlier referral with transplantation prior to excessive transfusions. Improvement is due to a better donor selection and modification of the conditioning; the best results have been published by the Minnesota group using CY 40 mg/kg, TBI 300 cGy with thymic shielding, FLU 140 mg/m^2, and ATG or T-cell depletion [21].

Cord Blood Transplantation

An entire chapter is dedicated to this question in this book (cf Umbilical cord blood transplantation for patients with acquired and inherited bone marrow failure syndromes on behalf of Eurocord, E Gluckman, AL Ruggeri, R Peffault de Latour). Briefly, HLA identical sibling cord blood transplantation, if available, is a good option in patients with FA where collection of cord blood from a healthy sibling can be anticipated before birth. If cord blood unit does not contain enough cells, addition of BM from the same donor gives excellent results. Use of an unrelated cord blood transplant for FA patients is controversial because of the high incidence of graft failure and transplant-related mortality. However, unrelated cord blood transplantation using high number of cells (more than 4×10^7 TNC/kg) with no more than 0–1 HLA mismatches in selected patients (children or/and negative recipient cytomegalovirus serology) is of interest.

FA Patients With MDS and AML

HSCT is the choice treatment in patients with FA and clonal evolution. However, while some situations are indications for rapid transplantation (overt AML, significant MDS, and cytogenetic/molecular abnormality of chromosome 3q, 7q, and/or RUNX1), it is still not clear whether other patients with a level of isolated clonal evolution, such as sole +1q clones, should receive immediate treatment. As far as cytoreduction pre-HSCT is concerned, all experts agree worldwide that toxicity is high in FA patients. However, certain specific subsets might benefit from pre-HSCT cytoreduction (i.e., MDS

with excess of blasts, AML, or BRCA2/FANCD1 patients) [22,23]. What remains clear is the risk of prolonged chemotherapy-associated aplasia, highlighting the importance of securing a donor before offering chemotherapy in these patients. Regarding HSCT per se, it should not be different than in patients without clonal evolution (outside patients with MDS with excess of blasts, AML, or BRCA2/FANCD1 discussed before). Alternative donors, that is, mismatched-, cord blood-, or haplo-HSCT, may be considered, given the poor prognosis of such patients according to the experience of each center.

POST-HCT MONITORING IN FA

Patients with FA require particular attention because of their sensitivities to toxic agents, various organ dysfunctions due to congenital malformations, and increased risk of developing malignancies. Patients with FA are at an increased risk of malignancy. The cumulative incidence of solid malignancy by age 50 years in the German Fanconi Anemia Registry is 28% [10,24]. A 4.4-fold higher rate of squamous cell carcinoma is reported in FA patients who undergo HSCT compared to patients with FA of the same age who have not received a HSCT [10,16]. In a background of competing causes of mortality, the cumulative incidence of squamous cell carcinoma is between 24% at 15 years [10,24] and 34% at 20 years in a more recent study [11] (Fig. 16.3). While patients with FA are inherently prone to develop tumors, pretransplant conditioning regimen and chronic GvHD are key factors in their development post-HSCT [11,25]. In the latter study, the use of peripheral blood stem cells has been shown to increase the risk of death at 1-year post-HSCT and this mortality was strongly associated with secondary malignancy in the background of cGVHD. Therefore oropharyngeal, dental, and gynecologic examinations and screening for malignancy have to be part of a long-term care that should be tailored to the individual FA patient [10].

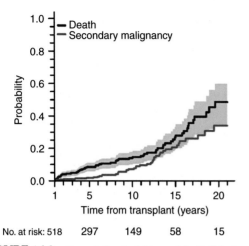

No. at risk: 518   297   149   58   15

FIGURE 16.3 Cumulative incidence of death *(blue)* and secondary cancer *(red)* in a 1-year survivor.

CONCLUSIONS

HSCT remains the only curative treatment option for FA in case of BMF or clonal evolution (i.e., MDS or AML). Sibling transplantation is the treatment of choice and overall survival is about 90% using the association of FLU and low-dose CY with BM as source of stem cells. Since 2000, due to better HLA typing and supportive care, matched unrelated transplantation has also drastically improved. The best results have been recently published by the Minnesota group using low-dose CY, FLU, TBI 300 cGy with thymic shielding, and ATG or T-cell depletion [21]. While related cord blood transplantation is associated with good results, unrelated cord blood are still associated with poor overall survival and should not be used outside prospective clinical trials. Given the fact that FA is characterized by nonhematologic disease manifestations and increased intrinsic susceptibility to both acute and late treatment-related toxicity, those patients should be carefully monitored long term notably because of the risk of secondary malignancy, for which early surgical intervention is mandatory. Prospective international clinical trials are urgently required to advance the management of these rare disorders and eventually improve outcomes.

All patients undergoing HSCT for FA should also receive routine monitoring of growth and development, cardiopulmonary, gastrointestinal, and renal function. Various nonspecific complications may occur after HCST for FA, such as, pulmonary disease, aseptic necrosis, or severe osteoporosis. Therefore monitoring of lung function test and bone-mineral density by dual energy X-ray absorptiometry using the age-adjusted Z-score is recommended. Even if less documented, patients with FA are exposed to endocrine dysfunction post-HSCT, as well as fertility problems and a comprehensive follow-up by a pediatric endocrinologist (considering, e.g., hypothyroidism, diabetes and metabolic syndrome, and growth hormone deficiency) is warranted. Patients should be encouraged to lead a healthy lifestyle, which includes a healthy diet, regular exercise, avoidance of alcohol, smoking and second-hand smoke, and limited sun exposure and use of sunscreen. Another important aspect of long-term care of FA patients post-HSCT is the ongoing assessment of adverse effects on psychosocial and mental health, quality of life, education, and employment.

References

[1] Dokal I, Vulliamy T. Inherited bone marrow failure syndromes. Haematologica 2010;95:1236–40.

[2] Moldovan GL, D'Andrea AD. How the Fanconi anemia pathway guards the genome. Annu Rev Genet 2009;43:223–49.

[3] Bogliolo M, Surralles J. Fanconi anemia: a model disease for studies on human genetics and advanced therapeutics. Curr Opin Genet Dev 2015;33:32–40.

[4] Kee Y, D'Andrea AD. Molecular pathogenesis and clinical management of Fanconi anemia. J Clin Invest 2012;122:3799–806.

[5] Kottemann MC, Smogorzewska A. Fanconi anaemia and the repair of Watson and Crick DNA crosslinks. Nature 2013;493:356–63.

[6] de Winter JP, Joenje H. The genetic and molecular basis of Fanconi anemia. Mutat Res 2009;668:11–9.

[7] Auerbach AD. Fanconi anemia and its diagnosis. Mutat Res 2009;668:4–10.

[8] Rosenberg PS, Greene MH, Alter BP. Cancer incidence in persons with Fanconi anemia. Blood 2003;101:822–6.

[9] Pinto FO, Leblanc T, Chamousset D, et al. Diagnosis of Fanconi anemia in patients with bone marrow failure. Haematologica 2009;94:487–95.

[10] MacMillan ML, Wagner JE. Haematopoeitic cell transplantation for Fanconi anaemia—when and how? Br J Haematol 2010;149:14–21.

[11] Peffault de Latour R, Porcher R, Dalle JH, et al. Allogeneic hematopoietic stem cell transplantation in Fanconi anemia: the European Group for Blood and Marrow Transplantation experience. Blood 2013;122:4279–86.

[12] Gluckman E, Devergie A, Dutreix J. Radiosensitivity in Fanconi anaemia: application to the conditioning regimen for bone marrow transplantation. Br J Haematol 1983;54:431–40.

[13] Guardiola P, Socie G, Li X, et al. Acute graft-versus-host disease in patients with Fanconi anemia or acquired aplastic anemia undergoing bone marrow transplantation from HLA-identical sibling donors: risk factors and influence on outcome. Blood 2004;103:73–7.

[14] Wagner JE, Eapen M, MacMillan ML, et al. Unrelated donor bone marrow transplantation for the treatment of Fanconi anemia. Blood 2007;109:2256–62.

[15] Locatelli F, Zecca M, Pession A, et al. The outcome of children with Fanconi anemia given hematopoietic stem cell transplantation and the influence of fludarabine in the conditioning regimen: a report from the Italian pediatric group. Haematologica 2007;92:1381–8.

[16] Rosenberg PS, Socie G, Alter BP, Gluckman E. Risk of head and neck squamous cell cancer and death in patients with Fanconi anemia who did and did not receive transplants. Blood 2005;105:67–73.

[17] Socie G, Devergie A, Girinski T, et al. Transplantation for Fanconi's anaemia: long-term follow-up of fifty patients transplanted from a sibling donor after low-dose cyclophosphamide and thoraco-abdominal irradiation for conditioning. Br J Haematol 1998;103:249–55.

[18] Pasquini R, Carreras J, Pasquini MC, et al. HLA-matched sibling hematopoietic stem cell transplantation for Fanconi anemia: comparison of irradiation and nonirradiation containing conditioning regimens. Biol Blood Marrow Transplant 2008;14:1141–7.

[19] Benajiba L, Salvado C, Dalle JH, et al. HLA-matched related-donor HSCT in Fanconi anemia patients conditioned with cyclophosphamide and fludarabine. Blood 2015;125:417–8.

[20] Nabhan SK, Bitencourt MA, Duval M, et al. Fertility recovery and pregnancy after allogeneic hematopoietic stem cell transplantation in Fanconi anemia patients. Haematologica 2010;95:1783–7.

[21] MacMillan ML, DeFor TE, Young JA, et al. Alternative donor hematopoietic cell transplantation for Fanconi anemia. Blood 2015;125:3798–804.

[22] Mitchell R, Wagner JE, Hirsch B, DeFor TE, Zierhut H, MacMillan ML. Haematopoietic cell transplantation for acute leukaemia and advanced myelodysplastic syndrome in Fanconi anaemia. Br J Haematol 2014;164:384–95.

[23] Talbot A, Peffault de Latour R, Raffoux E, et al. Sequential treatment for allogeneic hematopoietic stem cell transplantation in Fanconi anemia with acute myeloid leukemia. Haematologica 2014;99:e199–200.

[24] Rosenberg PS, Alter BP, Ebell W. Cancer risks in Fanconi anemia: findings from the German Fanconi Anemia Registry. Haematologica 2008;93:511–7.

[25] Deeg HJ, Socie G. Malignancies after hematopoietic stem cell transplantation: many questions, some answers. Blood 1998;91:1833–44.

17

Ribosomopathies and the Quality Control of Ribosome Assembly

A.J. Warren

Cambridge Institute for Medical Research, Cambridge, United Kingdom;
The Department of Haematology, University of Cambridge, Cambridge,
United Kingdom; Wellcome Trust–Medical Research Council Stem Cell Institute,
University of Cambridge, Cambridge, United Kingdom

INTRODUCTION

The ribosomopathies are a heterogeneous group of disorders characterized by bone marrow failure, physical anomalies, and cancer predisposition that are caused by mutations in multiple components of the apparatus involved in ribosome assembly. Ribosomes are essential macromolecular machines that catalyze protein synthesis in all kingdoms of life. In eukaryotes, the large (60S) and small (40S) ribosomal subunits are assembled from 4 ribosomal RNA (rRNA) species, 80 ribosomal proteins, and 76 small nucleolar RNAs in a complex process that is facilitated by around 200 ancillary factors [1]. Preribosome particles are released from the nucleolus into the nucleus and subsequently exported to the cytoplasm, where a small complement of assembly factors is removed and recycled in an ordered series of linked steps to generate translation-competent ribosomal subunits [2]. How defects in a ubiquitous and essential macromolecule, such as the ribosome, cause tissue-specific anomalies and cancer predisposition remains a key, unanswered question.

This review focuses on the clinical features, genetics, and molecular pathophysiology of the two human ribosomopathies, Diamond–Blackfan anemia (DBA) and Shwachman–Diamond syndrome (SDS). Elucidation of the molecular basis of these fascinating disorders is providing important new insights into the mechanism and quality control of ribosome assembly.

DIAMOND–BLACKFAN ANEMIA

Clinical Features

In 1938, Louis Diamond and Kenneth Blackfan described four children with a congenital hypoplastic anemia associated with physical anomalies [3]. This disorder is now recognized as Diamond–Blackfan anemia (DBA; OMIM

Congenital and Acquired Bone Marrow Failure
http://dx.doi.org/10.1016/B978-0-12-804152-9.00017-8

TABLE 17.1   Diagnostic Criteria for Diamond–Blackfan Anemia (DBA) [6]

	Supporting criteria	
Diagnostic criteria	Major criteria	Minor criteria
Age < 1 year	Pathogenic mutations	Elevated red cell adenosine deaminase
Normochromic, macrocytic anemia	Positive family history	Congenital anomalies
Reticulocytopenia		Elevated fetal hemoglobin
Normocellular bone marrow with selective reduction in erythroid precursors		No evidence of other congenital anemia

#105650) or congenital pure red cell aplasia. DBA is a rare, clinically and genetically heterogeneous disease, with an incidence of 2–7 per million live births. Individuals with DBA classically present within the first year of life (median 2 months) with normochromic, macrocytic anemia, reticulocytopenia, and reduced or absent erythroid precursors in the bone marrow [4]. Around 50% have associated short stature and physical anomalies including craniofacial (Pierre–Robin syndrome and cleft palate), thumb, cardiac, and urogenital defects. Importantly, DBA is associated with a 30- to 40-fold increased risk of AML, osteosarcoma, or colon cancer [5].

Diagnostic Criteria

The diagnostic criteria for DBA are summarized in Table 17.1 [6]. The differential diagnosis includes other congenital anemias; transient erythroblastopenia of childhood; viral infections, such as, parvovirus B19, HIV, and infectious mononucleosis; and exposure to drug and toxins.

Genetics

Familial and sporadic forms of DBA are recognized. Following the seminal report of *RPS19* mutations in a DBA patient carrying a constitutional X;19 chromosomal translocation [7], 25% of patients have been found to carry mutations in *RPS19* that encodes ribosomal protein uS19 [8] (the new system for naming ribosomal proteins [9] is used throughout the text).

Subsequently, DBA has been shown to be associated with haploinsufficiency of multiple genes encoding components of both the small and large ribosomal subunits (Table 17.2), leading to the concept of DBA as a disorder of ribosome biogenesis or ribosomopathy. De novo mutations are found in around two-thirds of cases, while in the remaining third the mutation is inherited as an autosomal dominant (AD) trait [10]. *RPL5* mutations are associated with craniofacial, thumb, and heart anomalies, while *RPL11* haploinsufficiency is associated with isolated thumb malformations [11]. Mutations in *RPS28*, *RPS26*, and its binding partner *TSR2*, have been identified in DBA associated with mandibulofacial dysostosis [12]. Interestingly, the yeast Tsr2 protein is a carrier (escortin) that specifically transfers ribosomal protein eS26 (also linked to DBA) from its importins to the 90S preribosome in the nucleolus [13]. The etiology of DBA extends beyond the ribosome to include mutations in the X-linked gene encoding the key hematopoietic transcription factor *GATA1* [14]. Indeed, disrupted *GATA1* translation has been proposed as a common mechanism linking the disparate ribosomal protein mutations in DBA [15].

Molecular Pathophysiology

How disrupted ribosome biogenesis promotes defective red cell maturation remains incompletely understood. The ribosomal stress model posits that a complex involving uL5 (RPL11), uL18 (RPL5), and 5S rRNA specifically binds to and inactivates the E3 ligase, MDM2,

TABLE 17.2   Genes Mutated in Human Ribosomopathies

Gene	Disease	Inheritance	Function	Clinical features	References
SBDS	SDS	AR	Release of eIF6 from late cytoplasmic 60S subunits	Neutropenia, exocrine pancreatic insufficiency, and metaphyseal chondrodysplasia	[16–20]
RPS7 *RPS10* *RPS17* *RPS19* *RPS24* *RPS26* *RPS27* *RPS28* *RPS29*	DBA	Sporadic, AD	Components of the 40S subunit	Macrocytic anemia, craniofacial abnormalities, thumb abnormalities, and short stature	[11] [21] [22] [7] [23] [24] [25] [12] [26]
RPL5 *RPL11* *RPL15* *RPL26* *RPL27* *RPL35A*			Components of the 60S subunit		[11] [11] [27] [28] [25] [24]
GATA1		X-linked	Transcription factor eS26 escortin		[14]
TSR2		X-linked			[12]
RPS14	5q– syndrome	Sporadic	Component of the 40S subunit	MDS and macrocytic anemia	[29]
RPL5, RPL10, and *RPL22*	Pediatric T-ALL	Sporadic	Components of 60S subunit	T-ALL	[30]
RPSA	Isolated congenital asplenia	AD	Component of 40S ribosomal subunit	Absence of spleen	[31]
BMS1	Aplasia cutis congenita	AD	Ribosomal GTPase	Skin agenesis on scalp vertex	[32]
EMG1	Bowen–Conradi syndrome	AR	Maturation of 40S ribosomal subunit	Growth retardation and psychomotor delay	[33]
CIRH1A (*UTP14*)	North American Indian childhood cirrhosis	AR	Maturation of 18S rRNA and rDNA transcription	Cirrhosis	[34]
RBM28	ANE syndrome	AR	Pre-rRNA processing, 60S maturation	Hair loss, microcephaly mental retardation, progressive motor retardation, and adrenal insufficiency	[35]
RPL10	Microcephaly	X-linked	Component of the 60S subunit	Microcephaly, growth retardation, and seizures	[36]
RPL10	Autism	X-linked		Autism	[37]
RMRP	Cartilage–hair hypoplasia	AR	RNA component of RNase MRP	Short limb dwarfism, metaphyseal dysplasia, hypoplastic anemia, defective B- and T cell-mediated immunity, and variable intestinal aganglionosis	[38]

(Continued)

TABLE 17.2   Genes Mutated in Human Ribosomopathies *(cont.)*

Gene	Disease	Inheritance	Function	Clinical features	References
DKC1	Dyskeratosis congenita	X-linked	Pseudouridine synthase	Bone marrow failure and mucocutaneous defects	[39]
RPS20	Familial colorectal cancer type X	AD	Component of 40S ribosomal subunit	Hereditary nonpolyposis colorectal carcinoma with no mismatch repair defects	[40]
RPS15, RPSA, and *RPS20*	Relapsed CLL	Sporadic	Component of 40S ribosomal subunit, binds MDM2	Adverse prognosis CLL after first line therapy	[41]

AD, Autosomal dominant; ANE, alopecia, neurologic defects, and endocrinopathy; AR, autosomal recessive; CLL, chronic lymphocytic leukemia; DBA, Diamond–Blackfan anemia; MDS, myelodysplastic syndrome; SDS, Shwachman–Diamond syndrome.

thereby inhibiting the ubiquitylation and degradation of the tumor suppressor, p53, to promote cell cycle arrest and apoptosis [42]. The p53 dependence of the erythroid phenotype in DBA is supported by the mouse genetic analysis [43,44]. Thus, p53 appears to have a key role in the quality control surveillance of the integrity of ribosome assembly pathway. However, much remains to be learned about the underlying molecular mechanisms driving this signaling pathway.

Management

Corticosteroids and red cell transfusion support are the mainstay of therapy in DBA [45]. The maintenance dose of corticosteroid is tapered to the minimal necessary to support transfusion independence. Iron chelation is required to avoid the complications of iron overload in chronically transfused patients but is contraindicated in pregnancy. Particularly if an HLA matched unaffected sibling is available, allogeneic hematopoietic stem cell transplantation (HSCT) should be considered in individuals who are corticosteroid-resistant or who progress to aplastic anemia, MDS, or AML.

A recent retrospective multicenter study showed a 5-year overall survival (OS) of 74.4% and a transplant-related mortality (TRM) of 25.6% in DBA patients undergoing HSCT [46].

No significant differences were observed between transplants from HLA matched unrelated donors (69.9% OS and 30% TRM) or matched sibling donors (80.4% OS and 20% TRM). The cumulative incidence of grade III–IV acute and chronic graft versus host disease (GvHD) was 24% and 21%, respectively, with no significant difference between HLA matched unrelated donor- and matched sibling donor transplants. OS was better for children transplanted under 10 years of age, while high-pretransplant ferritin and acute GvHD were negative prognostic factors. Gender or malformations did not affect transplant outcome.

Interestingly, the branched chain amino acid, L-leucine, conferred transfusion independence in a DBA patient [47] and improved the anemia in murine- and zebrafish-disease models, potentially through activation of mTOR signaling [48,49]. Clinical trials of L-leucine for the treatment of DBA are currently underway.

It is important to note that 20–30% of DBA patients may undergo spontaneous hematologic remission [4], although the mechanism remains unclear. In pregnancy, almost half of DBA patients show transient worsening of anemia and there is an increased frequency of maternal complications including preeclampsia, fetal loss, intrauterine death, premature delivery, intrauterine growth retardation, and congenital anomalies [50,51].

SHWACHMAN–DIAMOND SYNDROME

Clinical Features

In 1961, Nezelof and Watchi first described "congenital lipomatosis of the pancreas" in two children with exocrine pancreatic insufficiency and leucopenia [52]. Shwachman–Diamond syndrome (SDS, OMIM #260400), as it was subsequently termed, is now recognized as an autosomal recessive (AR) disorder caused by mutations in the eponymous SBDS gene [16] that is named after the American physician Harry Shwachman, the British ophthalmologist Martin Bodian, and the American pediatrician Louis Diamond who later reported the syndrome in 1964 [53,54]. The associated skeletal abnormalities [55,56] define a third key feature of the syndrome.

SDS is rare, with an estimated prevalence of 1 in 77,000 births [57]. The most common clinical features of SDS are exocrine pancreatic insufficiency with malabsorption, malnutrition, and poor growth; skeletal abnormalities; and neutropenia and bone marrow failure with increased risk of progression to myelodysplastic syndrome (MDS) and acute myeloid leukemia (AML). However, the phenotypic spectrum in SDS is broad, with almost half of the 37 individuals in the North American SDS registry confirmed genetically to have SDS lacking the classic combination of clinical manifestations [58]. The clinical diagnosis of SDS is confirmed by identifying biallelic pathogenic mutations in the SBDS gene. However, as 10% of SDS patients lack biallelic SBDS mutations [59,60], mutations in another gene may be linked to this disorder.

Hematology

The clinical features of SDS have been defined in eight cohort studies, with the inclusion of genotyping in three [58,59,61–66]. Almost all affected individuals have persistent or intermittent neutropenia [58]. Patients may present with or progress to severe aplastic anemia with pancytopenia [67,68]. Although defective neutrophil chemotaxis has been described in some patients [69], the clinical significance of this observation is unclear. B- and T-cell numbers may be altered and immunoglobulin levels reduced [70]. The bone marrow is generally hypocellular, with left-shifted granulopoiesis [63]. Interestingly, the 10% of patients with a clinical diagnosis of SDS who carried a wild type SBDS gene had a more severe hematopoietic but milder pancreatic phenotype [59].

Predisposition to MDS/AML is increased in SDS patients. However, the definition of MDS varies between reports, limiting the precision of the available data. In the largest study to date involving a French cohort of 102 SDS patients, the cumulative incidence of MDS/AML was 18.8% at 20 years and 36.1% at 30 years of age [68]. Acquired clonal cytogenetic abnormalities, particularly monosomy 7, i(7q10) and 7q deletion, are common [71]. Interestingly, the i(7q10) and del(20)(q) abnormalities may fluctuate over time and appear to be associated with a reduced risk of transformation to MDS/AML [72], although this needs to be confirmed in a larger cohort. The favorable prognosis of del(20)(q) may reflect the reduced dosage of the EIF6 gene [73] (discussed further).

Solid Tumors

There are limited case reports of solid tumors in SDS patients, including bilateral breast cancer [74], dermatofibrosarcoma [75], CNS lymphoma [76], pancreaticoduodenal carcinoma [77], pancreatic ductal adenocarcinoma [78], and Non-Hodgkin's lymphoma [79].

Gastrointestinal Tract

SDS is the second commonest cause of congenital exocrine pancreatic insufficiency in children after cystic fibrosis. Pancreatic insufficiency commonly presents with steatorrhea within the first year. Histologic analysis reveals fatty replacement of the acinar cells but relative preservation of ductal architecture and the islets of Langerhans [61]. The presence of parotid acinar cell dysfunction suggests that there may be a generalized acinar cell defect in SDS [80]. However, up to 50% of patients show age-related improvements in pancreatic function, leading to pancreatic sufficiency [62].

Hepatopathy in SDS is associated with a good prognosis [81]. While the elevated transaminases and hepatomegaly resolve over time, cholestasis may persist. Hepatic microcysts may develop in older patients.

Skeletal

Patients with SDS exhibit a range of skeletal abnormalities including metaphyseal chondrodysplasia, rib cage dysplasia, and osteopenia [82]. At the most severe end of the spectrum, SBDS mutations are associated with neonatal respiratory failure in the context of severe neonatal spondylometaphyseal dysplasia Sedaghatian-type with marked platyspondyly, lacy iliac crests, and severe metaphyseal dysplasia [83]. SBDS mutations have also been identified in patients presenting with asphyxiating thoracic dystrophy (Jeune syndrome) [84]. Low-turnover osteoporosis associated with low-bone mineral density and vertebral compression fractures has also been reported [85].

Neurocognitive

Neurocognitive and behavioral problems are recognized in SDS [86–88]. About 20% of patients demonstrate intellectual disability, most commonly perceptual reasoning. This may be related to structural brain changes identified by magnetic resonance imaging, such as reduced global brain volume (both gray and white matter) and significantly smaller age- and head size-adjusted areas of the posterior fossa, vermis,

corpus callosum, and pons [89]. Impaired verbal, perceptual, and memory skills and altered executive functions are associated with alterations in white matter connectivity [90].

Cardiac

Cardiomyopathy and congenital cardiac abnormalities are common in SDS [91]. Twelve of the 102 (11%) patients in the French severe chronic neutropenia registry had cardiac abnormalities, including 6 with cardiomyopathy (CMP) and 6 with congenital heart disease (CHD). CMP was diagnosed in the first year of life in three cases and in the second decade of life in the remaining three, while all CHD patients were diagnosed in the first year of life. Myocardial fibrosis has also been reported [92,93].

Miscellaneous

Additional findings include ichthyosis, eczema, oral disease with delayed dental development, [94] and congenital gastrointestinal, neurologic, kidney/urinary tract, eye, and ear anomalies [58]. Severe myopathy in a newborn requiring assisted ventilation has recently been reported [95]. Consistent with the observed phenotypic variability among affected sibs [64], no genotype–phenotype correlations have been observed with specific SBDS variant alleles [66].

Diagnosis

Consensus guidelines for the diagnosis and surveillance of SDS patients have been published [96] and are summarized in Tables 17.3 and 17.4. The diagnosis of SDS relies on clinical findings, particularly pancreatic dysfunction and characteristic hematologic problems, and is confirmed by the detection of biallelic pathogenic SBDS variants. The differential diagnosis includes cystic fibrosis, Pearson syndrome, Fanconi anemia, Johanson–Blizzard syndrome, DBA, celiac disease, and AR severe congenital neutropenia due to G6PC3 deficiency.

TABLE 17.3 Diagnostic Criteria for SDS

Exocrine pancreatic insufficiency

Reduced levels of pancreatic enzymes adjusted to age (serum trypsinogen, serum isoamylase, fecal elastase, and serum lipase)

Tests that support the diagnosis but require corroboration:

Abnormal 72-h fecal fat analysis

Reduced levels of fat-soluble vitamins (A, D, E, or K)

Evidence of pancreatic lipomatosis by ultrasound, CT, or MRI scanning

AND

Hematologic abnormalities (on at least two occasions over at least 3 months):

Neutropenia $< 1.5 \times 10^9 \, L^{-1}$

Anemia or thrombocytopenia below the age-adjusted norm

Tests that support the diagnosis but require corroboration:

Elevated hemoglobin F

Macrocytosis

TABLE 17.4 SDS Surveillance

	Frequency
Hematology	
FBC	3–6 monthly
Bone marrow examination with cytogenetics	1–3 yearly
Gastroenterology	
Vitamins A, D, E, and K and prothrombin time	6 monthly
Liver function tests	
Skeletal system, growth	
Height, weight, and head circumference	Yearly
Skeletal survey	As clinically indicated, during rapid growth stages
Densitometry	Before puberty, during puberty, then as clinically indicated
Oral and dental review	At least yearly
Development and neuropsychological evaluation	Infancy/preschool and at ages 6–8 years, 11–13 years, and 15–17 years

Clinical Management

Multidisciplinary care is strongly recommended. The mainstay of treatment is oral pancreatic enzyme replacement and fat-soluble vitamins for exocrine pancreatic insufficiency, but it is important to note that the steatorrhea may remit spontaneously in up to 50% of older patients [62]. Patients with learning difficulties may require specific educational and psychological support. Treatment with granulocyte-colony stimulating factor may be appropriate in patients with severe recurrent infections associated with neutropenia. HSCT should be considered for the treatment of severe pancytopenia or in the context of MDS or AML [97,98]. The largest retrospective multicenter study of allogeneic stem cell transplantation in 26 SDS patients with severe bone marrow dysfunction showed an incidence of grade III and IV acute GvHD of 24% and of chronic GvHD 29%. After a median follow-up of 1.1 years, the TRM was 35.5% while the OS was 64.5%, suggesting some benefit for SDS patients with severe bone marrow abnormalities [99]. Of note, 100% survival and far lower rates of GvHD were achieved in a more recent, though smaller study, using a reduced intensity-conditioning regimen along with Campath [98].

Genetics

In 2003, the Rommens group identified biallelic mutations in the SBDS gene on chromosome 7 in the majority of SDS patients using a positional cloning strategy [16]. Although most parents of children with SDS are carriers, about 10% of SBDS mutations arise de novo [100]. In more than 90% of affected individuals, one of three common pathogenic SBDS variants is detected in exon 2 (183_184TA>CT, 258+2T>C, or the combination of 183_184TA>CT + 258+2T>C on one allele) as a result of gene conversion between the SBDS gene and an unprocessed pseudogene that lies in a paralogous duplicon located 5.8 Mb distally. While 89% of affected individuals carried at least one converted allele, 60% carried two. The 258+2T>C mutation is predicted to disrupt the donor splice site of intron 2, while the dinucleotide alteration 183–184TA→CT introduces an in-frame stop codon (K62X). In addition, more than 40 novel sequence variants have been described in individuals who are compound heterozygotes with one of the three common gene conversion alleles [16,17,66,68,82,101–103]. Loss of the SBDS protein is lethal early in mammalian development [104], consistent with the observation that no individuals have been identified who are homozygous for the dinucleotide variant

(183–184TA→CT) that introduces an in-frame stop codon (K62X) [16]. Thus, the SBDS gene is essential and the SDS phenotype is associated with the expression of at least one hypomorphic allele [101].

SBDS Protein Structure

The SBDS gene encodes a widely expressed and highly conserved protein of 250 amino acids with orthologs in eukaryotes and archaea, but not eubacteria [16]. Consistent with an evolutionarily conserved function, the human and archaeal SBDS proteins have the same overall tripartite architecture [17,101,105–107]. Compared with its archaeal ortholog, the human SBDS protein carries additional unstructured N- (residues S2–V15) and C- (residues N238–E250) terminal extensions. The human SBDS N-terminal domain I (residues S2–S96), termed the FYSH (Fungal, Yhr087w, Shwachman) domain [101], has striking structural homology to a single-domain yeast protein, Yhr087w (Rtc3), implicated in translational regulation in response to high-glucose stress [108]. SBDS domain I consists of a twisted five-stranded antiparallel β-sheet with four α-helices on one face. The central domain (residues D97–A170) comprises a three-helical, right-handed twisted bundle, while the C-terminal domain (residues H171–E250) has a ferredoxin-like fold, the most common RNA-binding motif in the translation system. The intriguing similarity between the C-terminal ferredoxin motif of SBDS and domain V of elongation factor 2 [101] initially suggested the hypothesis that SBDS might bind to the ribosome in the vicinity of the GTPase center or functionally interact with a translational GTPase in some aspect of ribosome biogenesis or translation [18].

The majority of the amino acids targeted by SDS-associated missense mutations map to SBDS domains I and II [101]. Using NMR spectroscopic analysis, disease-associated SBDS missense mutations were classified into two discrete categories [17]. Class A mutations affect the stability or fold of the SBDS protein, while class B mutations alter surface epitopes without perturbing SBDS protein stability or fold.

Function of SBDS in Large Ribosomal Subunit Maturation

Compelling genetic, biochemical, and structural data in diverse species have revealed a specific role for the SBDS protein at a late cytoplasmic step during maturation of the large ribosomal subunit. Specifically, SBDS cooperates with the GTPase elongation factor-like 1 (EFL1, yeast Efl1) to catalyze the release and nuclear recycling of the shuttling ribosome assembly factor eukaryotic initiation factor 6 (eIF6, yeast Tif6) [17–20]. Highly conserved in archaea and all eukaryotes, eIF6 was originally identified as an antiassociation factor that inhibits joining of the small and large ribosomal subunits [109–111]. Essential in yeast [112,113] and mammals [114], eIF6 is required in the nucleus for pre-rRNA processing and for export of the pre-60S subunit to the cytoplasm [115], where eIF6 must be removed to allow ribosomal subunit joining and the activation of translation [116]. Both in archaea and eukaryotes, eIF6 consists of 5 copies of a repeating α/β subdomain of about 45 residues with an internal fivefold axis of pseudosymmetry called a pentein [117]. By binding to the sarcin–ricin loop (SRL) and to ribosomal proteins uL14 and eL24 on the intersubunit face of the large ribosomal subunit, eIF6 acts as a physical barrier to ribosomal subunit joining [118,119]. Hence, eIF6 removal effectively licenses the productive engagement of late cytoplasmic 60S with 40S subunits to activate translation.

The critical insight into the function of SBDS came from a reverse genetics approach in baker's yeast (Saccharomyces cerevisiae) [18]. Multiple gain-of-function TIF6 alleles suppressed the pre-60S nuclear export defect and cytoplasmic mislocalization of Tif6 in cells deleted either for the SBDS ortholog Sdo1 or for Efl1, or indeed for both Sdo1 and Efl1. Strikingly, the 28 independent amino acids in the Tif6 protein targeted by mutations in cells deleted for Sdo1 or Efl1 delineated the surface of the Tif6 protein that was predicted [18] and later confirmed by crystallography [119] to interact with the 60S ribosomal subunit. These experiments strongly suggested that in yeast Tif6 was the direct target of both Sdo1 and Efl1.

EFL1 is a cytoplasmic GTPase that is homologous to the ribosomal translocase elongation factor 2 (EF-2) in eukaryotes and to prokaryotic EF-G [120]. Similar to EF-G and EF-2, EFL1 has an overall five-domain architecture that is distinguished from other ribosomal translocases by an insertion of variable length within domain II. In archaea, the conservation of SBDS and eIF6, but the absence of EFL1, suggests that the archaeal EF-G homolog may function both in biogenesis (eIF6 release) and during translation as an elongation factor.

Following on from the genetic experiments in yeast, biochemical experiments were required to validate the hypothesis that eIF6 was indeed the direct target of SBDS and EFL1. Indeed, recombinant human SBDS and EFL1 catalyzed the release of endogenous eIF6 from cytoplasmic pre-60S ribosomal particles purified from Sbds-deleted mouse liver [17]. Removal of eIF6 was strictly dependent on the presence of SBDS and on GTP binding and hydrolysis by EFL1. Although SBDS stimulated 60S-dependent GTP hydrolysis by EFL1, informative disease-associated SBDS missense variants revealed that the critical role of the SBDS protein was to couple EFL1–GTP hydrolysis to eIF6 release.

Parallel experiments in the ancient eukaryote Dictyostelium discoideum confirmed the generality of the proposal that SBDS functions in 60S subunit maturation [19]. As the Dictyostelium SBDS ortholog is essential, temperature-sensitive self-splicing intein technology was introduced to this organism for the first time to generate a conditional SBDS mutant. At the restrictive temperature, ribosomal subunit joining was dramatically impaired in SBDS mutant cells due to impaired release and recycling of eIF6. Remarkably, wild type human SBDS, but not SDS-associated disease variants, complemented the growth and ribosome assembly defects in mutant *Dictyostelium* cells. Endogenous *Dictyostelium* SBDS and EFL1 proteins directly interacted on the nascent 60S ribosomal subunit in vivo and together, human SBDS and EFL1 catalyzed GTP-dependent release of endogenous eIF6 from purified *Dictyostelium* pre-60S ribosomal subunits. This study also demonstrated for the first time that SDS patient-derived lymphoblasts harbored ribosomal subunit joining defects whose magnitude was inversely proportional to the level of residual SBDS protein expression. Taken together, these data from diverse species including *Dictyostelium*, yeast, and mice through to SDS patient-derived cells strongly support the hypothesis that SDS is a ribosomopathy that is caused by a specific defect in 60S ribosomal subunit maturation.

Alternative Models for eIF6 Release

An alternative model for the mechanism of eIF6 release posits that RACK1 (receptor for activated C kinase 1) recruits PKCβII to the ribosome where it phosphorylates eIF6 residue S235 to promote its eviction [116]. However, a critical role for PKCβII in eIF6 release seems incompatible with the viability and normal growth of PKCβ null mutant mice [121]. In yeast, the phylogenetically conserved eIF6 residues 1–224 are sufficient for eIF6 recycling in vivo [20] and ex vivo biochemical experiments using purified mammalian pre-60S subunits show that eIF6 release depends on SBDS and EFL1, but not phosphorylation of eIF6 residue S235 [17]. Furthermore, binding of RACK1 to the head of the 40S subunit [122] is difficult to rationalize in the proposed RACK1–PKCβII model. By contrast, there is compelling evidence that the yeast RACK1 homolog, Asc1, is involved in polypeptide quality control [123].

In yeast, it was proposed that phosphorylation of Tif6 residues S174 and S175 by the casein kinase 1 homolog Hrr25 regulates Tif6 nuclear recycling and cell viability [124]. However, these findings have been challenged by the observation that cells expressing a *TIF6–S174A, S175A* double mutant as the sole source of *TIF6* show no loss-of-fitness compared with wild type [20]. Further genetic clarification of the in vivo functional importance of eIF6 phosphorylation in yeast and mammalian cells is required.

Putative Extraribosomal Functions of the SBDS Protein

A diverse array of extraribosomal functions has been suggested for the SBDS protein, including roles in bone marrow niche-induced oncogenesis [125], chemotaxis [126], mitosis [127], cellular stress responses [128], Rac2-mediated monocyte migration [129], Fas ligand-induced apoptosis [130,131], the regulation of lysosome homeostasis [132], mitochondrial function [133], the generation of reactive oxygen species [134], and rRNA processing [135]. However, none of these proposed functions for SBDS has yet been shown to be direct. It seems more likely that the large number of putative extraribosomal functions proposed for SBDS reflect the multiple potential downstream consequences of the primary defect ribosome assembly.

Subcellular fractionation, immunoblotting, and immunofluorescence experiments in multiple species, including human cells, demonstrate that SBDS is predominantly a cytoplasmic protein, consistent with its essential role during late cytoplasmic 60S ribosomal subunit maturation [17–20,128]. Indeed, recent biochemical data indicate that (late) integration of ribosomal protein uL16 into cytoplasmic ribosomal subunits is a prerequisite for Sdo1 recruitment to late 60S ribosomal subunits in vivo [20]. The functional significance of the proposed nucleolar enrichment of SBDS [136] remains unclear, as no direct evidence of a role for SBDS in rRNA processing has been demonstrated.

Visualizing SBDS and EFL1 Bound to the Ribosome

Despite establishing the requirement for SBDS and EFL1 in eIF6 release through genetic and biochemical approaches, the precise mechanism by which SBDS and EFL1 cooperate on the 60S subunit to evict eIF6 still remained unknown. To begin to address this issue, the latest advances in cryoelectron microscopy (cryo-EM) were applied to determine the structures of three independent complexes containing eIF6–SBDS, eIF6–SBDS–EFL1, and SBDS–EFL1 bound to the 60S subunit [20].

Interestingly, the structures reveal a physical link on the ribosome between two proteins [SBDS and uL16 (RPL10)] that are mutated in inherited (SDS) and sporadic [pediatric T-cell acute lymphoblastic leukemia (T-ALL)] forms of leukemia. The N-terminal domain of the SBDS protein binds to the ribosomal P-site on the intersubunit face of the 60S subunit in close proximity to a conserved essential internal loop of ribosomal protein uL16 that is targeted by mutations in T-ALL [30]. The N-terminus of the SBDS protein (residues S2–V15) interacts with components of the peptidyl-transferase center (PTC), with the six most N-terminal residues extending into the upper part of the ribosomal peptide exit tunnel. SBDS domain III interacts with the SRL and P-stalk base in a similar manner to domain V of prokaryotic EF-G and yeast EF-2. Thus, SBDS effectively interrogates all the active sites of the 60S subunit (P-site, PTC, the entrance to the polypeptide exit tunnel, and the binding site for the translational GTPases) (Fig. 17.1).

EFL1 binds in the GTPase center on the 60S intersubunit face, predominantly interacting with SBDS and eIF6 (Fig. 17.2). EFL1 adopts two distinct conformations depending on the presence or absence of eIF6, undergoing an arc-like interdomain movement in which domains I–II, and IV pivot (by ~20 Å and ~10 Å,

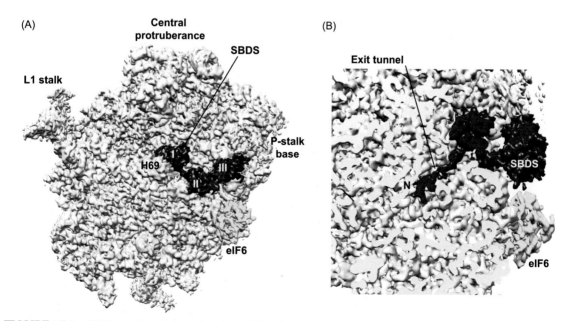

FIGURE 17.1 **SBDS proofreads the active sites of the ribosome.** (A) Crown view and (B) transverse section of the cryo-EM map of the 60S–eIF6–SBDS complex. The 60S ribosomal subunit is shown in *cyan*, eIF6 in *yellow*, and SBDS in *red*. SBDS domains I, II, and III are indicated.

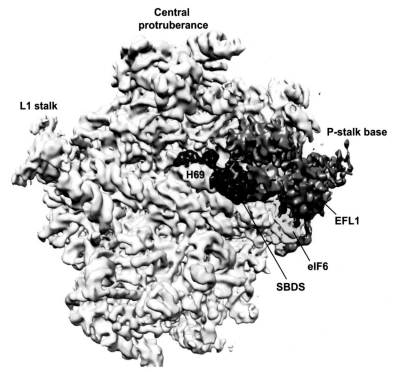

FIGURE 17.2   **Structure of the 60S–eIF6–SBDS–EFL1 complex.** The *60S* subunit is *cyan*, SBDS is *red*, EFL1 is *magenta*, and eIF6 is *yellow*. N, amino-terminus of SBDS.

respectively) around domains III and V. In the extended (accommodated) conformation, EFL1 competes with eIF6 for an overlapping binding site on the SRL, suggesting a mechanism for eIF6 displacement (Fig. 17.3). SBDS undergoes a marked conformational change on EFL1 binding. SBDS domain III rotates by 180 degrees away from the P-stalk base (closed state) toward helix 69 (open state), while SBDS domain I remains anchored in the P-site. The dramatic conformational change in SBDS observed by cryo-EM is consistent with the global dynamic motion of the SBDS protein off the ribosome revealed by solution NMR relaxation studies [17]. The identical fold of SBDS domain III and EFL1 domain V [101] permits both proteins to bind consecutively to a common site on the ribosome at base of the P-stalk base.

Interpreting Pathologic SBDS Missense Variants

Genetics and cryo-EM have provided unprecedented new insight into the consequences of disease-associated SBDS missense variants in the context of the ribosome [20]. Pathologic SBDS missense mutations disrupt rRNA contacts that are critical for 60S binding or for stabilizing key conformational states. For example, the highly conserved, disease-related residue K151 stabilizes the open SBDS conformation by interacting with the tip of helix 69 on the 60S subunit. The in vivo functional defect associated with disease-related variants that alter the length of the flexible linker (residues K90–R100) between SBDS domains I and II likely reflects the importance of this linker for conformational switching of SBDS on the ribosome.

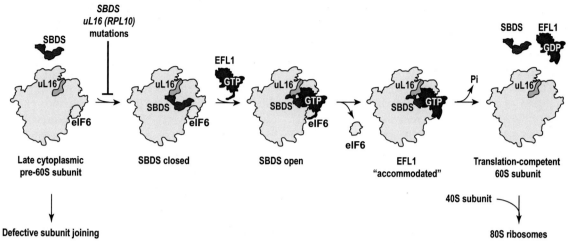

FIGURE 17.3   Proposed mechanism for eIF6 release by SBDS and the GTPase *EFL1*. SBDS binds in the "closed" conformation to a late cytoplasmic *pre-60S* particle that carries *uL16*. Binding of EFL1–GTP promotes the "open" SBDS conformation. EFL1–GTP adopts the "accommodated" state, which promotes eIF6 release by competing for an overlapping binding site on the ribosome. *GTP* hydrolysis destabilizes *EFL1* binding to the ribosome, promoting SBDS and *EFL1* dissociation and the formation of a translation-competent 60S subunit. Mutations in SBDS (in SDS) or *uL16* (in T-ALL) impair eIF6 release, causing defective ribosomal subunit joining.

Interpreting uL16 (RPL10) Mutations in Pediatric T-ALL

Disease mutations that target human ribosomal protein uL16 are found in around 5% of pediatric T-ALL [30]. Expression of disease-related *uL16* alleles in yeast and human lymphoid cells impaired the release and recycling of the ribosome ancillary factors Tif6 and Nmd3. In eukaryotes, nuclear export of the large ribosomal subunit from the nucleus depends on the shuttling adapter Nmd3, which provides the leucine-rich nuclear export signal that is recognized by the Crm1 export receptor to assemble a 60S subunit export complex with RanGTP [137–140]. The three recurrent T-ALL associated uL16 missense variants (R98S, R98C, and H123P) localize to a β-hairpin at the base of the conserved P-site loop that interacts closely with the N-terminal domain of SBDS [20]. As mutations in the uL16 P-loop reduce Sdo1 binding to the 60S subunit in vitro [141] and in vivo [20], T-ALL associated mutations likely impair eIF6 recycling indirectly as a consequence of reduced SBDS binding to the ribosome.

Proposed Mechanism of eIF6 Release by SBDS and the GTPase EFL1

Although there still remains much to be learnt, a conformational switching model has been proposed for the mechanism of eIF6 release by SBDS and EFL1 that is based on existing structural, biochemical, and genetic data [20] (Fig. 17.3). SBDS is likely recruited to a late cytoplasmic eIF6-bound pre-60S particle after the P-stalk base has been assembled and uL16 loaded. EFL1 binds to SBDS and eIF6 in the GTPase center of the 60S particle in a low-affinity GTP-bound state, inducing a 180-degree rotational displacement of SBDS domain III away from the P-stalk base (closed state) toward helix 69 (open state). In the open state, SBDS drives the conformational equilibrium of EFL1 toward a high-affinity conformation (the accommodated state) that competes more efficiently with eIF6 for an overlapping binding site on the SRL. In the accommodated state, EFL1–GTP hydrolysis is stimulated by the SRL, driving the conformational equilibrium of EFL1 toward a low-affinity ribosome-binding state, thereby promoting dissociation of both EFL1 and SBDS from the ribosome. Thus, an allosteric cascade, involving large-scale movements in SBDS and EFL1, links the conserved P-site loop of uL16 with eIF6.

The mechanism of eIF6 eviction that has emerged from recent cryo-EM data is reminiscent of bacterial ribosome recycling by RRF and EF-G [142–146], as previously hypothesized [17]. In the absence of EFL1, SBDS binds to the 60S subunit in an extended orientation that differs completely from any previously observed for tRNA, but which is similar to the orientation of RRF (also originally proposed as a tRNA mimic) bound to the bacterial 50S subunit.

In summary, the SBDS protein appears to have multiple roles during maturation of the large ribosomal subunit. Domain I masks and proofreads the structural integrity of the peptide exit tunnel and the PTC; domains I and II bind to the 60S subunit and mediate dynamic interdomain motion that is coupled to EFL1 conformational switching; in the closed state, SBDS domain III initially interrogates the translational GTPase-binding site at the P-stalk base and subsequently, in the open state, promotes EFL1 conformational switching.

SBDS and the Quality Control of Ribosome Assembly

Why evolve such a complex molecular mechanism for the final step in maturation of the nascent 60S subunit? As eIF6 acts as a physical barrier to ribosomal subunit joining, its release effectively gates the entry of mature 60S subunits into the actively translating pool. By proofreading the peptide exit tunnel, the P-site, the GTPase center, and the SRL, SBDS and EFL1 provide an elegant mechanism to couple eIF6 release to a final quality control assessment of the structural integrity of the key functional sites on the nascent 60S subunit. Taken together with the structures of the 60S–eIF6 [118,119,147], 60S–Nmd3 [148], and 60S–Arx1–Alb1–Rei1 [149,150]

complexes, the 60S–eIF6–SBDS–EFL1 structures support the hypothesis that cytoplasmic pre-60S assembly factors actively proofread the functional sites of the nascent ribosome. Interestingly, Rei1 inserts a C-terminal segment deeply into the polypeptide tunnel exit, forming specific contacts along most its entire length [151]. As SBDS inserts six-N-terminal residues into the proximal part of the polypeptide exit tunnel, Rei1 and SBDS together appear to proofread the entire length of the ribosomal tunnel during assembly. Interrogation of the SRL by EFL1 provides an elegant mechanism that couples proofreading of the structural integrity of the SRL to the activation of EFL1–GTP hydrolysis, proposed as the final step in the EFL1 catalytic cycle. Thus, SBDS and EFL1 emerge as critical gatekeepers that regulate the entry of functionally competent ribosomes into active translation.

How Do We Explain the SDS Phenotype?

How does the retention of eIF6 on nascent 60S ribosomal subunits promote the SDS phenotype? The reduced availability of free 60S subunits likely affects general protein translation when the requirements for protein synthesis are high at specific timepoints during development [152,153]. However, deregulated translation of specific mRNAs may also contribute to the SDS phenotype. Supporting this hypothesis, translational reinitiation of mRNAs encoding the indispensible regulators of granulocytic differentiation, C/EBPα and -β, is defective in SBDS-deficient cells [154]. In functional experiments, dysregulated translation of C/EBPα and -β mRNAs reduced MYC expression and cellular proliferation. An important next step will be to determine the functional importance of defective translational reinitiation in vivo in relevant SDS animal models.

The SDS phenotype may reflect relative stress sensitivities of particular cell types to perturbed ribosome assembly at particular timepoints during development. Although the p53 dependence of the phenotypic abnormalities associated with defective ribosome biogenesis is supported by genetic data in multiple animal models [155], the functional relevance of this pathway to SDS pathogenesis remains to be more fully clarified [152,156,157].

Dameshek's riddle encapsulates an intriguing paradox: how do defects in the translational machinery associated with a hypoproliferative state (bone marrow failure) promote the transition to a hyperproliferative disorder (leukemia) [158]? Cells harboring a defect in ribosome assembly appear to be under strong selective pressure to select for mutations that suppress the underlying growth defect and reverse the impaired proliferation [159]. A striking example of this is the recurrent del(20)(q) abnormality observed in the bone marrow of SDS patients. Microarray based comparative genomic hybridization of bone marrow cells revealed that the proximal del(20)(q) breakpoint always results in loss of the EIF6 gene, an event that is associated with a lower risk of transformation to MDS/AML [72,73]. A lower dose of eIF6 protein may promote the formation of more actively translating 80S ribosomes in SBDS-deficient hematopoietic stem- and progenitor cells, potentially reducing the selective pressure on cells to acquire additional, potentially oncogenic, suppressor mutations. On the other hand, although an *NMD3–Y379D* mutant suppressed the proliferation defect of the T-ALL-associated *uL16–R98S* mutant in yeast, translational fidelity defects, including −1 programmed frame shifting, were still clearly detected [160]. Indeed, the reduced stability of four mRNAs containing a −1 programmed frame-shifting signal that encode proteins implicated in telomere maintenance (Est1, Est2, Stn1, and Cdc13), was associated with a 25% reduction in yeast telomeric DNA repeat sequences in *uL16–R98S* mutant cells, supporting the presence of a telomere maintenance defect. This example illustrates that although cells may successfully bypass the proliferation defect caused by defective ribosome biogenesis, they

may still be at a long-term disadvantage due to engagement of defective ribosomes in active translation.

ADDITIONAL RIBOSOMOPATHIES

The growing number of human disorders associated with mutations in genes encoding ribosomal proteins or ribosome assembly factors, is summarized in Table 17.2. Similar to DBA, several of the inherited ribosomopathies are characterized by increased cancer risk. Furthermore, somatically acquired mutations in ribosomal proteins are also increasingly implicated in hematologic malignancy and cancer. Thus, haploinsufficiency of *RPS14* is associated with the 5q− syndrome subtype of MDS [29], while 10% of pediatric T-ALL harbor mutations in *RPL10*, *RPL5*, and *RPL22* [30]. Heterozygous deletions of *eL22* on chromosome 1p were identified in an additional 10% of T-ALL patients [161], while *RPL5* and *RPL22* are significantly mutated in 12 tumor types reported in the Cancer Genome Atlas [162]. Recently, 19.5% of relapsed chronic lymphocytic leukemia cases were found to harbor somatic missense mutations in *RPS15* [41]. Supporting an evolutionarily conserved role for ribosomal proteins as tumor suppressors, heterozygous loss-of-function mutations in multiple ribosomal proteins promote the development of peripheral nerve sheath tumors in zebrafish [163], while deletion of *RPS6* induced hematopoietic proliferation in *Drosophila* [164].

CONCLUSIONS

Elucidation of the molecular basis of the ribosomopathies continues to provide key insights into disease pathogenesis and unexpected new perspectives on the fundamental mechanisms of ribosome assembly and its quality control. How defects in the assembly of a ubiquitous, essential macromolecular machine, such as the ribosome, can cause human diseases with such striking tissue specificity remains a key unanswered question for the field. Recent advances in single-particle cryo-EM [165], together with new RNA–protein crosslinking methods [166–168] are likely to provide further significant mechanistic insights into basic aspects of ribosome biogenesis. In parallel, the development of robust animal models will be critical to illuminate disease pathophysiology. A critical challenge for the future is to translate new molecular insights into novel targeted therapeutics that will improve the quality of life and long-term outlook for patients affected by the ribosomopathies.

Acknowledgments

We apologize to colleagues whose work was not cited in this review due to space constraints. Work in the author's lab is supported by a Specialist Programme from Bloodwise (12048 to A.J.W.), the UK Medical Research Council (MRC) (MC_U105161083 to A.J.W. and U105115237 to R.R.K.), a Wellcome Trust strategic award to the Cambridge Institute for Medical Research (100140), a core support grant from the Wellcome Trust and MRC to the Wellcome Trust–Medical Research Council Cambridge Stem Cell Institute, Ted's Gang, and the Cambridge National Institute for Health Research Biomedical Research Centre.

References

[1] Nerurkar P, Altvater M, Gerhardy S, et al. Eukaryotic ribosome assembly and nuclear export. Int Rev Cell Mol Biol 2015;319:107–40.

[2] Lo KY, Li Z, Bussiere C, Bresson S, Marcotte EM, Johnson AW. Defining the pathway of cytoplasmic maturation of the 60S ribosomal subunit. Mol Cell 2010;39(2):196–208.

[3] Diamond L, Blackfan K. Hypoplastic anemia. Am J Dis Child 1938;56:4.

[4] Lipton JM, Atsidaftos E, Zyskind I, Vlachos A. Improving clinical care and elucidating the pathophysiology of Diamond Blackfan anemia: an update from the Diamond Blackfan Anemia Registry. Pediatr Blood Cancer 2006;46(5):558–64.

[5] Vlachos A, Rosenberg PS, Atsidaftos E, Alter BP, Lipton JM. Incidence of neoplasia in Diamond Blackfan anemia: a report from the Diamond Blackfan Anemia Registry. Blood 2012;119(16):3815–9.

[6] Vlachos A, Ball S, Dahl N, et al. Diagnosing and treating Diamond Blackfan anaemia: results of an international consensus conference. Br J Haematol 2008;142(6):859–76.

[7] Draptchinskaia N, Gustavsson P, Andersson B, et al. The gene encoding ribosomal protein S19 is mutated in Diamond–Blackfan anaemia. Nat Genet 1999;21(2):169–75.

[8] Willig TN, Draptchinskaia N, Dianzani I, et al. Mutations in ribosomal protein S19 gene and Diamond Blackfan anemia: wide variations in phenotypic expression. Blood 1999;94(12):4294–306.

[9] Ban N, Beckmann R, Cate JH, et al. A new system for naming ribosomal proteins. Curr Opin Struct Biol 2014;24:165–9.

[10] Boria I, Garelli E, Gazda HT, et al. The ribosomal basis of Diamond–Blackfan anemia: mutation and database update. Hum Mutat 2010;31(12):1269–79.

[11] Gazda HT, Sheen MR, Vlachos A, et al. Ribosomal protein L5 and L11 mutations are associated with cleft palate and abnormal thumbs in Diamond–Blackfan anemia patients. Am J Hum Genet 2008;83(6):769–80.

[12] Gripp KW, Curry C, Olney AH, et al. Diamond-Blackfan anemia with mandibulofacial dystostosis is heterogeneous, including the novel DBA genes TSR2 and RPS28. Am J Med Genet A 2014;164A(9):2240–9.

[13] Schutz S, Fischer U, Altvater M, et al. A RanGTP-independent mechanism allows ribosomal protein nuclear import for ribosome assembly. Elife 2014;3:e03473.

[14] Sankaran VG, Ghazvinian R, Do R, et al. Exome sequencing identifies GATA1 mutations resulting in Diamond–Blackfan anemia. J Clin Invest 2012;122(7):2439–43.

[15] Ludwig LS, Gazda HT, Eng JC, et al. Altered translation of GATA1 in Diamond–Blackfan anemia. Nat Med 2014;20(7):748–53.

[16] Boocock GR, Morrison JA, Popovic M, et al. Mutations in SBDS are associated with Shwachman–Diamond syndrome. Nat Genet 2003;33(1):97–101.

[17] Finch AJ, Hilcenko C, Basse N, et al. Uncoupling of GTP hydrolysis on eIF6 release on the ribosome causes Shwachman–Diamond syndrome. Genes Dev 2011;25(9):917–29.

[18] Menne TF, Goyenechea B, Sanchez-Puig N, et al. The Shwachman–Bodian–Diamond syndrome protein mediates translational activation of ribosomes in yeast. Nat Genet 2007;39(4):486–95.

[19] Wong CC, Traynor D, Basse N, Kay RR, Warren AJ. Defective ribosome assembly in Shwachman–Diamond syndrome. Blood 2011;118(16):4305–12.

[20] Weis F, Giudice E, Churcher M, et al. Mechanism of eIF6 release from the nascent 60S ribosomal subunit. Nat Struct Mol Biol 2015;22(11):914–9.

[21] Doherty L, Sheen MR, Vlachos A, et al. Ribosomal protein genes RPS10 and RPS26 are commonly mutated in Diamond–Blackfan anemia. Am J Hum Genet 2010;86(2):222–8.

[22] Cmejla R, Cmejlova J, Handrkova H, Petrak J, Pospisilova D. Ribosomal protein S17 gene (RPS17) is mutated in Diamond–Blackfan anemia. Hum Mutat 2007;28(12):1178–82.

[23] Gazda HT, Grabowska A, Merida-Long LB, et al. Ribosomal protein S24 gene is mutated in Diamond–Blackfan anemia. Am J Hum Genet 2006;79(6):1110–8.

[24] Farrar JE, Nater M, Caywood E, et al. Abnormalities of the large ribosomal subunit protein, Rpl35a, in Diamond–Blackfan anemia. Blood 2008;112(5):1582–92.

[25] Wang R, Yoshida K, Toki T, et al. Loss of function mutations in RPL27 and RPS27 identified by whole-exome sequencing in Diamond–Blackfan anaemia. Br J Haematol 2015;168(6):854–64.

[26] Mirabello L, Macari ER, Jessop L, et al. Whole-exome sequencing and functional studies identify RPS29 as a novel gene mutated in multicase Diamond–Blackfan anemia families. Blood 2014;124(1):24–32.

[27] Landowski M, O'Donohue MF, Buros C, et al. Novel deletion of RPL15 identified by array-comparative genomic hybridization in Diamond–Blackfan anemia. Hum Genet 2013;132(11):1265–74.

[28] Gazda HT, Preti M, Sheen MR, et al. Frameshift mutation in p53 regulator RPL26 is associated with multiple physical abnormalities and a specific pre-ribosomal RNA processing defect in Diamond–Blackfan anemia. Hum Mutat 2012;33(7):1037–44.

[29] Ebert BL, Pretz J, Bosco J, et al. Identification of RPS14 as a 5q– syndrome gene by RNA interference screen. Nature 2008;451(7176):335–9.

[30] De Keersmaecker K, Atak ZK, Li N, et al. Exome sequencing identifies mutation in CNOT3 and ribosomal genes RPL5 and RPL10 in T-cell acute lymphoblastic leukemia. Nat Genet 2013;45(2):186–90.

[31] Bolze A, Mahlaoui N, Byun M, et al. Ribosomal protein SA haploinsufficiency in humans with isolated congenital asplenia. Science 2013;340(6135):976–8.

[32] Marneros AG. BMS1 is mutated in aplasia cutis congenita. PLoS Genet 2013;9(6):e1003573.

[33] Armistead J, Khatkar S, Meyer B, et al. Mutation of a gene essential for ribosome biogenesis, EMG1, causes Bowen–Conradi syndrome. Am J Hum Genet 2009;84(6):728–39.

[34] Chagnon P, Michaud J, Mitchell G, et al. A missense mutation (R565W) in cirhin (FLJ14728) in North American Indian childhood cirrhosis. Am J Hum Genet 2002;71(6):1443–9.

[35] Nousbeck J, Spiegel R, Ishida-Yamamoto A, et al. Alopecia, neurological defects, and endocrinopathy syndrome caused by decreased expression of RBM28, a nucleolar protein associated with ribosome biogenesis. Am J Hum Genet 2008;82(5):1114–21.

[36] Brooks SS, Wall AL, Golzio C, et al. A novel ribosomopathy caused by dysfunction of RPL10 disrupts neurodevelopment and causes X-linked microcephaly in humans. Genetics 2014;198(2):723–33.

[37] Klauck SM, Felder B, Kolb-Kokocinski A, et al. Mutations in the ribosomal protein gene RPL10 suggest a novel modulating disease mechanism for autism. Mol Psychiatry 2006;11(12):1073–84.

[38] Ridanpaa M, van Eenennaam H, Pelin K, et al. Mutations in the RNA component of RNase MRP cause a pleiotropic human disease, cartilage–hair hypoplasia. Cell 2001;104(2):195–203.

[39] Heiss NS, Knight SW, Vulliamy TJ, et al. X-linked dyskeratosis congenita is caused by mutations in a highly conserved gene with putative nucleolar functions. Nat Genet 1998;19(1):32–8.

[40] Nieminen TT, O'Donohue MF, Wu Y, et al. Germline mutation of RPS20, encoding a ribosomal protein, causes predisposition to hereditary nonpolyposis colorectal carcinoma without DNA mismatch repair deficiency. Gastroenterology 2014;147(3):595–8.

[41] Ljungstrom V, Cortese D, Young E, et al. Whole-exome sequencing in relapsing chronic lymphocytic leukemia: clinical impact of recurrent RPS15 mutations. Blood 2016;127(8):1007–16.

[42] Zhang Y, Lu H. Signaling to p53: ribosomal proteins find their way. Cancer Cell 2009;16(5):369–77.

[43] McGowan KA, Li JZ, Park CY, et al. Ribosomal mutations cause p53-mediated dark skin and pleiotropic effects. Nat Genet 2008;40(8):963–70.

[44] Jaako P, Debnath S, Olsson K, et al. Disruption of the 5S RNP–Mdm2 interaction significantly improves the erythroid defect in a mouse model for Diamond–Blackfan anemia. Leukemia 2015;29(11):2221–9.

[45] Vlachos A, Muir E. How I treat Diamond–Blackfan anemia. Blood 2010;116(19):3715–23.

[46] Fagioli F, Quarello P, Zecca M, et al. Haematopoietic stem cell transplantation for Diamond Blackfan anaemia: a report from the Italian Association of Paediatric Haematology and Oncology Registry. Br J Haematol 2014;165(5):673–81.

[47] Pospisilova D, Cmejlova J, Hak J, Adam T, Cmejla R. Successful treatment of a Diamond–Blackfan anemia patient with amino acid leucine. Haematologica 2007;92(5):e66–7.

[48] Jaako P, Debnath S, Olsson K, Bryder D, Flygare J, Karlsson S. Dietary L-leucine improves the anemia in a mouse model for Diamond–Blackfan anemia. Blood 2012;120(11):2225–58.

[49] Payne EM, Virgilio M, Narla A, et al. L-leucine improves the anemia and developmental defects associated with Diamond–Blackfan anemia and del(5q) MDS by activating the mTOR pathway. Blood 2012;120(11): 2214–24.

[50] Ball SE, McGuckin CP, Jenkins G, Gordon-Smith EC. Diamond–Blackfan anaemia in the U.K.: analysis of 80 cases from a 20-year birth cohort. Br J Haematol 1996;94(4):645–53.

[51] Faivre L, Meerpohl J, Da Costa L, et al. High-risk pregnancies in Diamond–Blackfan anemia: a survey of 64 pregnancies from the French and German registries. Haematologica 2006;91(4):530–3.

[52] Nezelof C, Watchi M. Lipomatous congenital hypoplasia of the exocrine pancreas in children. Arch Fr Pediatr 1961;18:1135–72.

[53] Bodian M, Sheldon W, Lightwood R. Congenital hypoplasia of the exocrine pancreas. Acta Paediatr 1964;53:282–93.

[54] Shwachman H, Diamond LK, Oski FA, Khaw KT. The syndrome of pancreatic insufficiency and bone marrow dysfunction. J Pediatr 1964;65:645–63.

[55] Burke V, Colebatch JH, Anderson CM, Simons MJ. Association of pancreatic insufficiency and chronic neutropenia in childhood. Arch Dis Child 1967;42(222): 147–57.

[56] Pringle EM, Young WF, Haworth EM. Syndrome of pancreatic insufficiency, blood dyscrasia and metaphyseal dysplasia. Proc R Soc Med 1968;61(8):776–8.

[57] Goobie S, Popovic M, Morrison J, et al. Shwachman–Diamond syndrome with exocrine pancreatic dysfunction and bone marrow failure maps to the centromeric region of chromosome 7. Am J Hum Genet 2001;68(4):1048–54.

[58] Myers KC, Bolyard AA, Otto B, et al. Variable clinical presentation of Shwachman–Diamond syndrome: update from the North American Shwachman–Diamond Syndrome Registry. J Pediatr 2014;164(4):866–70.

[59] Hashmi SK, Allen C, Klaassen R, et al. Comparative analysis of Shwachman–Diamond syndrome to other inherited bone marrow failure syndromes and genotype-phenotype correlation. Clin Genet 2011;79(5): 448–58.

[60] Woloszynek JR, Rothbaum RJ, Rawls AS, et al. Mutations of the SBDS gene are present in most patients with Shwachman–Diamond syndrome. Blood 2004;104(12):3588–90.

[61] Aggett PJ, Cavanagh NP, Matthew DJ, Pincott JR, Sutcliffe J, Harries JT. Shwachman's syndrome. A review of 21 cases. Arch Dis Child 1980;55(5):331–47.

[62] Mack DR, Forstner GG, Wilschanski M, Freedman MH, Durie PR. Shwachman syndrome: exocrine pancreatic dysfunction and variable phenotypic expression. Gastroenterology 1996;111(6):1593–602.

[63] Smith OP, Hann IM, Chessells JM, Reeves BR, Milla P. Haematological abnormalities in Shwachman–Diamond syndrome. Br J Haematol 1996;94(2):279–84.

[64] Ginzberg H, Shin J, Ellis L, et al. Shwachman syndrome: phenotypic manifestations of sibling sets and isolated cases in a large patient cohort are similar. J Pediatr 1999;135(1):81–8.

[65] Cipolli M, D'Orazio C, Delmarco A, Marchesini C, Miano A, Mastella G. Shwachman's syndrome: pathomorphosis and long-term outcome. J Pediatr Gastroenterol Nutr 1999;29(3):265–72.

[66] Kuijpers TW, Alders M, Tool AT, Mellink C, Roos D, Hennekam RC. Hematologic abnormalities in Shwachman Diamond syndrome: lack of genotype-phenotype relationship. Blood 2005;106(1):356–61.

[67] Kuijpers TW, Nannenberg E, Alders M, Bredius R, Hennekam RC. Congenital aplastic anemia caused by mutations in the SBDS gene: a rare presentation of Shwachman–Diamond syndrome. Pediatrics 2004;114(3):e387–91.

[68] Donadieu J, Fenneteau O, Beaupain B, et al. Classification of and risk factors for hematologic complications in a French national cohort of 102 patients with Shwachman–Diamond syndrome. Haematologica 2012;97(9):1312–9.

[69] Stepanovic V, Wessels D, Goldman FD, Geiger J, Soll DR. The chemotaxis defect of Shwachman–Diamond Syndrome leukocytes. Cell Motil Cytoskeleton 2004;57(3):158–74.

[70] Dror Y, Ginzberg H, Dalal I, et al. Immune function in patients with Shwachman–Diamond syndrome. Br J Haematol 2001;114(3):712–7.

[71] Dror Y, Durie P, Ginzberg H, et al. Clonal evolution in marrows of patients with Shwachman–Diamond syndrome: a prospective 5-year follow-up study. Exp Hematol 2002;30(7):659–69.

[72] Pressato B, Valli R, Marletta C, et al. Deletion of chromosome 20 in bone marrow of patients with Shwachman–Diamond syndrome, loss of the EIF6 gene and benign prognosis. Br J Haematol 2012;157(4):503–5.

[73] Valli R, Pressato B, Marletta C, et al. Different loss of material in recurrent chromosome 20 interstitial deletions in Shwachman–Diamond syndrome and in myeloid neoplasms. Mol Cytogenet 2013;6(1):56.

[74] Singh SA, Vlachos A, Morgenstern NJ, et al. Breast cancer in a case of Shwachman Diamond syndrome. Pediatr Blood Cancer 2012;59(5):945–6.

[75] Sack JE, Kuchnir L, Demierre MF. Dermatofibrosarcoma protuberans arising in the context of Shwachman–Diamond syndrome. Pediatr Dermatol 2011;28(5): 568–9.

[76] Sharma A, Sadimin E, Drachtman R, Glod J. CNS lymphoma in a patient with Shwachman Diamond syndrome. Pediatr Blood Cancer 2014;61(3):564–6.

[77] Nakaya T, Kurata A, Hashimoto H, et al. Young-age-onset pancreatoduodenal carcinoma in Shwachman–Diamond syndrome. Pathol Int 2014;64(2):75–80.

[78] Dhanraj S, Manji A, Pinto D, et al. Molecular characteristics of a pancreatic adenocarcinoma associated with Shwachman–Diamond syndrome. Pediatr Blood Cancer 2013;60(5):754–60.

[79] Verbrugge J, Tulchinsky M. Lymphoma in a case of Shwachman–Diamond syndrome: PET/CT findings. Clin Nucl Med 2012;37(1):74–6.

[80] Stormon MO, Ip WF, Ellis L, Schibli S, Rommens JM, Durie PR. Evidence of a generalized defect of acinar cell function in Shwachman–Diamond syndrome. J Pediatr Gastroenterol Nutr 2010;51(1):8–13.

[81] Toiviainen-Salo S, Durie PR, Numminen K, et al. The natural history of Shwachman–Diamond syndrome-associated liver disease from childhood to adulthood. J Pediatr 2009;155(6):807–11.

[82] Makitie O, Ellis L, Durie PR, et al. Skeletal phenotype in patients with Shwachman–Diamond syndrome and mutations in SBDS. Clin Genet 2004;65(2):101–12.

[83] Nishimura G, Nakashima E, Hirose Y, et al. The Shwachman–Bodian–Diamond syndrome gene mutations cause a neonatal form of spondylometaphysial dysplasia (SMD) resembling SMD Sedaghatian type. J Med Genet 2007;44(4):e73.

[84] Keogh SJ, McKee S, Smithson SF, Grier D, Steward CG. Shwachman–Diamond syndrome: a complex case demonstrating the potential for misdiagnosis as asphyxiating thoracic dystrophy (Jeune syndrome). BMC Pediatr 2012;12:48.

[85] Toiviainen-Salo S, Mayranpaa MK, Durie PR, et al. Shwachman–Diamond syndrome is associated with low-turnover osteoporosis. Bone 2007;41(6):965–72.

[86] Kent A, Murphy GH, Milla P. Psychological characteristics of children with Shwachman syndrome. Arch Dis Child 1990;65(12):1349–52.

[87] Perobelli S, Nicolis E, Assael BM, Cipolli M. Further characterization of Shwachman–Diamond syndrome: psychological functioning and quality of life in adult and young patients. Am J Med Genet A 2012;158A(3):567–73.

[88] Kerr EN, Ellis L, Dupuis A, Rommens JM, Durie PR. The behavioral phenotype of school-age children with shwachman diamond syndrome indicates neurocognitive dysfunction with loss of Shwachman–Bodian–Diamond syndrome gene function. J Pediatr 2010;156(3):433–8.

[89] Toiviainen-Salo S, Makitie O, Mannerkoski M, Hamalainen J, Valanne L, Autti T. Shwachman–Diamond syndrome is associated with structural brain alterations on MRI. Am J Med Genet A 2008;146A(12):1558–64.

[90] Perobelli S, Alessandrini F, Zoccatelli G, et al. Diffuse alterations in grey and white matter associated with cognitive impairment in Shwachman–Diamond syndrome: evidence from a multimodal approach. Neuroimage Clin 2015;7:721–31.

[91] Hauet Q, Beaupain B, Micheau M, et al. Cardiomyopathies and congenital heart diseases in Shwachman–Diamond syndrome: a national survey. Int J Cardiol 2013;167(3):1048–50.

[92] Nivelon JL, Michiels R, Martres-Lassauniere MN, Passavy-Trouche AM, Petit A, Lorenzini JL. Myocardial fibrosis in Shwachman's syndrome: pathogenic discussion of cardiac complications. Pediatrie 1978;33(5):461–9.

[93] Savilahti E, Rapola J. Frequent myocardial lesions in Shwachman's syndrome. Eight fatal cases among 16 Finnish patients. Acta Paediatr Scand 1984;73(5):642–51.

[94] Ho W, Cheretakis C, Durie P, Kulkarni G, Glogauer M. Prevalence of oral diseases in Shwachman–Diamond syndrome. Spec Care Dentist 2007;27(2):52–8.

[95] Topa A, Tulinius M, Oldfors A, Hedberg-Oldfors C. Novel myopathy in a newborn with Shwachman–Diamond syndrome and review of neonatal presentation. Am J Med Genet A 2016;170(5):1155–64.

[96] Dror Y, Donadieu J, Koglmeier J, et al. Draft consensus guidelines for diagnosis and treatment of Shwachman–Diamond syndrome. Ann NY Acad Sci 2011;1242:40–55.

[97] Sauer M, Zeidler C, Meissner B, et al. Substitution of cyclophosphamide and busulfan by fludarabine, treosulfan and melphalan in a preparative regimen for children and adolescents with Shwachman–Diamond syndrome. Bone Marrow Transplant 2007;39(3):143–7.

[98] Bhatla D, Davies SM, Shenoy S, et al. Reduced-intensity conditioning is effective and safe for transplantation of patients with Shwachman–Diamond syndrome. Bone Marrow Transplant 2008;42(3):159–65.

[99] Cesaro S, Oneto R, Messina C, et al. Haematopoietic stem cell transplantation for Shwachman–Diamond disease: a study from the European Group for Blood and Marrow Transplantation. Br J Haematol 2005;131(2):231–6.

[100] Steele L, Rommens JM, Stockley T, Baskin B, Ray P. De Novo Mutations Causing Shwachman-Diamond Syndrome and a Founder Mutation in SBDS in the French Canadian Population. J Investig Genomics 2014;1(2):4.

[101] Shammas C, Menne TF, Hilcenko C, et al. Structural and mutational analysis of the SBDS protein family. Insight into the leukemia-associated Shwachman-Diamond Syndrome. J Biol Chem 2005;280(19):19221–9.

[102] Nicolis E, Bonizzato A, Assael BM, Cipolli M. Identification of novel mutations in patients with Shwachman–Diamond syndrome. Hum Mutat 2005;25(4):410.

[103] Erdos M, Alapi K, Balogh I, et al. Severe Shwachman-Diamond syndrome phenotype caused by compound heterozygous missense mutations in the SBDS gene. Exp Hematol 2006;34(11):1517–21.

[104] Zhang S, Shi M, Hui CC, Rommens JM. Loss of the mouse ortholog of the Shwachman–Diamond syndrome gene (Sbds) results in early embryonic lethality. Mol Cell Biol 2006;26(17):6656–63.

[105] Savchenko A, Krogan N, Cort JR, et al. The Shwachman–Bodian–Diamond syndrome protein family is involved in RNA metabolism. J Biol Chem 2005;280(19):19213–20.

[106] Ng CL, Waterman DG, Koonin EV, et al. Conformational flexibility and molecular interactions of an archaeal homologue of the Shwachman–Bodian–Diamond syndrome protein. BMC Struct Biol 2009;9:32.

[107] de Oliveira JF, Sforca ML, Blumenschein TM, et al. Structure, dynamics, and RNA interaction analysis of the human SBDS protein. J Mol Biol 2010;396(4): 1053–69.

[108] Gomar-Alba M, Jimenez-Marti E, del Olmo M. The Saccharomyces cerevisiae Hot1p regulated gene YHR087W (HGI1) has a role in translation upon high glucose concentration stress. BMC Mol Biol 2012;13:19.

[109] Valenzuela DM, Chaudhuri A, Maitra U. Eukaryotic ribosomal subunit anti-association activity of calf liver is contained in a single polypeptide chain protein of Mr = 25,500 (eukaryotic initiation factor 6). J Biol Chem 1982;257(13):7712–9.

[110] Russell DW, Spremulli LL. Identification of a wheat germ ribosome dissociation factor distinct from initiation factor eIF-3. J Biol Chem 1978;253(19):6647–9.

[111] Russell DW, Spremulli LL. Purification and characterization of a ribosome dissociation factor (eukaryotic initiation factor 6) from wheat germ. J Biol Chem 1979;254(18):8796–87800.

[112] Sanvito F, Piatti S, Villa A, et al. The beta4 integrin interactor p27(BBP/eIF6) is an essential nuclear matrix protein involved in 60S ribosomal subunit assembly. J Cell Biol 1999;144(5):823–37.

[113] Si K, Maitra U. The Saccharomyces cerevisiae homologue of mammalian translation initiation factor 6 does not function as a translation initiation factor. Mol Cell Biol 1999;19(2):1416–26.

[114] Gandin V, Miluzio A, Barbieri AM, et al. Eukaryotic initiation factor 6 is rate-limiting in translation, growth and transformation. Nature 2008;455(7213):684–8.

[115] Basu U, Si K, Warner JR, Maitra U. The Saccharomyces cerevisiae TIF6 gene encoding translation initiation factor 6 is required for 60S ribosomal subunit biogenesis. Mol Cell Biol 2001;21(5):1453–62.

[116] Ceci M, Gaviraghi C, Gorrini C, et al. Release of eIF6 (p27BBP) from the 60S subunit allows 80S ribosome assembly. Nature 2003;426(6966):579–84.

[117] Groft CM, Beckmann R, Sali A, Burley SK. Crystal structures of ribosome anti-association factor IF6. Nat Struct Biol 2000;7(12):1156–64.

[118] Gartmann M, Blau M, Armache JP, Mielke T, Topf M, Beckmann R. Mechanism of eIF6-mediated inhibition of ribosomal subunit joining. J Biol Chem 2010;285(20):14848–51.

[119] Klinge S, Voigts-Hoffmann F, Leibundgut M, Arpagaus S, Ban N. Crystal structure of the eukaryotic 60S ribosomal subunit in complex with initiation factor 6. Science 2012;334(6058):941–8.

[120] Senger B, Lafontaine DL, Graindorge JS, et al. The nucle(ol)ar Tif6p and Efl1p are required for a late cytoplasmic step of ribosome synthesis. Mol Cell 2001;8(6):1363–73.

[121] Bansode RR, Huang W, Roy SK, Mehta M, Mehta KD. Protein kinase C deficiency increases fatty acid oxidation and reduces fat storage. J Biol Chem 2008;283(1): 231–6.

[122] Khatter H, Myasnikov AG, Natchiar SK, Klaholz BP. Structure of the human 80S ribosome. Nature 2015;520(7549):640–5.

[123] Kuroha K, Akamatsu M, Dimitrova L, et al. Receptor for activated C kinase 1 stimulates nascent polypeptide-dependent translation arrest. EMBO Rep 2010;11(12):956–61.

[124] Basu U, Si K, Deng H, Maitra U. Phosphorylation of mammalian eukaryotic translation initiation factor 6 and its Saccharomyces cerevisiae homologue Tif6p: evidence that phosphorylation of Tif6p regulates its nucleocytoplasmic distribution and is required for yeast cell growth. Mol Cell Biol 2003;23(17): 6187–99.

[125] Raaijmakers MH, Mukherjee S, Guo S, et al. Bone progenitor dysfunction induces myelodysplasia and secondary leukaemia. Nature 2010;464(7290): 852–7.

[126] Wessels D, Srikantha T, Yi S, Kuhl S, Aravind L, Soll DR. The Shwachman–Bodian–Diamond syndrome gene encodes an RNA-binding protein that localizes to the pseudopod of Dictyostelium amoebae during chemotaxis. J Cell Sci 2006;119(Pt. 2):370–9.

[127] Austin KM, Gupta ML, Coats SA, et al. Mitotic spindle destabilization and genomic instability in Shwachman–Diamond syndrome. J Clin Invest 2008;118(4):1511–8.

[128] Ball HL, Zhang B, Riches JJ, et al. Shwachman–Bodian–Diamond syndrome is a multi-functional protein implicated in cellular stress responses. Hum Mol Genet 2009;18(19):3684–95.

[129] Leung R, Cuddy K, Wang Y, Rommens J, Glogauer M. Sbds is required for Rac2-mediated monocyte migration and signaling downstream of RANK during osteoclastogenesis. Blood 2011;117(6):2044–53.

[130] Rujkijyanont P, Watanabe K, Ambekar C, et al. SBDS-deficient cells undergo accelerated apoptosis through the Fas-pathway. Haematologica 2008;93(3):363–71.

[131] Watanabe K, Ambekar C, Wang H, Ciccolini A, Schimmer AD, Dror Y. SBDS-deficiency results in specific hypersensitivity to Fas stimulation and accumulation of Fas at the plasma membrane. Apoptosis 2009;14(1):77–89.

[132] Vitiello SP, Benedict JW, Padilla-Lopez S, Pearce DA. Interaction between Sdo1p and Btn1p in the *Saccharomyces cerevisiae* model for Batten disease. Hum Mol Genet 2010;19(5):931–42.

[133] Henson AL, Moore JBt, Alard P, Wattenberg MM, Liu JM, Ellis SR. Mitochondrial function is impaired in yeast and human cellular models of Shwachman Diamond syndrome. Biochem Biophys Res Commun 2013;437(1):29–34.

[134] Ambekar C, Das B, Yeger H, Dror Y. SBDS-deficiency results in deregulation of reactive oxygen species leading to increased cell death and decreased cell growth. Pediatr Blood Cancer 2010;55(6):1138–44.

[135] Miluzio A, Beugnet A, Volta V, Biffo S. Eukaryotic initiation factor 6 mediates a continuum between 60S ribosome biogenesis and translation. EMBO Rep 2009;10(5):459–65.

[136] Austin KM, Leary RJ, Shimamura A. The Shwachman–Diamond SBDS protein localizes to the nucleolus. Blood 2005;106(4):1253–8.

[137] Ho JH, Kallstrom G, Johnson AW. Nmd3p is a Crm1p-dependent adapter protein for nuclear export of the large ribosomal subunit. J Cell Biol 2000;151(5): 1057–66.

[138] Gadal O, Strauss D, Kessl J, Trumpower B, Tollervey D, Hurt E. Nuclear export of 60s ribosomal subunits depends on Xpo1p and requires a nuclear export sequence-containing factor, Nmd3p, that associates with the large subunit protein Rpl10p. Mol Cell Biol 2001;21(10):3405–15.

[139] Thomas F, Kutay U. Biogenesis and nuclear export of ribosomal subunits in higher eukaryotes depend on the CRM1 export pathway. J Cell Sci 2003;116(Pt. 12): 2409–19.

[140] Trotta CR, Lund E, Kahan L, Johnson AW, Dahlberg JE. Coordinated nuclear export of 60S ribosomal subunits and NMD3 in vertebrates. EMBO J 2003;22(11): 2841–51.

[141] Sulima SO, Gulay SP, Anjos M, et al. Eukaryotic RPL10 drives ribosomal rotation. Nucleic Acids Res 2014;42(3):2049–63.

[142] Lancaster L, Kiel MC, Kaji A, Noller HF. Orientation of ribosome recycling factor in the ribosome from directed hydroxyl radical probing. Cell 2002;111(1):129–40.

[143] Wilson DN, Schluenzen F, Harms JM, et al. X-ray crystallography study on ribosome recycling: the mechanism of binding and action of RRF on the 50S ribosomal subunit. EMBO J 2005;24(2):251–60.

[144] Gao N, Zavialov AV, Ehrenberg M, Frank J. Specific interaction between EF-G and RRF and its implication for GTP-dependent ribosome splitting into subunits. J Mol Biol 2007;374(5):1345–58.

[145] Weixlbaumer A, Petry S, Dunham CM, Selmer M, Kelley AC, Ramakrishnan V. Crystal structure of the ribosome recycling factor bound to the ribosome. Nat Struct Mol Biol 2007;14(8):733–7.

[146] Seshadri A, Singh NS, Varshney U. Recycling of the posttermination complexes of *Mycobacterium smegmatis* and *Escherichia coli* ribosomes using heterologous factors. J Mol Biol 2010;401(5):854–65.

[147] Greber BJ, Boehringer D, Godinic-Mikulcic V, et al. Cryo-EM structure of the archaeal 50S ribosomal subunit in complex with initiation factor 6 and implications for ribosome evolution. J Mol Biol 2012;418(3–4):145–60.

[148] Sengupta J, Bussiere C, Pallesen J, West M, Johnson AW, Frank J. Characterization of the nuclear export adaptor protein Nmd3 in association with the 60S ribosomal subunit. J Cell Biol 2010;189(7): 1079–86.

[149] Greber BJ, Boehringer D, Montellese C, Ban N. Cryo-EM structures of Arx1 and maturation factors Rei1 and Jjj1 bound to the 60S ribosomal subunit. Nat Struct Mol Biol 2012;19(12):1228–33.

[150] Bradatsch B, Leidig C, Granneman S, et al. Structure of the pre-60S ribosomal subunit with nuclear export factor Arx1 bound at the exit tunnel. Nat Struct Mol Biol 2012;19(12):1234–41.

[151] Greber BJ, Gerhardy S, Leitner A, et al. Insertion of the biogenesis factor Rei1 probes the ribosomal tunnel during 60S maturation. Cell 2016;164(1–2):91–102.

[152] Provost E, Wehner KA, Zhong X, et al. Ribosomal biogenesis genes play an essential and p53-independent role in zebrafish pancreas development. Development 2012;139(17):3232–41.

[153] Bielczyk-Maczynska E, Lam Hung L, Ferreira L, et al. The ribosome biogenesis protein nol9 is essential for definitive hematopoiesis and pancreas morphogenesis in zebrafish. PLoS Genet 2015;11(12):e1005677.

[154] In K, Zaini MA, Muller C, Warren AJ, von Lindern M, Calkhoven CF. Shwachman–Bodian–Diamond syndrome (SBDS) protein deficiency impairs translation re-initiation from C/EBPalpha and C/EBPbeta mRNAs. Nucleic Acids Res 2016;44(9):4134–46.

[155] Barlow JL, Drynan LF, Trim NL, Erber WN, Warren AJ, McKenzie AN. New insights into 5q– syndrome as a ribosomopathy. Cell Cycle 2010;9(24):4286–93.

[156] Zambetti NA, Bindels EM, Van Strien PM, et al. Deficiency of the ribosome biogenesis gene Sbds in hematopoietic stem and progenitor cells causes neutropenia

in mice by attenuating lineage progression in myelocytes. Haematologica 2015;100(10):1285–93.

[157] Tourlakis ME, Zhang S, Ball HL, et al. In vivo senescence in the Sbds-deficient murine pancreas: cell-type specific consequences of translation insufficiency. PLoS Genet 2015;11(6):e1005288.

[158] Dameshek W. Riddle: what do aplastic anemia, paroxysmal nocturnal hemoglobinuria (PNH) and "hypoplastic" leukemia have in common? Blood 1967;30(2):251–4.

[159] De Keersmaecker K, Sulima SO, Dinman JD. Ribosomopathies and the paradox of cellular hypo- to hyperproliferation. Blood 2015;125(9):1377–82.

[160] Sulima SO, Patchett S, Advani VM, De Keersmaecker K, Johnson AW, Dinman JD. Bypass of the pre-60S ribosomal quality control as a pathway to oncogenesis. Proc Natl Acad Sci USA 2014;111(15):5640–5.

[161] Rao S, Lee SY, Gutierrez A, et al. Inactivation of ribosomal protein L22 promotes transformation by induction of the stemness factor, Lin28B. Blood 2012;120(18):3764–73.

[162] Kandoth C, McLellan MD, Vandin F, et al. Mutational landscape and significance across 12 major cancer types. Nature 2013;502(7471):333–9.

[163] MacInnes AW, Amsterdam A, Whittaker CA, Hopkins N, Lees JA. Loss of p53 synthesis in zebrafish tumors with ribosomal protein gene mutations. Proc Natl Acad Sci USA 2008;105(30):10408–13.

[164] Watson KL, Konrad KD, Woods DF, Bryant PJ. Drosophila homolog of the human S6 ribosomal protein is required for tumor suppression in the hematopoietic system. Proc Natl Acad Sci USA 1992;89(23):11302–6.

[165] Bai XC, McMullan G, Scheres SH. How cryo-EM is revolutionizing structural biology. Trends Biochem Sci 2015;40(1):49–57.

[166] Granneman S, Petfalski E, Swiatkowska A, Tollervey D. Cracking pre-40S ribosomal subunit structure by systematic analyses of RNA–protein cross-linking. EMBO J 2010;29(12):2026–36.

[167] Kudla G, Granneman S, Hahn D, Beggs JD, Tollervey D. Cross-linking, ligation, and sequencing of hybrids reveals RNA–RNA interactions in yeast. Proc Natl Acad Sci USA 2011;108(24):10010–5.

[168] Spitale RC, Crisalli P, Flynn RA, Torre EA, Kool ET, Chang HY. RNA SHAPE analysis in living cells. Nat Chem Biol 2013;9(1):18–20.

Dyskeratosis Congenita

M. Ayas*, S.O. Ahmed**

*Pediatric Hematology–Oncology and Stem Cell Transplantation, King Faisal Specialist Hospital and Research Center, Riyadh, Saudi Arabia; **Adult Hematology and Bone Marrow Transplantation, Oncology Center, King Faisal Specialist Hospital and Research Center, Riyadh, Saudi Arabia

BACKGROUND

Dyskeratosis congenita (DC), also known eponymously as Zinsser–Cole–Engman syndrome after the three physicians who separately described the clinical features in the early 1900s, is a rare inherited multisystem disorder characterized by mucocutaneous features of reticulated skin pigmentation, oral leukoplakia, nail dystrophy, and progressive bone marrow failure that is often the main cause of death. The disorder can affect other organs including, but not limited to, the lungs, liver, genitourinary, and skeletal systems and is associated with a predisposition to the development of cancers.

PATHOBIOLOGY

The process of understanding and diagnosing DC has undergone a paradigmatic shift over the last decade, with recognition that abnormalities of telomere maintenance, resulting from one of a number of mutations, underlie the syndrome.

Telomeres are repetitive nucleotide sequences that cap the ends of chromosomes and shorten with successive cell division; these sequences are replenished by functioning telomerase, which is a ribonucleoprotein complex. Heiss et al. identified a gene coding for dyskerin, DKC1, as being mutated in X-linked DC in 1998 [1]. Dyskerin is a nucleolar protein and is also a component of the telomerase complex. The fact that DC was caused by mutations affecting telomerase function was further uncovered in a number of families with autosomal dominant (AD) inheritance of DC with mutations in the telomerase-RNA component, TERC [2]. Mutations affecting the protein component of the telomerase enzyme, TERT, were subsequently described in patients with acquired and inherited bone marrow failure syndromes [3,4]. In 2008, mutations in TINF2 affecting shelterin—a protein complex that protects the telomeres by controlling the synthesis of telomeric DNA—were found to be implicated in AD DC by more than one group [5,6]. Mutations affecting NOP10—a small nucleolar ribounucleoprotein component of the

Congenital and Acquired Bone Marrow Failure
http://dx.doi.org/10.1016/B978-0-12-804152-9.00018-X

225

TABLE 18.1 List of Genes That Have Been Identified in Patients and Families With Dyskeratosis Congenita (DC) [7]

Gene	Chromosomal location	Inheritance
DKC1	Xq28	X-linked recessive
TINF2	17p13.1	AD
TERC	3q26.2	AD
TERT	5p15.33	AD/AR
NOP10	15q14	AR
NHP2	5q35.3	AR
USB1	16q21	AR
TCAB1	17p13.1	AR
CTC1	17p13.1	AR
RTEL1	20q13.3	AR

AD, Autosomal dominant; AR, autosomal recessive.

telomerase complex—were described in a large consanguineous Saudi Arabian family leading to autosomal recessive (AR) DC; homozygous individuals had shortened telomeres and reduced TERC levels. Currently some 10 genes have been implicated with either X-linked-, AD-, or AR-inheritance patterns, with a tendency for X-linked forms to have the mildest and the AR forms, the most severe phenotype (Table 18.1) [7,8].

Mutations affecting telomerase and the accelerated telomere loss that ensues lead to the clinical manifestations seen in DC: (1) depletion of the stem cell compartment that leads to bone marrow failure, (2) accelerated cell senescence causing mucocutaneous abnormalities, and (3) the chromosomal instability that predisposes to other malignancies [9]. Cancer susceptibility in patients with DC is, therefore, part of the spectrum of the disease. In this respect, DC closely resembles Fanconi anemia (FA) in both the rates and types of neoplastic events [10]. The crude rate of malignancy in DC has been reported to be approximately 10% [11]. In one study, the actuarial risk of cancer was 40% by age 50 years, and more than 60% by age 68 years [12]. The most

frequent solid tumors are head- and neck squamous cell carcinomas and the incidence of acute myeloid leukemia is reported to be 10% between the ages of 30 and 40 years. The general cancer predisposition may well also vary according to the underlying mutation.

CLINICAL FEATURES

The clinical presentation of DC can be highly variable. While manifestations of DC often appear and are diagnosed in childhood, some cases may only be diagnosed when they present with a severe bone marrow failure syndrome; a malignancy, such as, acute leukemia; or pulmonary abnormalities. Diagnosis requires the presence of two out of four major features (bone marrow failure, nail dystrophy, abnormal skin pigmentation, and leukoplakia) and at least two other systemic features (Table 18.2). Bone marrow failure occurs in over 90% of patients by the age of 40 and is the main cause of mortality [13,14].

DKC1 mutations are associated with X-linked childhood DC and the affected males usually demonstrate the classical mucocutaneous features [1]. A less common but more severe variant of DC, the Hoyeraal–Hreidarsson syndrome (including mucocutaneous findings, severe marrow failure along with growth retardation, immune deficiency, and cerebellar hypoplasia) has also been associated with other mutations of the DKC1 and more rarely of the TINF2 gene [15]. Another severe manifestation, Revesz syndrome, in which classic DC features occur along with exudative retinopathy, is caused by mutations in TINF2 and is transmitted in an AD fashion [5,16]. AD DC is heterogeneous and mutations in *TERC*, *TERT*, and *TINF1* can cause telomeropathies in both children and adults [17–19].

The lung has been well documented as a target organ for telomeropathies. Idiopathic pulmonary fibrosis is the classic lung disease and accounts for 65% of telomere-mediated lung

TABLE 18.2 Clinical Features of DC Syndromes

Hematologic	**Bone marrow failure**
	Isolated cytopenias
	Macrocytosis
	Elevated fetal hemoglobin
Dermatologic (90%)	**Nail dystrophy**
	Abnormal skin pigmentation
	Thinning/early graying of hair
	Hyperhidrosis and hyperkera-
	tosis of soles and palms
	Adermatoglyphia
	Acrocyanosis
Oral	Leukoplakia
	Erythematous patches
	Brown/black patches
	Short tooth roots
	Enlarged dental pulp chambers
	Increased rate of dental decay
Ophthalmic	**Epiphora**
	Conjunctivitis
	Blepharitis
	Strabismus
	Cataracts
	Ectropion
	Entropion
	Sparse eyelashes
	Optic nerve atrophy
	Retinal vessel fragility and
	hemorrhages
	Exudative retinopathy
Neurologic	**Learning disability**
	Ataxia
	Cerebellar hypoplasia
	Microcephaly
	Deafness
Pulmonary disease	**Pulmonary fibrosis**
	Pulmonary vasculopathy
Muskuloskeltal	**Short stature**
	Osteoporosis
	Long bone fractures
	Avascular necrosis
	Scoliosis
	Mandibular hypoplasia
Gastrointestinal	Esophageal strictures
	Hepatomegaly
	Cirrhosis
	Peptic ulceration
	Enteropathy
Genitourinary	Hypogonadism
	Undescended testes
	Hypospadias
	Phimosis
	Urethral stenosis
	Horseshoe kidney
Cancer predisposi-	Myelodysplastic syndrome
tion	Acute myeloid leukemia
	Solid cancers (oropharynx,
	gastrointestinal)
Other	Intrauterine growth retardation

Features in bold occur at a frequency of ≥20% [13,14].

pathology. TERT, TERC, PARN, and RTEL1 have been closely associated with the development idiopathic pulmonary fibrosis [17,20]. Pulmonary fibrosis is also well described in patients with the classic DC. It is usually age related in both types (idiopathic and in patients with DC) and is considered a frequent cause of mortality in adults with DC. Smoking has been blamed as a likely cofactor in accelerating the development of the lung fibrosis at least in patients with the idiopathic form.

Liver involvement is also part of the telomere syndromes, more often in the form of cryptogenic cirrhosis that may be the only manifestation of the disease or be associated with pulmonary injury and marrow failure; TERC and TERT are the genes most frequently involved [10].

DIAGNOSIS

In making the diagnosis, the importance of a careful personal and family history and detailed clinical examination cannot be overstated. Investigations may reveal a pancytopenia that usually develops within the first two decades of life; less frequently, patients may have an isolated cytopenia. Often macrocytosis and fetal hemoglobin may be elevated. Imaging may reveal cerebellar hypoplasia, pulmonary fibrosis, renal and liver abnormalities, and osteopenia.

Telomere length can be measured in leukocyte subsets by flow cytometry and fluorescence in situ hybridization (flow-FISH) and compared to a population control [21]. Telomere length in patients with DC appears to a fundamental measure of clinical severity [22]. Interestingly, unlike patients with DC, patients with non-DC-inherited bone marrow failure syndromes usually tend to have the telomere lengths within the normal range, albeit shorter than in unaffected individuals [23].

With the advent of next-generation sequencing technologies, the underlying mutation can now be more rapidly identified in patients and families. Many labs can now offer a next-generation sequencing-based bone marrow failure gene panel that can be customized to test for a variable number of gene mutations implicated in inherited bone marrow failure syndromes; such panels have been shown to be efficient and accurate in detecting mutations in inherited bone marrow failure syndromes [24–26].

MANAGEMENT

Given the multisystem nature of the disease, patients may be managed in conjunction with, or with consultation of, pediatricians, hematologists, dermatologists, dentists, and oncologists. Patients and their families will need to be seen by a medical geneticist for confirmation of a diagnosis, identification of mutations, and appropriate counseling. Given the rarity of the disease, patients should ideally be managed by a multidisciplinary team experienced in the management of DC.

Management recommendations should include general measures, such as, dental hygiene, skin care, and avoidance of smoking and alcohol given the propensity for the lungs and liver to be affected by the disease and risk of cancer.

Due to the wide clinical spectrum of DC, therapy should be tailored according to the manifestations and the underlying genetic abnormality. Hematopoietic stem-cell transplantation (HCT) becomes inevitable when hematopoiesis is

exhausted and patients become pancytopenic or when there is evidence of clonal/leukemic evolution. Until that time, some therapeutic modalities have been shown, at least transiently, to ameliorate the hematopoietic defect in DC patients.

Androgens

Androgens have been used to treat inherited bone marrow failures even in the current era of HCT, particularly, when a suitable donor is not readily available or when the patient is not an HCT candidate. The use of androgen in DC per se is less common than in FA but there are data to suggest that these patients are particularly sensitive to androgen therapy even with lower doses. A recent observational cohort study evaluated hematologic response and side effects of androgen therapy in 16 patients with DC, with untreated DC patients serving as controls; 70% of treated-DC patients had a hematologic response with red blood cell and/or platelet transfusion independence, suggesting that androgen therapy for the pancytopenia in DC patients may be a viable option while awaiting HCT. Of note is that in this study, telomere length decline in androgen-treated patients was as expected by age [17,27].

How androgens exactly stimulate hematopoiesis is not well known but it has been suggested that they increase erythropoietin production, which in turn stimulates erythropoietic stem cells and, to a lesser extent, myeloid progenitor cells in the bone marrow [28]. More recent studies suggest that androgens, such as testosterone, do not increase erythropoietin levels but rather work at the level of the erythropoietin receptor to elicit a hematologic response [29]. Others have suggested that in view of the telomerase defect observed in DC patients, some of the effects of androgens may be achieved via upregulation of telomerase activity by slowing the rate of telomere attrition and enhancing cell regeneration. Interestingly this hypothesis may suggest that androgens could improve not only hematopoiesis but also other organ dysfunction [30].

Immune Suppressive Therapy

Immune suppressive therapy is probably not an appropriate option for patients with DC but there have been some reported observations in adult patients with TERT- and TERC mutations, which have shown blood count improvement after immune suppressive therapy suggesting that such genetic mutations may act as risk factors in stem-cell susceptibility to immune attacks [17,31].

Growth Factors

Although the clinical availability of recombinant hematopoietic growth factors was initially thought to be a breakthrough in the treatment of bone marrow failure syndromes, congenital neutropenia is probably the only marrow failure syndrome where long-term administration of granulocyte colony-stimulating factor was associated with a maintained increase in absolute neutrophil count and a reduction of severe bacterial infections. On the whole, results have generally been disappointing in other constitutional marrow failures. In patients with DC, one report described excellent neutrophil and hemoglobin responses with granulocyte-macrophage colony-stimulating factor and erythropoietin [32]. If the decision is made to treat with growth factors, it is strongly recommended to monitor patients for clonal aberrations prior to and during long-term treatment.

Hematopoietic Cell Transplantation

As is true in the majority of hereditary bone marrow failure syndrome, the only curative modality, (up until now) for bone marrow failure in patients with DC is allogeneic HCT. Overall, the outcome of HCT has been poor. A recent systematic review of literature of HCT in patients with DC revealed 5- and 10-year survival estimates of 57 and 23%, respectively. However, different factors can affect outcome [33].

Choice of a Suitable Conditioning Regimen

Despite many similarities, the heterogeneity of the bone marrow failure syndromes precludes the implementation of general rules and guidelines when deciding about the process of HCT; hence what is now considered as the standard conditioning for patients with FA, may not necessarily be the best for DC patients. In FA patients who suffer from abnormal DNA damage repair mechanisms, the sensitivity of FA cells to alkylating agents and radiation has been clearly established and thus, the use of lower doses of alkylating agents and radiation conditioning has been pivotal to avoid undue transplant toxicity [18]. In patients with DC, on the other hand, such evidence supporting the use of reduced intensity conditioning (RIC) transplants is lacking. However, as many series reported increased pulmonary and hepatic toxicities with conventional intensity regimens, it has been proposed that the use of RIC may induce milder toxicity and better survival, presumably, by reducing the need for extensive cellular repair due to higher doses of chemotherapy and limiting the consequences of defective telomere functioning in tissues of DC patients undergoing HCT. Given the rarity of the disease, the results of RIC in the literature are rather sketchy [34–37]. Nishio et al. reported a favorable outcome in three DC patients who had HCT from mismatched-related bone marrow (two) and unrelated BM donor (one); patients engrafted successfully and were alive at 10, 66, and 72 months after transplantation [35]. Dietz et al. reported six patients who underwent allogeneic HCT using nonmyeloablative conditioning regimen for DC patients, graft sources included (1) related PBSCs, (2) unrelated BM, and (3) unrelated double umbilical cord blood; and with a median follow-up of 26.5 months, four patients are alive, three of whom were recipients of unrelated grafts [37]. A report from the Eastern

Mediterranean Blood and Marrow Transplantation Registry described nine patients with a homogeneous donor source (all underwent matched related HCT); out of the eight who received RIC, seven were reported to be alive and hematologically stable at a median follow-up of 61 months (0.8–212 months); one patient was reported to have developed liver cirrhosis [19]. On a larger scale, Barbaro et al. reported in their recent review that RIC did not significantly improve survival but there was a trend toward less pulmonary disease in RIC recipients [33].

It remains to be said that although some advantages in survival and incidence of complications have been reported with RIC, most of the reports are of small series and with many complicating cofounding factors, such as, different donor sources and different comorbid conditions before HCT, which precludes making solid recommendations regarding the use of RIC in DC patients [30]. Thus, concerns over graft failure remain realistic. A recent study in patients with DC demonstrated that skeletal stem cells within the bone marrow stromal population (known as bone marrow-derived mesenchymal stem cells) are responsible for the creation of the hematopoietic environment and may contribute (along with the defect in hematopoietic stem cells) to the DC phenotype [38]. This may further emphasize the role of adequate marrow ablation in creating the necessary environment for the donor cells and may account for the primary and secondary graft failures noted in several reports [38]. Interestingly, in one study from Center for International Blood and Marrow Transplant Research (CIBMTR) on 34 patients with DC, cyclophosphamide was used alone at 200 mg/kg in 10 patients (8 matched related and 2 matched unrelated) and was found to be associated with the longest survival [39].

Choice of a Suitable Donor

MATCHED RELATED DONORS

One major drawback of interpreting data from the relatively larger series for patients with DC is the great heterogeneity and diversity of the conditioning regimens employed by different centers. In general, however, the reported outcomes of matched related donor HCT remain suboptimal. In the cohort of patients reported by the CIBMTR, 18 received grafts from related donors and only 8 of them were reported to be alive at the time of publication [39]. Others had marginally better survival; seven of the nine patients in the study by Ayas et al. were reported to be alive (median follow up of 61 months) [19]. Advancements in supportive care and early recognition of potential complications peculiar to DC patients may play a pivotal role in improving outcome; two studies showed that patients who underwent HCT more recently (after 2000) had a statistically improved survival suggesting that improved supportive care is an essential prognostic factor [33,39]. However, although the 5-year survival was better in one study, the long-term survival was similar and pulmonary disease was one of the leading causes of death.

Additionally, most of the existing reports on matched related HCT in patients with DC do not refer to the carrier status of the donor and thus, its impact on the eventual outcome is not well delineated. Should the carrier status be determined in related donors for DC patients undergoing related HCT? They probably should, as concerns are raised about the possibility of future hematologic deterioration even in carriers. One report indicates that one silent carrier has subsequently developed mild thrombocytopenia and a hypocellular marrow [22]. Therefore, it is generally already accepted that silent carriers should be avoided as donors and, as some families may have unknown gene mutations, it is strongly recommended that lymphocyte telomere length in the donor be determined before donation [40]. Furthermore, as the study by Balakumaran et al. suggests that the issue may not only be in the hematopoietic cells, it can be speculated that carriers are not suitable donors as their shorter telomeres of the SSCs/BMSCs may fail to produce the necessary healthy hematopoietic environment after transplant and hence, contribute to graft failure [38].

MATCHED UNRELATED DONORS

Thus far, the results of unrelated donor HCT have been disappointing. The largest review of outcome of unrelated donor HCT in DC patient is probably the CIBMTR report cited earlier; out of the 34 DC patients, 16 had unrelated donors and only 6 were alive at the time of publication. Of note, graft failure was the most common cause of early death; all but one of those who developed graft failure (four primary and six secondary) were recipients of mismatched related or unrelated donor transplants [39].

It is noteworthy here that even in unrelated donor HCT, the telomere length in the donor leukocyte might actually play a crucial role in transplant outcome. A recent report on 330 patients who received HCT for severe aplastic anemia (235 acquired and 95 hereditary) demonstrated that longer leukocyte telomere length of the donor was associated with significantly increased 5-year survival. None of the patient in the study had DC but these findings may be particularly pertinent in DC, where telomeropathy is the underlying pathology of the disease [25,41].

UNRELATED CORD BLOOD TRANSPLANTATION

It appears that cord blood transplantation is associated with dismal outcome for patients with DC. In a study by the EBMT group about outcome after cord blood transplantation in non-FA hereditary bone marrow failures, eight patients had DC, (two related and six unrelated), all patients died except one, who is alive at 126 months after an HLA matched sibling unrelated cord blood transplantation. In the group given unrelated grafts, three had been conditioned with conventional intensity conditioning with cyclophosphamide (120 mg/kg), busulphan (16 mg/kg), and antithymocyte globulin; three had RIC with fludarabine and reduced

doses of busulphan or cyclophosphamide. All six died of transplant-related toxicity [42].

HAPLOIDENTICAL TRANSPLANTATION

In the absence of a suitable matched related or unrelated donor, the use of haploidentical donors is gaining momentum in hereditary bone marrow failure syndromes; encouraging data have been reported in FA [43]. Recently one DC patient was reported to be alive and transfusion independent 26 months after T-cell depleted haploidentical HCT from a family donor [44].

Does Transplantation Accelerate Pulmonary Complications?

Mortality due to pulmonary complications has been reported as a major cause of late death in the literature of DC patients after HCT even for those who underwent RIC [33]; the impact of donor source on the incidence of such complications is not clear either. Many patients who died of pulmonary complications were reported not to have lung disease at transplantation and consequently many transplanters feel pulmonary complications may indeed be accelerated after HCT [39,41]. Hence, it is strongly recommended that preparative regimens should be carefully selected to minimize the risk of pulmonary toxicity. Thorough and regular pre- and posttransplant-pulmonary function assessment should be done on these patients for early detection of pulmonary insufficiency.

Conditioning and Cancer Risk

As outlined earlier, patients with DC have a significantly higher risk of developing solid and hematologic malignancies. Whereas in FA patients it has been suggested that better prevention, treatment of GVHD, and avoiding the use radiation-containing regimens might reduce the risk of late tumor development; such data are lacking however in patients with DC. It might be safer, however, to recommend that radiation be avoided in the conditioning of DC patients.

CONCLUSIONS

Long strides have been made in understanding the biology and in the diagnosis of DC. The classically described clinical features combined with the current state-of-the-art of genomic technologies should allow specialists to accurately diagnose the disease and the underlying mutations.

Our knowledge reservoir as related to definitive management, however, remains deficient especially regarding HCT, as many of the available studies have limited power to evaluate the impact of different key factors on outcome due to small sample sizes. More work is clearly needed; larger collaborative studies with well-defined clinical and laboratory characterization of the subtype of DC and a more uniform approach to the choice of conditioning regimens and donors are warranted to be able make solid recommendations. Meanwhile, however, stem-cell transplantation remains the ultimate and only curative modality for marrow failure/clonal evolution in this disorder. Conditioning regimens should be selected carefully and preferably of lesser intensity. Matched related or unrelated donors are both viable options but telomere-length measurements should be included in the assessment of any potential donor.

References

[1] Heiss NS, Knight SW, Vulliamy TJ, et al. X-linked dyskeratosis congenita is caused by mutations in a highly conserved gene with putative nucleolar functions. Nat Genet 1998;19(1):32–8.

[2] Vulliamy T, Marrone A, Goldman F, et al. The RNA component of telomerase is mutated in autosomal dominant dyskeratosis congenita. Nature 2001;413(6854):432–5.

[3] Vulliamy TJ, Walne A, Baskaradas A, Mason PJ, Marrone A, Dokal I. Mutations in the reverse transcriptase component of telomerase (TERT) in patients with bone marrow failure. Blood Cells Mol Dis 2005;34(3):257–63.

[4] Yamaguchi H, Calado RT, Ly H, et al. Mutations in TERT, the gene for telomerase reverse transcriptase, in aplastic anemia. New Engl J Med 2005;352(14):1413–24.

[5] Savage SA, Giri N, Baerlocher GM, Orr N, Lansdorp PM, Alter BP. TINF2, a component of the shelterin telomere protection complex, is mutated in dyskeratosis congenita. Am J Human Genet 2008;82(2):501–9.

[6] Walne AJ, Vulliamy T, Beswick R, Kirwan M, Dokal I. TINF2 mutations result in very short telomeres: analysis of a large cohort of patients with dyskeratosis congenita and related bone marrow failure syndromes. Blood 2008;112(9):3594–600.

[7] Dokal I, Vulliamy T, Mason P, Bessler M. Clinical utility gene card for: dyskeratosis congenita—update 2015. Eur J Human Genet 2015;23(4).

[8] Walne AJ, Vulliamy T, Marrone A, et al. Genetic heterogeneity in autosomal recessive dyskeratosis congenita with one subtype due to mutations in the telomerase-associated protein NOP10. Human Mol Genet 2007;16(13):1619–29.

[9] Calado RT, Cooper JN, Padilla-Nash HM, et al. Short telomeres result in chromosomal instability in hematopoietic cells and precede malignant evolution in human aplastic anemia. Leukemia 2012;26(4):700–7.

[10] Calado RT, Regal JA, Kleiner DE, et al. A spectrum of severe familial liver disorders associate with telomerase mutations. PloS One 2009;4(11):e7926.

[11] Vulliamy T, Dokal I. Dyskeratosis congenita. Semin Hematol 2006;43(3):157–66.

[12] Alter BP, Giri N, Savage SA, Rosenberg PS. Cancer in dyskeratosis congenita. Blood 2009;113(26):6549–57.

[13] Dokal I. Dyskeratosis congenita. Hematology Am Soc Hematol Educ Program 2011;2011:480–6.

[14] Dokal I. Dyskeratosis congenita in all its forms. Br J Haematol 2000;110(4):768–79.

[15] Hoyeraal HM, Lamvik J, Moe PJ. Congenital hypoplastic thrombocytopenia and cerebral malformations in two brothers. Acta Paediatr Scand 1970;59(2):185–91.

[16] Revesz T, Fletcher S, al-Gazali LI, DeBuse P. Bilateral retinopathy, aplastic anaemia, and central nervous system abnormalities: a new syndrome? J Med Genet 1992;29(9):673–5.

[17] Townsley DM, Dumitriu B, Young NS. Bone marrow failure and the telomeropathies. Blood 2014;124(18):2775–83.

[18] Gluckman E. Radiosensitivity in Fanconi anemia: application to the conditioning for bone marrow transplantation. Radiother Oncol 1990;18(Suppl. 1):88–93.

[19] Ayas M, Nassar A, Hamidieh AA, et al. Reduced intensity conditioning is effective for hematopoietic SCT in dyskeratosis congenita-related BM failure. Bone Marrow Transplant 2013;48(9):1168–72.

[20] Stuart BD, Choi J, Zaidi S, et al. Exome sequencing links mutations in PARN and RTEL1 with familial pulmonary fibrosis and telomere shortening. Nat Genet 2015;47(5):512–7.

[21] Alter BP, Baerlocher GM, Savage SA, et al. Very short telomere length by flow fluorescence in situ hybridization identifies patients with dyskeratosis congenita. Blood 2007;110(5):1439–47.

[22] Alter BP, Rosenberg PS, Giri N, Baerlocher GM, Lansdorp PM, Savage SA. Telomere length is associated with disease severity and declines with age in dyskeratosis congenita. Haematologica 2012;97(3):353–9.

[23] Alter BP, Giri N, Savage SA, Rosenberg PS. Telomere length in inherited bone marrow failure syndromes. Haematologica 2015;100(1):49–54.

[24] Ghemlas I, Li H, Zlateska B, et al. Improving diagnostic precision, care and syndrome definitions using comprehensive next-generation sequencing for the inherited bone marrow failure syndromes. J Med Genet 2015;52(9):575–84.

[25] Ayas M. Unrelated hematopoietic cell transplantation in aplastic anemia: there is more to a successful outcome than meets the eye. JAMA Oncol 2015;1(8):1164–5.

[26] Saudi Mendeliome Group. Comprehensive gene panels provide advantages over clinical exome sequencing for Mendelian diseases. Genome Biol 2015;16(1):1–14.

[27] Khincha PP, Wentzensen IM, Giri N, Alter BP, Savage SA. Response to androgen therapy in patients with dyskeratosis congenita. Br J Haematol 2014;165(3):349–57.

[28] Shahidi NT. A review of the chemistry, biological action, and clinical applications of anabolic-androgenic steroids. Clin Ther 2001;23(9):1355–90.

[29] Maggio M, Snyder PJ, Ceda GP, et al. Is the haematopoietic effect of testosterone mediated by erythropoietin? The results of a clinical trial in older men. Andrology 2013;1(1):24–8.

[30] de la Fuente J, Dokal I. Dyskeratosis congenita: advances in the understanding of the telomerase defect and the role of stem cell transplantation. Pediatr Transplant 2007;11(6):584–94.

[31] Comoli P, Basso S, Huanga GC. Intensive immunosuppression therapy for aplastic anemia associated with dyskeratosis congenita: report of a case. Int J Hematol 2005;82(1):35–7.

[32] Erduran E, Hacisalihoglu S, Ozoran Y. Treatment of dyskeratosis congenita with granulocyte-macrophage colony-stimulating factor and erythropoietin. J Pediatric Hematol Oncol 2003;25(4):333–5.

[33] Barbaro P, Vedi A. Survival after hematopoietic stem cell transplant in patients with dyskeratosis congenita: systematic review of the literature. Biol Blood Marrow Transplant 2016;22(7):1152–8.

[34] Kharfan-Dabaja MA, Otrock ZK, Bacigalupo A, Mahfouz RA, Geara F, Bazarbachi A. A reduced intensity conditioning regimen of fludarabine, cyclophosphamide, antithymocyte globulin, plus 2 Gy TBI facilitates successful hematopoietic cell engraftment in an adult with dyskeratosis congenita. Bone Marrow Transplant 2012;47(9):1254–5.

[35] Nishio N, Takahashi Y, Ohashi H, et al. Reduced-intensity conditioning for alternative donor hematopoietic stem cell transplantation in patients with dyskeratosis congenita. Pediatr Transplant 2011;15(2):161–6.

[36] Dror Y, Freedman MH, Leaker M, et al. Low-intensity hematopoietic stem-cell transplantation across human leucocyte antigen barriers in dyskeratosis congenita. Bone Marrow Transplant 2003;31(10):847–50.

[37] Dietz AC, Orchard PJ, Baker KS, et al. Disease-specific hematopoietic cell transplantation: nonmyeloablative conditioning regimen for dyskeratosis congenita. Bone Marrow Transplant 2011;46(1):98–104.

[38] Balakumaran A, Mishra PJ, Pawelczyk E, et al. Bone marrow skeletal stem/progenitor cell defects in dyskeratosis congenita and telomere biology disorders. Blood 2015;125(5):793–802.

[39] Gadalla SM, Sales-Bonfim C, Carreras J, et al. Outcomes of allogeneic hematopoietic cell transplantation in patients with dyskeratosis congenita. Biol Blood Marrow Transplant 2013;19(8):1238–43.

[40] Bessler M, Du HY, Gu B, Mason PJ. Dysfunctional telomeres and dyskeratosis congenita. Haematologica 2007;92(8):1009–12.

[41] Gadalla SM, Wang T, Haagenson M, et al. Association between donor leukocyte telomere length and survival after unrelated allogeneic hematopoietic cell transplantation for severe aplastic anemia. JAMA 2015;313(6):594–602.

[42] Bizzetto R, Bonfim C, Rocha V, et al. Outcomes after related and unrelated umbilical cord blood transplantation for hereditary bone marrow failure syndromes other than Fanconi anemia. Haematologica 2011;96(1):134–41.

[43] Thakar MS, Bonfim C, Sandmaier BM, et al. Cyclophosphamide-based in vivo T-cell depletion for HLA-haploidentical transplantation in Fanconi anemia. Pediatr Hematol Oncology 2012;29(6):568–78.

[44] Algeri M, Comoli P, Strocchio L, et al. Successful T-cell-depleted haploidentical hematopoietic stem cell transplantation in a child with dyskeratosis congenita after a fludarabine-based conditioning regimen. J Pediatr Hematol Oncology 2015;37(4):322–6.

Amegakaryocytic Thrombocytopenia

D. Meyran*, T. Leblanc*, S. Giraudier**, J.H. Dalle*

*Hematology–Immunology Pediatric Department, Robert–Debre Hospital, Paris, France; **Hematology Laboratory, Henri–Mondor Hospital, Paris–Est–Creteil University, Paris, France

INTRODUCTION

Congenital amegakaryocytic thrombocytopenia (CAMT) is a rare autosomal recessive bone marrow failure syndrome characterized by severe reduction or absence of megakaryocytes in bone marrow. The diagnosis must be considered in any young patient with a history of early bruising or bleedings with nonimmune thrombocytopenia and no specific physical abnormalities. The diagnosis is confirmed by blood sample and bone marrow examination, determination of platelet volume, dosage of thrombopoietin (TPO), and the identification of mutations in the c-MPL gene. Some patients with mild thrombocytopenia may be missed and only diagnosed later. However, all published patients were diagnosed and reported before adulthood.

Molecular Pathogenesis

CAMT is due to mutations on the c-MPL gene resulting in an abnormal TPO receptor without the binding capacity for TPO. This results in a failure of megakaryopoiesis even if the TPO serum level is markedly elevated [1–10].

As megakaryopoiesis mainly depends on TPO pathway, inadequate binding of TPO to the TPO receptor induces a severe reduction of megakaryocyte number into the bone marrow. However, megakaryopoiesis is not as dependent on TPO as erythropoiesis is on erythropoietin. In fact, megakaryopoiesis may be induced also by the synergistc action of IL-3, IL-6, and SCF. This is in keeping with the observation that mice inactivated for TPO or its receptor, MPL, may still have a residual megakaryopoiesis. This probably also explains why in some CAMT patients, a low and residual megakaryopoiesis persists. TPO receptor is a member of the cytokine receptor superfamily and its functions are not restricted to megakaryopoiesis but also involve other hematopoietic cells and tissues (CD34+, bone marrow, spleen, and fetal liver), thus ensuring a pivotal role in hematopoietic stem cell homeostasis, which explains why associated CAMT

Congenital and Acquired Bone Marrow Failure
http://dx.doi.org/10.1016/B978-0-12-804152-9.00019-1

mutations can lead to full blown marrow failure with pancytopenia [11–13].

More than 36 different mutations have been described since the first one reported by Kenji Hiara [14] in 1999. Some studies suggested a genotype/phenotype correlation in which null mutations result in severe early thrombocytopenia and rapid progression to pancytopenia (CAMT-I), while missense mutations have led to improvement in platelet counts early in childhood and delayed evolution to aplastic anemia (CAMT-II) [15]. These correlations were partly confirmed by other groups [7,8,16]. However, Savoia in 2007 in a study about five patients did not confirm this finding [9]. Actually, the total number of reported patients is too small to determine whether the risk of aplastic anemia or malignant evolution significantly differs between the two groups. At the molecular level, the vast majority of CAMT patients are due to homozygous or compound heterozygous mutations in MPL [2]. CAMT-associated truncating mutations in MPL principally lead to defective MPL presentation on the cell surface that is responsible for unresponsiveness to TPO [17,18].

Clinical Manifestation and Diagnosis

The clinical presentation is variable but usually hemorrhagic manifestations are severe, specifically when the first symptoms occur during neonatal period. During childbirth, these patients could develop central nervous system hemorrhage with life-threatening prognosis. Infants and children present with hemorrhagic symptoms even before learning to walk, with sometime unusual localizations. Buchanan hemorrhagic score (namely established for idiopathic thrombocytopenic purpura) may be variable from low grade with few purpuric lesions up to grade 5 with life-threatening hemorrhages. However, no hemorrhagic manifestations are pathognomonic of CAMT. The clinical presentation can often be mistaken for fetal/neonatal alloimmune thrombocytopenia or idiopathic thrombocytopenic purpura. Due to treatment failure a central thrombocytopenia is suspected. Diagnosis requires the association of thrombocytopenia, reduced or absent bone marrow megakaryocytes, highly elevated TPO serum level, and eventual progression to bone marrow failure. Identification of mutations in c-MPL gene confirms the diagnosis since now no other genes have been involved in this disease. However, some patients present with morphologic amegakaryocytosis without c-MPL gene mutation. They are considered as CAMT-III.

Some patients with mild thrombocytopenia may be missed during the first years of life and only diagnosed later. However, all published patients were diagnosed and reported before adulthood.

Thrombocytopenia progressed to pancytopenia at a median age of 39 months in 68% of reported cases up to 2011 [15]. The projected median age for aplastic anemia was 5 years. However progression to aplastic anemia is seen in over 90% of cases by the age of 13 years.

Depending on the authors, CAMT must be divided in two or three different subgroups based on the clinical course [6]: type I is characterized by persistent severe thrombocytopenia (below $50 \times 10^9 \text{ L}^{-1}$) and early evolution to central pancytopenia by 2 years of life; type II is characterized by a transient amelioration of platelet count up to $50 \times 10^9 \text{ L}^{-1}$ but by a subsequent fall, with mean onset of marrow failure at 5 years of age; and type III with ineffective megakaryopoiesis with no genetic defects in the c-MPL gene but is associated to amegakaryocytosis and high TPO serum level.

CAMT is also classified as preleukemic disease with a median age of 17 years to overt leukemia. However, reported cases of malignant evolution in CAMT are rare and the risk of malignant transformation seems to be very low as compared to other bone marrow failure syndromes. This may be due to the fact that hematopoietic stem cell transplantation (HSCT) actually prevents the risk of leukemic evolution.

To note, although malformations do not classically belong to the clinical picture of CAMT, a few patients have been reported with associated congenital abnormalities, such as septal defect of the heart, eye anomalies, and cerebral malformations including cortical dysplasia, lissencephaly, hypoplastic cerebellar vermis with communication 4th ventricle, and cisterna magna. Indeed the role for TPO in the brain development brain is currently discussed [6,15] and in these patients no MPL mutations were demonstrated.

When the clinical features are typical and amegakaryocytosis is demonstrated, there is nearly no differential diagnosis for CAMT. Thrombocytopenia with absent radii, which is systematically associated with bilateral and symmetric radial hypoplasia with normal thumbs, is easy to diagnose. Thrombocytopenia associated with radioulnar synostosis is a very rare condition [19]. Other inherited bone marrow failure (IBMF) syndrome, such as Fanconi anemia, dyskeratosis congenita, and Shwachman–Diamond syndrome may also be considered. Nevertheless, these syndromes usually do not associate a very severe thrombocytopenia to normal counts for other lineages; the distinction may be more difficult at the aplastic stage.

Wiskott–Aldrich syndrome with microthrombocytes [low mean platelet volume (MPV) as opposite to CAMT where MPV is normal] associated to immune deficiency and skin lesions may be discussed in male patients but megakaryocytic lineage is actually normal on bone marrow aspiration.

To note, very rare cases of alloimmune amegakaryocytosis have been published in infants [20] although this picture has not been reported in recent times. More recently, in neonates in whom a CAMT was discussed, 22q22.11 was demonstrated. Lastly, Noonan patients may present with amekaryocytic thrombocytopenia [21–23]

Laboratory Features

Thrombocytopenia is usually severe, the median platelets counts on the first diagnosis of thrombocytopenia is 17×10^9 L^{-1} with a 4–96 range [15]. Platelets are normal in size (normal MPV) and on blood films [24] and the expression of platelets surface glycoproteins is also normal [9]. There is no other cytopenia at diagnosis.

Bone marrow is usually normocellular with absent or severely reduced number of megakaryocytes; other lineages are normal at diagnosis. When still present, the megakaryocytes are often reported as small and immature. To note, as megakaryocyte may be normally present on the first bone marrow aspiration [15], this examination has to be repeated during evolution when diagnosis is highly suspected.

TPO levels are very high in CAMT patients and higher than those observed in other hypomegakaryocytosis whatever the etiology.

Some other diagnostic tools have been reported but are not currently used in clinical practice [15] In fact, diagnosis is based on isolated severe thrombopenia with normal platelet size and with low number (or absence) of megakaryocytes in the bone marrow. TPO level is measured to confirm the decreased megakaryopoiesis but the highest diagnostic levels is achieved by sequencing the entire coding region of c-MPL gene.

Treatment and Supportive Care

Treatment options in CAMT are highly restricted. Patients with CAMT do not respond to immunoglobulins, corticosteroids, splenectomy, or androgens and to date there is no reported benefit associated with the use of TPO receptor agonists in CAMT. Supportive care is limited to transfusion of blood products for long-life duration.

As for other chronic transfusion programs with a possible curative treatment by HSCT, the following rules have to be respected:

- Transfusions with compatible irradiated blood products.
- Limit platelet transfusions to clinical hemorrhagic symptoms. Do not perform transfusion only for low platelet count without clinical manifestations.

Due to the risk of malignant transformation, a bone marrow examination associated with karyotype monitoring has to be performed once a year.

If the patient develops pancytopenia, prophylactic therapy against infectious disease (bacterial and fungal) have to be discussed in order to avoid severe infections that may impair the result or even contraindicate subsequent HSCT.

HSCT

As for other IBMF syndromes, HSCT represents the only curative option.

A retrospective study was recently conducted by EBMT (Fahd et al., CIBMTR Tandem Meeting, 2014, Dallas, TX, USA); 63 patients (30 males/33 females) were identified. They received 80 HSCT from June 1987 to January 2013 (11 and 6 patients underwent a second or a third HSCT, respectively for primary graft rejection). The median age was 1.3 years (0.12–12.7 years) and 7 years (0.3–17.7 years) at CAMT diagnosis and HSCT, respectively. More than 80% of patients received myeloablative conditioning regimen for the first HSCT. None of conditioning regimens were TBI-based. For the first HSCT, 40% of patients underwent transplantation from sibling donor and 40% from unrelated donor. Seventeen percent of HSCT were performed from unrelated cord blood. The 5-year OS was 76.6% without any difference regarding the type of donor and stem-cell source. Transplant-related mortality (TRM) was about 13% at 3 years.

As other chronically transfused patients, CAMT subjects present rather a high risk of graft rejection as 11 and 6 patients (out of 63) received a second and a third HSCT, respectively. Seven out of 10 patients died after a first HSCT from TRM. OS was not significantly different according to donor type and stem-cell source although there was a trend toward higher OS and lower TRM for patients transplanted from sibling donor and with bone marrow as source of cells.

Overall CAMT patients have an increased risk for transplant-related toxicity due to prior infections, in case of long duration of aplasia, and due to iron overload consequent to red blood cell transfusions when patients develop pancytopenia.

As described by Alter et al., CAMT patients also have an increased risk of clonal evolution overtime [1]. Therefore HSCT should be performed as early as possible during childhood but probably not before 1 year of age in order to avoid excessive toxicity during infancy. As in other nonmalignant diseases, HSCT from alternative donor [defined as less than 10/10 allelic HLA-matched compatibility (HLA-A,-B,-C, -DQ, and -DRB1 typing at the allele level)] or unrelated cord blood has to be considered with caution because of higher expected graft failure and higher TRM than from sibling donor (defined as a family donor who shares the same parental haplotypes as the patient). These HSCT from alternative donor should be performed only at an experimental center and needs to be registered in the international registry dedicated to CAMT or to IBMF, as it appears it is very unlikely to have an international HSCT protocol for this very rare IBMF.

Source of Stem Cells

Bone marrow has been shown to be superior to peripheral blood as a stem-cell source in patients with acquired aplastic anemia undergoing matched sibling or unrelated transplants [25,26]. Bone marrow turned out to be superior to peripheral blood also in HSCT in FA patients

[27]. Also our EBMT study in CAMT patients showed a trend toward a better outcome when bone marrow was used as a source of cells. Based on these findings it can be inferred that in CAMT also peripheral blood stem cells should be avoided because it is likely to be associated with a higher risk of extensive chronic graft versus host disease (GvHD) compared with bone marrow. In general, these results lead us to recommend bone marrow as the preferred choice of stem cells for all children with IBMF.

Cell Dose

On a similar basis, it seems logical to recommend a cell dose greater than 3×10^8 total nucleated cells/kg of recipient body weight for bone marrow stem cells and greater than 3×10^7 total nucleated cells/kg of recipient body weight before freezing for cord stem cells [28].

Graft Versus Host Disease Prophylaxis

GvHD is one of the most significant complications in nonmalignant diseases and must be avoided. GvHD prophylaxis should include cyclosporin A plus methotrexate. Cyclosporin A should be gradually reduced until it can be discontinued, in the absence of chronic GvHD, 6–12 months after HSCT. These latter recommendations are derived from what has been already published in acquired severe aplastic anemia from sibling donors [29] and unrelated transplantation [30], using bone marrow as a stem-cell source. Pretransplantation serotherapy (either antithymoglobulin or alemtuzumab) is recommended for unrelated donor HSCT, but not for matched sibling donor HSCT because of the associated immunosuppression in the early months post-HSCT, which has the potential for higher TRM and mortality due to opportunistic infections.

References

[1] Alter BP. Diagnosis, genetics, and management of inherited bone marrow failure syndromes. Hematology Am Soc Hematol Educ Program 2007;29–39.

[2] Ballmaier M, Germeshausen M. Congenital amegakaryocytic thrombocytopenia: clinical presentation, diagnosis, and treatment. Semin Thromb Hemost 2011;37(6):673–81.

[3] Freedman MH, Estrov Z. Congenital amegakaryocytic thrombocytopenia: an intrinsic hematopoietic stem cell defect. Am J Pediatr Hematol Oncol 1990;12(2):225–30.

[4] Geddis AE. Congenital amegakaryocytic thrombocytopenia and thrombocytopenia with absent radii. Hematol Oncol Clin North Am 2009;23(2):321–31.

[5] Geddis AE. Congenital amegakaryocytic thrombocytopenia. Pediatr Blood Cancer 2011;57(2):199–203.

[6] King S, Germeshausen M, Strauss G, Welte K, Ballmaier M. Congenital amegakaryocytic thrombocytopenia: a retrospective clinical analysis of 20 patients. Br J Haematol 2005;131(5):636–44.

[7] Passos-Coelho JL, Sebastiao M, Gameiro P, Reichert A, Vieira L, Ferreira I, et al. Congenital amegakaryocytic thrombocytopenia—report of a new c-mpl gene missense mutation. Am J Hematol 2007;82(3):240–1.

[8] Rose MJ, Nicol KK, Skeens MA, Gross TG, Kerlin BA. Congenital amegakaryocytic thrombocytopenia: the diagnostic importance of combining pathology with molecular genetics. Pediatr Blood Cancer 2008;50(6):1263–5.

[9] Savoia A, Dufour C, Locatelli F, Noris P, Ambaglio C, Rosti V, et al. Congenital amegakaryocytic thrombocytopenia: clinical and biological consequences of five novel mutations. Haematologica 2007;92(9):1186–93.

[10] Stoddart MT, Connor P, Germeshausen M, Ballmaier M, Steward CG. Congenital amegakaryocytic thrombocytopenia (CAMT) presenting as severe pancytopenia in the first month of life. Pediatr Blood Cancer 2013;60(9):E94–6.

[11] Kobayashi M, Laver JH, Kato T, Miyazaki H, Ogawa M. Thrombopoietin supports proliferation of human primitive hematopoietic cells in synergy with steel factor and/or interleukin-3. Blood 1996;88(2):429–36.

[12] Kaushansky K. The molecular mechanisms that control thrombopoiesis. J Clin Invest 2005;115(12):3339–47.

[13] Yoshihara H, Arai F, Hosokawa K, Hagiwara T, Takubo K, Nakamura Y, et al. Thrombopoietin/MPL signaling regulates hematopoietic stem cell quiescence and interaction with the osteoblastic niche. Cell Stem Cell 2007;1(6):685–97.

[14] Ihara K, Ishii E, Eguchi M, Takada H, Suminoe A, Good RA, et al. Identification of mutations in the c-mpl gene in congenital amegakaryocytic thrombocytopenia. Proc Natl Acad Sci USA 1999;96(6):3132–6.

[15] Ballmaier M, Germeshausen M. Advances in the understanding of congenital amegakaryocytic thrombocytopenia. Br J Haematol 2009;146(1):3–16.

[16] Muraoka K, Ishii E, Ihara K, Imayoshi M, Miyazaki S, Hara T, et al. Successful bone marrow transplantation

in a patient with c-mpl-mutated congenital amegakaryocytic thrombocytopenia from a carrier donor. Pediatr Transplant 2005;9(1):101–3.

[17] Ballmaier M, Germeshausen M, Schulze H, Cherkaoui K, Lang S, Gaudig A, et al. c-Mpl mutations are the cause of congenital amegakaryocytic thrombocytopenia. Blood 2001;97(1):139–46.

[18] Varghese LN, Zhang JG, Young SN, Willson TA, Alexander WS, Nicola NA, et al. Functional characterization of c-Mpl ectodomain mutations that underlie congenital amegakaryocytic thrombocytopenia. Growth Factors 2014;32(1):18–26.

[19] Thompson AA, Nguyen LT. Amegakaryocytic thrombocytopenia and radio-ulnar synostosis are associated with HOXA11 mutation. Nature Genet 2000;26(4): 397–8.

[20] Bizzaro N, Dianese G. Neonatal alloimmune amegakaryocytosis. Case report. Vox Sang 1988;54(2):112–4.

[21] Christensen RD, Wiedmeier SE, Yaish HM. A neonate with congenital amegakaryocytic thrombocytopenia associated with a chromosomal microdeletion at 21q22.11 including the gene RUNX1. J Perinatol 2013;33(3):242–4.

[22] Christensen RD, Yaish HM, Leon EL, Sola-Visner MC, Agrawal PB. A de novo T73I mutation in PTPN11 in a neonate with severe and prolonged congenital thrombocytopenia and Noonan syndrome. Neonatology 2013;104(1):1–5.

[23] Evans DG, Lonsdale RN, Patton MA. Cutaneous lymphangioma and amegakaryocytic thrombocytopenia in Noonan syndrome. Clin Genet 1991;39(3):228–32.

[24] Drachman JG. Inherited thrombocytopenia: when a low platelet count does not mean ITP. Blood 2004;103(2):390–8.

[25] Eapen M, Le Rademacher J, Antin JH, Champlin RE, Carreras J, Fay J, et al. Effect of stem cell source on outcomes after unrelated donor transplantation in severe aplastic anemia. Blood 2011;118(9): 2618–21.

[26] Bacigalupo A, Socie G, Schrezenmeier H, Tichelli A, Locasciulli A, Fuehrer M, et al. Bone marrow versus peripheral blood as the stem cell source for sibling transplants in acquired aplastic anemia: survival advantage for bone marrow in all age groups. Haematologica 2012;97(8):1142–8.

[27] Peffault de Latour R, Porcher R, Dalle JH, Aljurf M, Korthof ET, Svahn J, et al. Allogeneic hematopoietic stem cell transplantation in Fanconi anemia: the European Group for Blood and Marrow Transplantation experience. Blood 2013;122(26):4279–86.

[28] Gluckman E, Ruggeri A, Volt F, Cunha R, Boudjedir K, Rocha V. Milestones in umbilical cord blood transplantation. Br J Haematol 2011;154(4):441–7.

[29] Locatelli F, Bruno B, Zecca M, Van-Lint MT, McCann S, Arcese W, et al. Cyclosporin A and short-term methotrexate versus cyclosporin A as graft versus host disease prophylaxis in patients with severe aplastic anemia given allogeneic bone marrow transplantation from an HLA-identical sibling: results of a GITMO/EBMT randomized trial. Blood 2000;96(5): 1690–7.

[30] Bacigalupo A, Socie G, Lanino E, Prete A, Locatelli F, Locasciulli A, et al. Fludarabine, cyclophosphamide, antithymocyte globulin, with or without low dose total body irradiation, for alternative donor transplants, in acquired severe aplastic anemia: a retrospective study from the EBMT-SAA Working Party. Haematologica 2010;95(6):976–82.

Severe Congenital Neutropenias and Other Rare Inherited Disorders With Marrow Failure

F. Fioredda*, P. Farruggia**, M. Miano*, C. Dufour*

*Hematology Unit, G. Gaslini Children's Hospital; Unità di Ematologia Istituto Giannina Gaslini, Genova, Italy; **Pediatric Hematology and Oncology Unit, A.R.N.A.S. Ospedale Civico, Palermo, Italy

SEVERE CONGENITAL NEUTROPENIA

Definition, Epidemiology, and Genetics

Neutropenia is a persistent decrease in circulating mature neutrophils, defined as severe if the absolute neutrophil count is below $0.5 \times 10^9 \, L^{-1}$ [1]. This condition may be acquired or inherited. Severe congenital neutropenia (SCN) is a heterogeneous group of disorders whose incidence depends on the broadness of the definition. Incidence for cases in which the neutropenia is the main element of the disease is reported as one per million births [2]. However, there are patients in which neutropenia is associated with extrahematologic abnormalities as in the case of glycogen storage disease 1b (GSD1b), Shwachman–Diamond syndrome, WHIM syndrome, and Barth disease. Incidence

of these neutropenias has been assessed at 10–15 new cases/million births [2]. Whatever the association, the triad: (1) profound neutropenia ($<0.2 \times 10^9 \, mm^{-3}$) occurring during the first weeks of life, (2) maturation arrest of granulopoiesis at the promyelo/myelocyte stage, and (3) high risk of death due to bacterial infections characterizes many forms of SCN [3]. SCN was first described as a granulocytosis by Dr. Rolf Kostmann in 1956 in his doctoral thesis, where he described a recessive disease in a North Swedish family, which was fatal within the first year of life because of infections of various kinds. The disease was characterized by a block at the promyelocyte stage in the bone marrow (BM) that impaired neutrophil maturation and exit of mature neutrophils from the marrow into the peripheral blood circulation [4]. Later on, this neutropenia has been attributed to mutations of the HCLS1-associated protein X1

TABLE 20.1   Severe Congenital Neutropenia (SCN) According to the OMIM Classification; More Than One-Third of SCN Patients Are Orphan of Any Mutation

Gene	Phenotype	Mode of inheritance	Location
ELANE	Severe congenital neutropenia, SCN1	Autosomal dominant	19p13.3
GFI1	Severe congenital neutropenia, SCN2	Autosomal dominant	1p22.1
HAX1	Severe congenital neutropenia, SCN 3	Autosomal recessive	1q21.3
G6PC3	Severe congenital neutropenia, SCN 4	Autosomal recessive	17q21.31
VPS45A	Severe congenital neutropenia, SCN 5	Autosomal recessive	1q21.2
JAGN1	Severe congenital neutropenia, SCN 6	Autosomal recessive	3p25.3
WAS	Severe congenital neutropenia, X-linked	X-linked	Xp11.23

ELANE, Elastase, neutrophil expressed; GFI1, growth factor independent 1, G6PC3, glucose-6-phosphastase subunit 3; HAX1, HCLS1-associated protein X1; JAGN1, jagunal homolog 1; VPS45A, vacuolar protein sorting 45A; WAS, Wiskott–Aldrich.

(HAX1) gene that are transmitted in an autosomal recessive fashion [5]. More frequently, SCN is an autosomal dominant or sporadic disorder caused by mutations in elastase, neutrophil expressed (ELANE), a gene encoding the elastase protein of neutrophils [6] that accounts for more than half of the SCN population in Europe and North America. Mutations in other genes, such as, growth factor independent (GFI)1, Wiskott–Aldrich (WAS) protein, glucose-6-phosphastase subunit 3 (G6PC3), and vacuolar protein sorting 45 (VPS45) are also involved although less frequently (Table 20.1) [7–11].

Bialleic mutation of the extracellular portion of granulocyte colony-stimulating factor (G-CSF) receptor (G-CSFR) makes it insensitive to the effect of endogenous G-CSF and may also cause SCN [12]. Recently, mutations in new genes, including JAG1 and TCIRG1, were reported as causative of SCN but nonetheless, more than one-third of SCN patients are orphan of any known mutation [13,14].

Physiopathologic Mechanism

Although the genetic diversity of SCN suggests that various mechanisms may be involved, the common final pathway is the apoptosis of the granulocytic precursor. The intracellular stress leading to apoptosis may be due to functional disturbances of different cellular compartments including endoplasmic reticulum (ER) or endosomal system, mitochondria, ribosomes, and glucose homeostasis [7–12]. In the case of neutropenia due to ELANE mutations, the genetic lesion causes the accumulation of misfolded proteins in the ER with consequent cellular stress and apoptosis of promyelocytes [15]. Lack of expression of HAX1 protein in hematopoietic and nonhematopoietic cells, as occurs in neutropenias due to mutations in this gene, destabilizes the mitochondrial membrane potential thus leading to the increased apoptosis [16]. A small subgroup of SCN patients have mutations in the transcriptional repressor oncoprotein GFI1 that plays a critical role in hematopoiesis. WAS protein's function is involved in regulation of the cytoskeleton, while VPS45 regulates endosomal trafficking [17,18]. G6PC3 mutations are associated with a disturbed glucose homeostasis and an increased ER stress causing neutrophil apoptosis [9]. In the rare cases in which neutropenia is due to TCIRG1 mutations, the damage originates from a disturbed marrow "microenvironment" rather than from a myeloid progenitor [13].

Diagnosis and Clinical Phenotype

The most common presentation of the disease is a severe infection occurring in the neonatal period, typically an omphalytis in the first months of life, associated with very low neutrophil count. Sometimes other abrnormalities can be present: facial dismorphysms, failure to thrive, cardiac- and renal malformation, or metabolic disturbances.

In most cases, and particularly, in those due to ELANE, HAX 1, G6PC3, WAS, GFI1, and JAGN1 mutations, the BM aspiration shows a complete absence or a significant reduction of the elements downstreamed to promyelocytes with a virtual absence of mature neutrophils [3]. In more rare cases, as those due to biallelic mutations of CSFR3, VPS 45, and TCGR1 genes, the maturation block at promyleocyte is not present [7–14,18].

The advent of the new genetic tools (next generation sequencing and whole exome sequencing) may facilitate diagnosis of genetic forms of SCN and may also help to identify the genetic origin of those forms that are currently yet gene orphan.

Deep neutropenia and related infectious risk are the main elements affecting survival and quality of life of patients. Even if on G-CSF, infections can be severe. They may include fulminant sepsis and/or localize in unusual sites, such as, umbilical cord at birth, skin, and soft-tissue of face and perineal region. Of note, infections occur more frequently and tend to be more severe as compared to acquired autoimmune neutropenias [19] mainly because in genetic SCN also, neutrophil functions, such as, killing and chemotaxis are impaired. This explains why complete recovery from infections is sometimes difficult, even when neutrophil count is normalized by growth factors [20].

Natural History

Before the 1990s, when treatment with G-CSF was unavailable, life expectancy of SCN patients was poor with 42% of subjects dying at a median of 2 years, mainly due to sepsis or pneumonia [21].

After the introduction of G-CSF in the treatment of SCN, the natural history of the disease radically changed although infectious deaths are still a major issue even on G-CSF [22–24]. Indeed cumulative risk of death for sepsis has been estimated around 15% after 20 years from the start of G-SCF by the Severe Chronic Neutropenia International Registry (SCNIR). The experience of the French Severe Chronic Neutropenia Registry (SNFR) reports 6 septic deaths in 101 SCN cases, but not all these patients were receiving G-CSF at the time of infection [25,26]. Lifetime G-CSF treatment effectively prevents the risk of lethal infections. Lack of adherence and/ or scarce compliance to treatment may be the reason of some deaths due to infections even in well-informed patients/patients' family [27]. To overcome poor compliance to daily administrations, the use of PEGylated G-CSF (pegfilgrastim) could be an option to pursue but within clinical trials, as this drug is not registered for SCN patients. Data on pegfilgratim in SCN in pediatric age is limited and rather controversial. In some patients, not eligible for classic G-CSF treatment, a better control of infections was achieved [28], although in other subjects an increase of pulmonary side effects was observed [29].

Another characteristic of SCN is the intrinsic tendency to clonal evolution to myelodysplasia/acute leukemia (MDS/AL) due to the instability of the precursors. According to the SCNIR and the SNFR the cumulative incidence of MDS/AL is 22 and 10.8%, respectively, after 15 years from the beginning of G-CSF therapy. The risk of transformation increases with median G-CSF doses higher than 8 µg/kg/day [25,26]. This is in line with the Swedish cohort analysis in which the transforming event occurred in 31% of the population after 15 years of G-CSF exposure at median doses greater than the standard 5 µg/kg/day [30].

Although over the last years, progress in the understanding of the clonal evolution has been achieved, the full process is not completely elucidated. Transformation indeed is a multistep pathway in which a number of factors play a role. In principle, some "constitutive mutations" are more prone than others to favor emergence of leukemic clones as it in the case of G214R or C151Y mutations of the ELANE gene [31].

Reduction of function of differentiating factors (LEF1 and C/EBPα), proven to occur in the "sick" hematopoietic progenitor cell, may favor the emergence of leukemic clones. Mutations in the external part of the G-CSFR could work as a precancer event, causing activation of STAT5, which in turn is responsible for reactive oxygen species production with consequent intracellular DNA damage [32].

A recent study showed that G-CSF induces proliferation with reduced myeloid differentiation of hematopoietic CD34+ cells coexpressing mutations of RUNX1 and CSF3R. On sequential analysis, RUNX1 mutations occurred as late events in leukemic transformation. This suggests that in the leukemogenic process of SCN, CSF3R mutation precedes mutations of the downstream hematopoietic transcription factor RUNX1 thus, pointing to lesions of these genes as markers for identifying SCN patients with a high risk of progression to leukemia or MDS [33].

Chromosomal markers, such as, monosomy 7, trisomy 8, and/or trisomy 21 may be acquired just before the appearance of MDS/AL and can be regarded as further contributing events toward clonal evolution [34].

Treatment

The only definitive cure of SCN is hematopoietic stem cell transplantation (HSCT). A recent collaborative study from EBMT and SCETIDE showed a 3-year overall survival (OS) and event-free survival of 82 and 71%, respectively [35]. Patients transplanted after the introduction of the high resolution HLA typing and at an age younger than 10 years have an OS of 85 and 87%, respectively. BM from an HLA matched donor,

either sibling or unrelated, is a preferred source of cells as compared to cord blood and peripheral stem cells because of the lower risk of chronic graft versus host disease (GvHD). Transplant-related mortality (which indicates deaths due to GvDH and infections) has a nonnegligible rate of 17% thus, raising the concern of recurrence to a potentially lethal procedure, in those patients who can be effectively and easily managed with G-CSF daily injections. Based on the available data, HSCT is a good option in SCN patients with absent or scarce response to a high dose of G-CSF (>10 mcg/kg), in subjects who have high infection rate and signs of transformation to MDS/AL, but it is not advisable in those patients who respond to standard dose (≤5 mcg/kg) of G-CSF [36].

Thus, most SCN patients do not need HSCT but can be easily and effectively treated with G-CSF [36]. The aimed dose of G-CSF should be individually tailored as the lowest, enabling an absolute neutrophil count that can protect from the infection (minimal effective dose) with none or minimal side effects (e.g., bone pain, splenomegaly, etc.) and lowering the proliferative stimulus [37]. In principle it should not exceed 5 mcg/kg/day although, patients are known to receive long-term dose up to 7.5 mcg/kg/day. However, it is important to outline that G-CSF chronically-treated patients need monitoring to survey neutrophil count and BM function. Guidelines on how to handle the long-term monitoring of these patients are available in literature and are reported in Table 20.2 [37]. This aim can be more efficiently pursued if the patient is followed in pediatric hematology centers experienced in neutropenias and marrow failure disorders.

In conclusion, even if the knowledge about this rare disease has greatly improved since the first description, a number of unsolved questions still remain open. Future researches and large registry studies will help to further define the best diagnostic and therapeutic management of patients affected with SCN

TABLE 20.2 Schedule of Controls in Different Types of Neutropenia

	Whole blood count	Biochemical[a] parameters	BM[b]	Abdomen ultrasound scan	Bone density	Further considerations
SCN	1–2 months	6 months	12 months[c]	12 months	12–24 months	If morphologic dysplasia or any abnormal cytogenetic clone during follow up occurs, repeat BM analysis every 2–3 months[d] If an isolated mutation of G-CSFR occurs, repeat BM analysis every 6 months[e]
CyN	On the basis of clinical symptoms	Not indicated	Not indicated	Not indicated	Not indicated	Not applicable
CyN	2–3 months	6 months	Not indicated	12 months	24 months	Not applicable
Autoimmune neutropenia	3 months		If clinical signs, BM aspiration suggested			If a spontaneous resolution of neutropenia does not occur, consider performing an enlarged panel of autoimmunity[f]
Autoimmune neutropenia G-CSF-treated	1 month	6 months		12 months	24 months	
Idiopathic neutropenia	3 months		Suggested at least once			If neutropenia is persistent and strongly suggestive of AN, repeat indirect antibodies against neutrophils If a spontaneous resolution of neutropenia does not occur, consider performing an enlarged panel of autoimmunity[f]
Idiopathic neutropenia G-CSF-treated	1 month	At least every 3 months	Periodic BM suggested	12 months	24 months	
GSD1b despite G-CSF	1–3 months	3 months	Not indicated	12 months	24 months	
Associated neutropenia	Timing of controls have to be established on the basis of the single disease					

ALP, Alkaline phosphatase; ALPS, autoimmune lymphoproliferative syndrome; AN, autoimmune neutropenia; BM, bone marrow; CyN, cyclic neutropenia, FISH, fluorescent in situ hybridization, G-CSFR, granulocyte colony-stimulating factor receptor; G-CSF, granulocyte-colony stimulating factor; GSD, glycogen storage disease, HbF, fetal hemoglobin, LDH, lactate dehydrogenase, MDS, myelodysplasia
[a] Liver and renal function test, ALP, LDH, uric acid, HbF, and urine analysis.
[b] Morphology, cytogenetics, FISH for chromosome 7 and 8, and immunophenotype. Trephine biopsy is recommended if MDS or hypocellular leukemia is suspected.
[c] G-CSFR mutation analysis on BM aspirate every 12 months.
[d] Morphology, cytogenetics, FISH for chromosome 7 and 8, and G-CSFR mutation analysis.
[e] Morphology, cytogenetics, and FISH for chromosome 7 and 8.
[f] Enlarged panel of autoimmunity has to be done periodically and includes tests for thyroiditis, coeliac disease, and ALPS.
From Fioredda F, Calvillo M, Bonanomi S. Congenital and acquired neutropenias consensus guidelines on therapy and follow-up in childhood from the Neutropenia Committee of the Marrow Failure Syndrome Group of the AIEOP. Am J Hematol 2012;87:238–43 [37].

OTHER RARE DISEASES

Osteopetrosis

Definition, Epidemiology, Genetics, and Physiopathologic Mechanism

Malignant inherited osteopetrosis, also called "marble bone disease" because of the typical increased density of bones, has been first described by Albert Schönberg, a German radiologist in 1904 [38]. The disease is caused by abnormal osteoclast differentiation and function and has a clinical phenotype with different degrees of severity: malignant, intermediate, and mild osteopetrosis. Around 70% of patients affected with osteopetrosis have a known mutation in 1 of the more than 10 genes that were shown, as of today, to be involved in the disease, whereas about 30% of subjects still remain gene orphan.

Modality of transmission can be autosomal recessive, dominant, or X-linked and the disease may present from neonatal age to adult life.

We will focus on the most severe form of autosomal recessive osteopetrosis (ARO), also named "malignant infantile," usually appearing early in life with a very severe phenotype and dismal prognosis if not recognized [39–41]. Incidence is difficult to evaluate because of frequent misdiagnosis, but it has been estimated to be 1 out of 250,000 births with a particularly high incidence reported in specific geographic regions (i.e., Costa Rica, Middle East, Chuvash Republic of Russia, and Northern Sweden) [39,42].

According to the Nosology Group of the International Skeletal Dysplasia Society, the malignant infantile form is mainly caused by mutations in of one of the following genes: TCIRG1, CLCN7, OSTM1, and SNX10 [43]. These genes code for enzymes that are critical for the degradation of mineral and organic bone matrix, which is the main function of the osteoclasts. The degradation process is mediated by the hydrochloric acid produced by osteoclasts. Hydrochloric acid secretion requires preliminary events including osteoclast "polarization," formation of a "ruffled border," predisposition of a "sealing zone," and a final step in which a niche (lacuna) is formed where the dissolution of bone mineral hydroxyapatite occurs.

The acid secretion by osteoclast also requires the action of two enzymes: ATPase (V-ATPase) and chloride channel 7 (CLCN-7). Homozygous mutation in genes coding these two proteins, TCIRG1 and CLCN-7, respectively account for 70% of severe malignant osteopetrosis phenotypes [44–46]. CLCN-7 is closely associated with another membrane protein, OSTM1. Mutations in the OSTM1 gene are found in gray-lethal mice and in a subset of ARO patients with neurologic involvement [47]. This is because OSTM is involved in lysosomal acidification of other cell types including neurons thus, explaining the storage and neurodegeneration in the central nervous system and in the retina that occurs in a subset of human ARO patients [48]. Another protein, sorting nexin 10 (SNX10), was shown to be involved in the development of ARO but the molecular mechanism by which the mutation can be linked to the disease is unclear yet.

Clinical Picture, Diagnosis, and Differential Diagnosis

First signs of the disease present at birth but sometimes they are rather ambiguous and become clear only at the end of the neonatal period. Patients show a generalized increase in bone density that results in macrocephaly, frontal bossing, exophthalmos/eye protrusion, and abnormally small jaw. These elements confer the affected subjects with a very typical facies.

The increased bone density is visible on standard X-ray scans even in neonates. Methaphysis of long bones may have a funnel-like appearance ("Erlenmeyer flask" deformity) and sometimes bone margins that are delimited by "lucent bands." "Bone in the bone" can be appreciated within the phalanges, vertebrae, and the iliac wings [42]. The expanding bone can narrow cranial foramina inducing compression of cranial nerves causing blindness, deafness, and facial palsy. Hearing loss is estimated to affect 78% of individuals with ARO [49]. Moreover, the skull changes can result in choanal stenosis and hydrocephalus [50]. Tooth eruption defects and severe dental caries are also common. Children with ARO are at risk of developing hypocalcaemia, with tetanic seizures and secondary hyperparathyroidism.

Reduction of the hematopoietic BM spaces generates BM failure with pancytopenia and extramedullary hematopoiesis that in turn causes hepato–splenomegaly thus, mimicking a myeloproliferative disorder. BM failure is one of the most important causes of mortality of ARO. Neuropathic ARO variant is characterized by seizures with normal calcium levels, developmental delay, hypotonia, retinal atrophy with absent evoked visual potentials, and sensor neural deafness due to neurodegeneration as seen in lysosomal storage diseases [51]. Brain magnetic resonance imaging shows delayed myelinization, diffuse progressive cortical and subcortical atrophy, and bilateral atrial subependymal heterotopias [51].

Also peripheral blood film may help to support diagnosis of ARO, by showing abnormalities of red cells (anisopoikylocyotosis, teardrop-shaped red cells), white cells, and platelets. Given the rarity and the heterogeneity of this disease, diagnosis is often delayed or incorrect. Differential diagnoses of ARO include, some forms of Budd–Chiari malformation, hereditary optic atrophy, and myeloproliferative or metabolic diseases. The correct clinical diagnosis can be made by radiography and by targeted molecular analysis.

Early diagnosis of the disease is fundamental to effectively apply therapeutic interventions that may prevent irreversible damage of central and peripheral nervous system.

Treatment

At present, no effective medical treatment for osteopetrosis exists. Stimulation of osteoclasts with calcium restriction, calcitrol, steroids, parathyroid hormone, and interferon has also been attempted without great success [52,53]. Low dose steroids may only temporary reduce transfusional need. If untreated, ARO has a mortality rate of 70% by 6 years of age due to BM failure

complications. HSCT is the only option capable to treat the BM failure but also to definitely resolve the bony disease. As osteoclasts are hematopoietically-derived cells, intrinsic osteoclasts defects can be treated by allogeneic graft [54–56]. Differentiation of normally functioning osteoclast from the "new" marrow can lead to bone remodeling, correction of BM failure, and reversal of extramedullary hematopoiesis. HSCT timing is critical to preserve the neurologic integrity of cranial nerves thus sparing eyesight and hearing ability and is reported to be optimal within the first 3 months of life. The two largest studies on outcome of HSCT in osteopetrosis have been published by EBMT and IBMDR and were composed of 122 and 193 subjects, respectively [57,58].

The 5-year OS of these cohorts was higher in HLA matched patients (62–73%) rather than in alternative donors (40–42%) with a difference that increases in the long-term follow up. According to the EBMT study, the survival of patients transplanted from mismatched donor is significantly lower if compared with that of those engrafted from matched related and unrelated donor. This data were not confirmed by the IBMDR study. This is not entirely unexpected because rejection, the primary cause of death, can occur very early after HSCT irrespective of donor type. The graft failure is probably due to architectural subversion of the bone which creates an unfavorable environment to host the infused marrow.

Crude mortality rate is about 50% within the first year after HSCT. Delayed hematopoietic reconstitution, venous occlusive disease, pulmonary hypertension, and hypercalcemic crisis are also HSCT-related problems that may affect posttransplant mortality [55,56].

These findings result from the association of the "constitutional" defect with the organ's drug toxicity (i.e., busulphan on lung function and/or endothelial tissue). Reduced intensity-conditioning regimen adopted with the aim to lower transplanted-related mortality seemed to have negatively affected the chance of engraftment [58]. Acute GvHD (grade II–IV) is more relevant in HLA matched sibling compared with alternative donor (15% vs. 38%; $p < 0.001$), while chronic GvDH is set around 8%, irrespective of the HLA matching. The inability to reverse complications and to correct the neurodegenerative damage have to be thoroughly weighted in the treatment-decision making process of this disorder [57–59]. In conclusion, HSCT in osteopetrosis remains a challenge: the prerequisites for favorable outcome are the earliest referral of the patient and the choice of the HLA matching of the donor. Outcome may be also favorably influenced by a strict follow up on possible liver- and lung post-HSCT complications [57,58].

Pearson Syndrome

Definition, Epidemiology, and Pathogenic Mechanism

Pearson syndrome (PS) is a rare sporadic, multisystemic mitochondrial disease with and estimated incidence of 1 case/million newborns [60], caused by large-scale rearrangements of mitochondrial DNA (mtDNA) leading to defects in the mitochondrial respiratory chain [61]. The mtDNA passes on to progeny via the cytoplasm and its random distribution during cell division is responsible for the presence of both normal and mutated copies in a single cell (heteroplasmy). A mitochondrial disease appears if the proportion of mutated mtDNA exceeds a given threshold. In PS the same 5.0 kb deletion constitutes the most common lesion [60,61] and no apparent correlation has been found between the size and site of mtDNA deletion and the phenotype [60–63].

Clinical Phenotype

Preterm delivery and birth weight under 2.5 kg is seen only in a minority of patients [60,63]. Acute fetal illness caused by severe anemia is frequent [60], whereas malformations are uncommon [60,64,65]. Splenic atrophy was reported in two patients in the original paper by Pearson [66], but no other case has been published. A postnatal failure to thrive is observed in at least two-thirds of the patients [60,67].

Exocrine pancreatic deficiency is present in 23–63% of cases [60,61,63,67], hepatomegaly occurs in about 70%, and splenomegaly in 30% of patients [60]. True hepatic failure is reported in up to one-third of the cases [60]. Kidney involvement, often progressing to renal failure, is reported in 20–45% of cases [60,61,63,67] and cardiac disease in 27–50% [60,63,68]. Endocrine system involvement, including hypoparathyroidism, hypoglycemia, type 1 insulin-dependent diabetes mellitus, and adrenal insufficiency, occurs in 10–20% of cases [60,67,69,70].

Neurologic and neuromuscular impairment of variable degree (hypotonia, developmental delay, ataxia, and seizures) is reported in more than half of the patients [60,63]. It usually manifests [60] very precociously, tends to progress to neurologic disability sometimes with a final evolution to Leigh syndrome [71], (that is characterized by dysphagia, hypotonia, ataxia, peripheral neuropathy, and ophthalmoparesis) or, more frequently, to Kearns–Sayre syndrome [60,63,72], (including progressive external ophthalmoplegia, pigmentary retinitis, ataxia, cardiac conduction abnormalities, and deafness). The neonatal cerebral ultrasound is usually uninformative [60] and neuroimaging (magnetic resonance imaging or computed tomography) is quite variable from completely normal to severely abnormal findings of white matter, basal ganglion, cerebellum, and brainstem [60,73,74]. Auditory evoked potentials can show a compromised brainstem [60].

Eye problems (ptosis and abnormalities of cornea and retina) have been reported in about 50% of the patients [60,67,75]. Hearing loss and skin abnormalities (hyperpigmentation, café au lait spots, etc.) are infrequent [60,63].

Anemia very often macrocytic and in a context of pancytopenia, is an early finding. Neutropenia is generally moderate (between 0.5 and 1×10^9 L^{-1}) [60,63]. Severe infections occur in about 80% of the patients [60]. The evolution to acute myeloid leukemia was observed in a single patient after HSCT [76].

Laboratory

Serum lactate and lactate/pyruvate ratio are increased sometimes with an intermittent pattern

TABLE 20.3   Differential Diagnosis With Diamond–
Blackfan Anemia (DBA)

	PS	DBA
Leucopenia and/or thrombocy-topenia	90%	Rare
Severe infections	80%	Rare
Hepato- and/or splenomegaly	75%	Rare
Neurologic symptoms	60%	Rare
Eye problems	50%	Rare
Pancreatic insufficiency	40%	No
↑ Serum lactate	90%	No
↑ Urine fumarate	90%	No
↑ Serum alanine	90%	No
Vacuoles in BM	80%	No
Sideroblasts in BM	80%	No
Steroid therapy efficacy	No	60%
Congenital malformations	Rare	>50%

BM, Bone marrow; PS, Pearson syndrome.

[77]. Urinary excretion of lactate and fumarate, and less frequently of malic [60] and methylglutaconic [60,78,79] acid, is increased. Increased alanine [60,63], probably reflecting the disruption of the Kreb's cycle [80] and reduced citrulline and arginine [81] plasma levels are frequently reported.

Fetal hemoglobin (HbF) and erythropoietin are elevated in almost all patients [60,82]. In a recent series of Diamond–Blackfan Anemia (DBA) patients [83], a small percentage was found to be affected by PS: adenosine deaminase, typically high in DBA, is of no help in differential diagnosis, as it is reported as high in about half of PS patients [60,82]. Table 20.3 indicates elements helpful to differentiate PS from DBA.

BM examination shows erythroid and myeloid precursor vacuolization, reduced progenitor growth and BM cellularity. Sideroblasts are seen on Perls staining in the vast majority of patients [60,63].

Treatment

Treatments are essentially supportive. Almost all children are transfused with packed red blood cells and/or platelets, usually starting in the first months of life [60] but transfusion independence becomes highly probable if the patient survives beyond the first 2–3 years of life [60,84]. Eythropoietin usually is not beneficial [60] and G-CSF was effective in some reports [69] but not in others [60]. Pancreatic extracts are efficient in children with exocrine deficiency [60] and bicarbonates can often control metabolic acidosis episodes [60,69]. Scanty data on HSCT are available; it has been reported that even if transplant is able to correct bone marrow failure and acidosis, it may conversely enhance the risk of development of secondary acute myeloid leukaemia [76].

Survival

Improvement of supportive therapy explains the higher survival rate of more recent series [60,63], however, survival to young adulthood has exceptionally been reported [63,83]. Causes of death are mainly sepsis, intractable metabolic acidosis, severe renal failure, and sometimes hepatocellular failure [60,63].

Congenital Dyserythropoietic Anemia

Congenital dyserythropoietic anemias (CDA) are a group of heterogeneous disorders characterized by anemia, ineffective erythropoiesis, and specific cytomorphologic features involving BM late erythroblasts. Apart from anemia that can have a variable grade of severity, patients usually show jaundice, hepatosplenomegaly, and a nonadequate reticulocyte response to the degree of anemia. Patients may also show some dysmorphisms involving fingers and nails [85,86]. BM smear shows megaloblastoid erythroid hyperplasia, nuclear chromatin bridges between erythroblasts and binuclear erythroblasts. Peripheral blood may also show aniso-poikilocytosis and basophilic stipplings.

CDA are grouped into three types: CDA I, II, and III [87,88]. In the last few years, genetic mutations explaining most cases of CDA have been identified.

CDA I is an autosomal recessive transmitted disorder and is characterized by macrocytic anemia and abnormal chromatin structure. CDAN1 [89] and C15ORF41 [90] gene mutations have been identified to be the cause of this form of CDA and both genes seem to be related to the cell proliferation and chromatin assembly that can lead to the impairment of terminal erythroid differentiation [91].

CDA II is an autosomal recessive transmitted disease and is characterized by bimultinuclearity of erythroblasts. It represents the most common type of CDA and is caused by the mutation of SEC23B gene [92], which encodes for the cytoplasmic protein COPII, involved in the deformation of cell membrane and intracellular trafficking of secretory cargo. Analysis of red cell membrane proteins represents a sensitive and specific test that identifies glycosylation abnormalities of this form.

CDA III is an autosomal dominant transmitted disorder characterized by large multinucleated erythroblasts and absence of splenomegaly. KIF3 gene mutation was shown to be associated with this form of CDA [93]. The protein encoded by this gene (MKLP1) is involved in cytokinesis; it connects the microtubule bundle and membranes at a cleavage plane and its impairment results in the accumulation of bimultinuclear cells [94,95].

The management of the disease is generally limited to blood transfusion and iron chelation according to the severity of the disease. Interferon alpha and splenectomy have also been used in the setting of CDA. In two reported cases, interferon improved hemoglobin levels and reduced iron overload [96]. The indication to splenectomy is still controversial, however, it has been reported to be useful in 19 out of 26 reported patients who showed an improved level of hemoglobin [97]. Stem cell transplantation represents the only curative option for this disease. Out of seven reported transplanted patients (six from matched sibiling and one from unrelated donor) all reached transfusion independency without major complication apart from two cases who experienced acute grade 3 and chronic GvHD, respectively [98–101]. Nonetheless, the cost–balance ratio of this procedure should be carefully evaluated and limited to heavily transfusion-dependent patients.

References

[1] Dinauer MC. The phagocyte system and disorders of granulopoiesis and granulocyte function. In: Nathan DG, Orkin SH, editors. Nathan and Oshi's hematology of infancy and childhood. Philadelphia: WBS Saunders Company; 2003. p. 923–1010.

[2] Donadieu J, Beaupain B, Mahlaoui N, et al. Epidemiology of congenital neutropenia. Hematol Oncol Clin North Am 2013;27:1–17.

[3] Welte K, Zeidler C, Dale DC. Severe congenital neutropenia. Semin Hematol 2006;43:189–95.

[4] Kostmann R. Infantile genetic agranulocytosis; agranulocytosis infantilis hereditaria. Acta Paediatr 1956;45(Suppl. 105):1–78.

[5] Klein C, Grudzien M, Appaswamy G, et al. HAX1 deficiency causes autosomal recessive severe congenital neutropenia (Kostmann disease). Nat Genet 2007;39:86–92.

[6] Horwitz M, Benson KF, Person RE, et al. Mutations in ELA2, encoding neutrophil elastase, define a 21-day biological clock in cyclic haematopoiesis. Nat Genet 1999;23:433–6.

[7] Xia J, Bolyard AA, Rodger E, et al. Prevalence of mutations in ELANE, GFI1, HAX1, SBDS, WAS and G6PC3 in patients with severe congenital neutropenia. Br J Haematol 2009;147:535–42.

[8] Person RE, Li FQ, Duan Z, et al. Mutations in proto-oncogene GFI1 cause human neutropenia and target ELA2. Nat. Genet 2003;34:308–12.

[9] Boztug K, Appaswamy G, Ashikov A, et al. A syndrome with congenital neutropenia and mutations in G6PC3. N Engl J Med 2009;360:32–43.

[10] Devriendt K, Kim AS, Mathijs G, et al. Constitutively activating mutation in WASP causes X-linked severe congenital neutropenia. Nat Genet 2001;27:313–7.

[11] Link DC. SNAREing a new cause of neutropenia. Blood 2013;121:5078–87.

[12] Tricot A, Järvinen PM, Arostegui JI, et al. Inherited bi allelic CSFR3 mutations in severe congenital neutropenia. Blood 2014;123:3811–7.

[13] Makaryan V1, Rosenthal EA, Bolyard AA, et al. TCIRG1-associated congenital neutropenia. Hum Mutat 2014;35:824–7.

[14] Boztug K, Järvinen PM, Salzer E, et al. JAGN1 deficiency causes aberrant myeloid cell homeostasis and congenital neutropenia. Nat Genet 2014;46:1021–7.

[15] Grenda DS, Murakami M, Ghatak J, et al. Mutations of the ELA2 gene found in patients with severe congenital neutropenia induce the unfolded protein response and cellular apoptosis. Blood 2007;110:4179–87.

[16] Skokowa J, Klimiankou M, Klimenkova O, et al. Interactions among HCLS1,HAX1 and LEF-1 proteins are essential for G-CSF-triggered granulopoiesis. Nat Med 2012;18:1550–9.

[17] Moulding DA, Blundell MP, Spiller DG, et al. Unregulated actin polymerization by WASp causes defects of mitosis and cytokinesis in X-linked neutropenia. J Exp Med 2007 3;204:2213–24.

[18] Vilboux T, Lev A, Malicdan MC, et al. A congenital neutrophil defect syndrome associated with mutations in VPS45. N Engl J Med 2013;369:54–65.

[19] Fioredda F, Calvillo M, Burlando O, et al. Infectious complications in children with severe congenital, autoimmune or idiopathic neutropenia: a retrospective study from the Italian Neutropenia Registry. Pediatr Infect Dis J 2013;32:410–2.

[20] Donini M, Fontana S, Savoldi G, et al. G-CSF treatment of severe congenital neutropenia reverses neutropenia but does not correct the underlying functional deficiency of the neutrophil in defending against microorganisms. Blood 2007;1(109):4716–23.

[21] Young NS, Alter BP. Kostmann's syndrome. In: Young NS, Alter BP, editors. Aplastic anemia: acquired and inherited. Philadelphia: Saunders; 1994. p. 391–4.

[22] Dale DC, Bonilla MA, Davis MW, et al. A randomized controlled phase III trial of recombinant human granulocyte colony-stimulating factor (filgrastim) for treatment of severe chronic neutropenia. Blood 1993;81:2496–502.

[23] Dale DC, Cottle TE, Fier CJ, et al. Severe chronic neutropenia: treatment and follow-up of patients in the Severe Chronic Neutropenia International Registry. Am J Hematol 2003;72:82–93.

[24] Lehrnbecher T, Welte K. Haematopoietic growth factors in children with neutropenia. Br J Haematol 2002;116:28–56.

[25] Rosenberg PS, Zeidler C, Bolyard AA, et al. Stable long-term risk of leukaemia in patients with severe congenital neutropenia maintained on G-CSF therapy. Br J Haematol 2010;150:196–9.

[26] Donadieu J, Leblanc T, Bader Meunier B, et al. Analysis of risk factors for myelodysplasias, leukemias and death from infection among patients with congenital neutropenia. Experience of the French Severe Chronic Neutropenia Study Group. Haematologica 2005;90:45–53.

[27] Fioredda F, Calvillo M, Lanciotti M. Lethal sepsis and malignant transformation in severe congenital neutropenia: report from the Italian Neutropenia Registry. Pediatr Blood Cancer 2015;62:1110–2.

[28] Fioredda F, Calvillo M, Lanciotti M. Pegfilgrastim in children with severe congenital neutropenia. Pediatr Blood Cancer 2010;54:465–7.

[29] Beaupain B, Leblanc T, Reman O, et al. Is pegfilgrastim safe and effective in congenital neutropenia? An analysis of the French Severe Chronic Neutropenia registry. Pediatr Blood Cancer 2009;53:1068–73.

[30] Carlsson G, Fasth A, Berglöf E, et al. Incidence of severe congenital neutropenia in Sweden and risk of evolution to myelodysplastic syndrome/leukaemia. Br J Haematol 2012;158:363–9.

[31] Makaryan V, Zeidler C, Bolyard AA, et al. The diversity of mutations and clinical outcomes for ELANE-associated neutropenia. Curr Opin Hematol 2015;22:3–11.

[32] Skokowa J, Welte K. Defective G-CSFR signaling pathways in congenital neutropenia. Hematol Oncol Clin North Am 2013;27(1):75–88.

[33] Skokowa J, Steinemann D, Katsman-Kuipers JE. Cooperativity of RUNX1 and CSF3R mutations in severe congenital neutropenia: a unique pathway in myeloid leukemogenesis. Blood 2014;123:2229–37.

[34] Touw IP. Game of clones: the genomic evolution of severe congenital neutropenia. Hematol Am Soc Hematol Educ Program 2015;1:1–7.

[35] Fioredda F, Iacobelli S, van Biezen A, et al. Stem cell transplantation in severe congenital neutropenia: an analysis from the European Society for Blood and MarrowTransplantation. Blood 2015;126:1885–92.

[36] Peffault de Latour R, Peters C, Gibson B, et al. Recommendations on hematopoietic stem cell transplantation for inherited bone marrow failure syndromes. Bone Marrow Transplant 2015;50:1168–72.

[37] Fioredda F, Calvillo M, Bonanomi S, et al. Congenital and acquired neutropenias consensus guidelines on therapy and follow-up in childhood from the Neutropenia Committee of the Marrow Failure Syndrome Group of the AIEOP (Associazione Italiana Emato-Oncologia Pediatrica). Am J Hematol 2012;87:238–43.

[38] Albert Schönberg H. Roentgenbilder einer seltenen Knochenerkrankung. Munch Med Wochenschr 1904;51:365–8.

[39] Loria-Cortes R, Quesada-Calvo E, Cordero-Chaverri C. Osteopetrosis in children: a report of 26 cases. J Pediatr 1977;91:43–7.

[40] Bollerslev J, Andersen PE Jr. Radiological, biochemical and hereditary evidence of two types of autosomal dominant osteopetrosis. Bone 1988;9:7–13.

[41] Stark Z, Savarirayan R. Osteopetrosis. Orphanet J Rare Dis 2009;4:5.

[42] Sobacchi C, Shulz A, Coxon PF, et al. Osteopetrosis: genetics, treatment and new insights into osteoclast function. Nature Rev 2013;9:522–36.

[43] Bonafe L, Cormier-Daire V, Hall C, et al. Nosology and classification of genetic skeletal disorders: 2015 revision. Am J Med Genet A 2015;167A(12):2869–92.

[44] Frattini A, Orchard PJ, Sobacchi C, et al. TCIRG1 subunit of the vacuolar proton pump are responsible for a subset of human autosomal recessive osteopetrosis. Nat Genet 2000;25:343–6.

[45] Kornak U, Kasper D, Bosl MR, et al. Loss of the ClC-7 chloride channelleads to osteopetrosis in mice and man. Cell 2001;104:205–15.

[46] Sobacchi C, Frattini A, Orchard P, et al. The mutationalspectrum of human malignant autosomal recessive osteopetrosis. Hum Mol Genet 2001;10:1767–73.

[47] Ramirez A, Faupel J, Goebel I, et al. Identification of a novel mutation in the coding region of the grey-lethal gene OSTM1 in human malignant infantile osteopetrosis. Hum Mutat 2004;23:471–6.

[48] Lange PF, Wartosch L, Jentsch TJ, Fuhrmann JC. CLC-7 requires Ostm1 as a beta-subunit to support bone resorption and lysosomal function. Nature 2006;440:220–3.

[49] Dozier TS, Duncan IM, Klein AJ, et al. Otologic manifestations of malignant osteopetrosis. Otol Neurotol 2005;26:762–76.

[50] Al-Tamimi YZ, Tyag,i AK, Chumas PD, et al. Patients with autosomal-recessive osteopetrosis presenting with hydrocephalusand hindbrain posterior fossa crowding. J Neurosurg Pediatr 2008;1:103–6.

[51] Steward CG. Neurological aspects of osteopetrosis. Neuropathol Appl Neurobiol 2003;29:87–97.

[52] Kocher MS, Kasser JR. Osteopetrosis. Am J Orthop 2003;32:222–8.

[53] Key L, Carnes D, Cole S, et al. Treatment of congenital osteopetrosis with high-dose calcitriol. N Engl J Med 1984;310(7):409–15.

[54] Coccia PF, Krivit W, Cervenka J, et al. Successful bone-marrow transplantation for infantile malignant osteopetrosis. N Engl J Med 1980;30:701–8.

[55] Sieff CA, Chessells JM, Levinsky RJ, et al. Allogeneic bone-marrow transplantation in infantile malignant osteopetrosis. Lancet 1983;1:437–41.

[56] Gerritsen EJ, Vossen JM, Fasth A, et al. Bone marrow transplantation for autosomal recessive osteopetrosis. A report from the Working Party on Inborn Errors of the European Bone Marrow Transplantation Group. J Pediatr 1994;125:896–902.

[57] Driessen GJ, Gerritsen EJ, Fischer, et al. Long-term outcome of haematopoietic stem cell transplantation in autosomal recessive osteopetrosis: an EBMT report. Bone Marrow Transplant 2003;32:657–63.

[58] Orchard PJ, Fasth AL, Le Rademacher J, et al. Hematopoietic stem cell transplantation for infantile osteopetrosis. Blood 2015;126:270–6.

[59] Steward CG. Neurological aspects of osteopetrosis. Neuropathol Appl Neurobiol 2003;29:87–97.

[60] Farruggia P, Di Cataldo A, Pinto RM, et al. Pearson syndrome: a retrospective cohort study from the Marrow Failure Study Group of A.I.E.O.P. (Associazione Italiana Emato-Oncologia Pediatrica). J Inherit Metab Dis Rep 2016;26:37–43.

[61] Rötig A, Bourgeron T, Chretien D, et al. Spectrum of mitochondrial DNA rearrangements in the Pearson marrow-pancreas syndrome. Hum Mol Genet 1995;4(8):1327–30.

[62] Topaloğlu R, Lebre AS, Demirkaya E, et al. Two new cases with Pearson syndrome and review of Hacettepe experience. Turk J Pediatr 2008;50(6):572–6.

[63] Broomfield A, Sweeney MG, Woodward CE, et al. Paediatric single mitochondrial DNA deletion disorders: an overlapping spectrum of disease. J Inherit Metab Dis 2015;38(3):445–57.

[64] Lohi O, Kuusela AL, Arola M. A novel deletion in a Pearson syndrome infant with hypospadias and cleft lip and palate. J Inherit Metab Dis 2005;28(6):1165–6.

[65] Gürgey A, Ozalp I, Rötig A, et al. A case of Pearson syndrome associated with multiple renal cysts. Pediatr Nephrol 1996;10(5):637–8.

[66] Pearson HA, Lobel JS, Kocoshis SA, et al. A new syndrome of refractory sideroblastic anemia with vacuolization of marrow precursors and exocrine pancreatic dysfunction. J Pediatr 1979;95(6):976–84.

[67] Atale A, Bonneau-Amati P, Rötig A, et al. Tubulopathy and pancytopaenia with normal pancreatic function: a variant of Pearson syndrome. Eur J Med Genet 2009;52(1):23–6.

[68] Krauch G, Wilichowski E, Schmidt KG, Mayatepek E. Pearson marrow-pancreas syndrome with worsening cardiac function caused by pleiotropic rearrangement of mitochondrial DNA. Am J Med Genet 2002;110(1):57–61.

[69] Seneca S, De Meirleir L, De Schepper J, et al. Pearson marrow pancreas syndrome: a molecular study and clinical management. Clin Genet 1997;51(5):338–42.

[70] Gürgey A, Ozalp I, Rötig A, et al. A case of Pearson syndrome associated with multiple renal cysts. Pediatr Nephrol 1996;10(5):637–8.

[71] Santorelli FM, Barmada MA, Pons R, Zhang LL, Di Mauro S. Leigh-type neuropathology in Pearson syndrome associated with impaired ATP production and a novel mtDNA deletion. Neurology 1996;47(5):1320–3.

[72] McShane MA, Hammans SR, Sweeney M, et al. Pearson syndrome and mitochondrial encephalomyopathy in a patient with a deletion of mtDNA. Am J Human Genet 1991;48(1):39–42.

[73] Lee HF, Lee HJ, Chi CS, Tsai CR, Chang TK, Wang CJ. The neurological evolution of Pearson syndrome: case report and literature review. Eur J Paediatr Neurol 2007;11(4):208–14.

[74] Morel AS, Joris N, Meuli R, et al. Early neurological impairment and severe anemia in a newborn with Pearson syndrome. Eur J Pediatr 2009;168(3):311–5.

[75] Cursiefen C, Küchle M, Scheurlen W, Naumann GO. Bilateral zonular cataract associated with the mitochondrial cytopathy of Pearson syndrome. Am J Ophthalmol 1998;125(2):260–1.

[76] Tumino M, Meli C, Farruggia P, et al. Clinical manifestations and management of four children with Pearson syndrome. Am J Med Genet A 2011;155A(12):3063–6.

[77] Yamada K, Toribe Y, Yanagihara K, Mano T, Akagi M, Suzuki Y. Diagnostic accuracy of blood and CSF lactate in identifying children with mitochondrial diseases affecting the central nervous system. Brain Dev 2012;34(2):92–7.

[78] Gibson KM, Bennett MJ, Mize CE, et al. 3-Methylglutaconic aciduria associated with Pearson syndrome and respiratory chain defects. J Pediatr 1992;121(6):940–2.

[79] Lichter-Konecki U, Trefz FK, Rötig A, Munnich A, Pfeil A, Bremer HJ. 3-Methylglutaconic aciduria in a patient with Pearson syndrome. Eur J Pediatr 1993;152(4):378.

[80] Zschocke J, Hoffmann GF, Burlina BA, et al. Vademecum metabolicum: manual of metabolic paediatrics. 2nd ed. Stuttgart, Germany: Schattauer Gmbh; 2004. 100.

[81] Crippa BL, Leon E, Calhoun A, Lowichik A, Pasquali M, Longo N. Biochemical abnormalities in Pearson syndrome. Am J Med Genet A 2015;167A(3):621–8.

[82] Superti-Furga A, Schoenle E, Tuchschmid P, et al. Pearson bone marrow-pancreas syndrome with insulin-dependent diabetes, progressive renal tubulopathy, organic aciduria and elevated fetal haemoglobin caused by deletion and duplication of mitochondrial DNA. Eur J Pediatr 1993;152(1):44–50.

[83] Gagne KE, Ghazvinian R, Yuan, et al. Pearson marrow pancreas syndrome in patients suspected to have Diamond–Blackfan anemia. Blood 2014;124(3):437–40.

[84] Muraki K, Nishimura S, Goto Y, Nonaka I, Sakura N, Ueda K. The association between haematological manifestation and mtDNA deletions in Pearson syndrome. J Inherit Metab Dis 1997;20(5):697–703.

[85] Shalev H, Tamary H, Shaft D, et al. Neonatal manifestations of congenital dyserythropoietic anemia typoe I. J Pediatr 1997;131:95–7.

[86] Wickramasinghe SN. Congenital dyserythropoietic anemias. Curr Opin Hematol 2000;7:71–8.

[87] Heimpel H, Wendt F. Congenital dyserythropoietic anemia with karyorrhexis and multinuclearity of erythroblasts. Helv Med Acta 1968;34(2):103–15.

[88] Wickramasinghe SN, Wood WG. Advances in the understanding of the congenital dyserythropoietic anemias. Br J Haematol 2005;131(4):431–46.

[89] Ahmed MR, Chehal A, Zahed L, et al. Linkage and mutational analysis of the CDAN1 gene reveals genetic heterogeneity in congenital dyserythropoietic anemia type I. Blood 2006;107(12):4968–9.

[90] Babbs C, Roberts NA, Sanchez-Pulido L, et al. Homozygous mutations in a predicted endonuclease are a novel cause of congenital dyserythropoietic anemia type I. Haematologica 2013;98(9):1383–7.

[91] Iolascon A, Heimpel H, Wahlin A, Tamary H. Congenital dyserythropoietic anemias: molecular insights and diagnostic approach. Blood 2013;122(13):2162–6.

[92] Schwarz K, Iolascon A, Verissimo F, et al. Mutations affecting the secretory COPII coat component SEC23B cause congenital dyserythropoietic anemia type II. Nat Genet 2009;41(8):936–40.

[93] Liljeholm M, Irvine AF, Vikberg AL, Norberg A, et al. Congenital dyserythropoietic anemia type III (CDA III) is caused by a mutation in kinesin family member, KIF23. Blood 2013;121(23):4791–9.

[94] Lind L, Sandström H, Wahlin A, et al. Localization of the gene for congenital dyserythropoietic anemia type III, CDAN3, to chromosome 15q21-q25. Hum Mol Genet 1995;4(1):109–12.

[95] Makyio H, Ohgi M, Takei T, et al. Structural basis for Arf6–MKLP1 complex formation on the Flemming body responsible for cytokinesis. EMBO J 2012;31(11):2590–603.

[96] Goede JS, Benz R, Fehr J, Schwarz K, Heimpel H. Congenital dyserythropoietic anemia type I with bone abnormalities, mutations of the CDAN I gene, and significant responsiveness to alpha-interferon therapy. Ann Hematol 2006;85:591–5.

[97] Iolascon A, Delaunay J, Wickramasinghe SN, et al. Natural history of congenital dyserythropoietic anemia type II. Blood 2001;98:1258–60.

[98] Ayas M, Al-Jefri A, Baothman A, et al. Transfusion-dependent congenital dyserythropoietic anemia type I successfully treated with allogeneic stem cell transplantation. Bone Marrow Transplant 2002;29:681–2.

[99] Unal S, Russo R, Gumruk F, et al. Successful hematopoietic stem cell transplantation in a patient with congenital dyserythropoietic anemia type II. Pediatr Transplant 2014;18(4):E130–3.

[100] Braun M, Wölfl M, Wiegering V, et al. Successful treatment of an infant with CDA type II by intrauterine transfusions and postnatal stem cell transplantation. Pediatr Blood Cancer 2014;61(4):743–5.

[101] Buchbinder D, Nugent D, Vu D, et al. Unrelated hematopoietic stem cell transplantation in a patient with congenital dyserythropoietic anemia and iron overload. Pediatr Transplant 2012;16(3):E69–73.

Bone Marrow Failure Syndromes in Children

S. Elmahdi, S. Kojima

Department of Pediatrics, Graduate School of Medicine, Nagoya University, Nagoya, Japan

INTRODUCTION

Childhood bone marrow failure syndromes (BMFs) represent a heterogeneous group of diseases characterized by impaired production of blood cells, which may be either inherited or acquired. The pathophysiology and treatment of pediatric BMFs are fundamentally the same as in adults. However, inherited BMFs are more frequent in pediatric populations, comprising 10–20% of all childhood BMFs. A definitive diagnosis of hypoplastic BMFs is clinically challenging, particularly in the absence of an identified chromosomal abnormality [1]. The most common and challenging clinical scenario is the differential diagnosis between aplastic anemia (AA) and refractory cytopenia of childhood (RCC), a provisional entity of the childhood myelodysplastic syndromes (MDS) [2]. This WHO classification has been proposed by the European Working Groups of MDS of Childhood (EWOG–MDS), but it is still not widely used outside of this group [2].

In contradiction to the previous era, in which outcomes in cases of severe BMFs were generally dismal and frequently fatal, we have reached a point where the majority of cases are cured. Over the last three decades, bone marrow transplantation (BMT) from an HLA matched family donor (MFD) has become a treatment of choice in severe forms of childhood BMFs [3–9]. Further, immunosuppressive therapy (IST) with antithymocyte globulin (ATG) and cyclosporine is indicated in children without an available MFD [6,10–13]. BMT from a matched but unrelated donor (MUD) is typically reserved for nonresponders to IST. Although there has been no improvement in the overall- and failure-free survival rates following IST in the last two decades, survival rates following BMT from an HLA-MUD have dramatically improved and are now comparable to those of BMT from a MFD. Based on these results, upfront BMT from an HLA-MUD is now considered in patients with childhood BMFs [8,14–17].

The present review aims to describe the incorporation of recent advances in biologic understanding with a particular focus on novel molecular findings in the diagnosis of childhood BMFs, as well as, the application of these achievements in management approaches for childhood BMFs.

Congenital and Acquired Bone Marrow Failure
http://dx.doi.org/10.1016/B978-0-12-804152-9.00021-X

NEXT GENERATION SEQUENCING FOR INHERITED BMFs

The differential diagnosis should always consider the inherited forms of childhood BMFs. Physical examination and a detailed family history are mandatory to establish the diagnosis of inherited BMFs. However, cryptic or atypical presentations and phenotypic overlap may obscure the correct diagnosis [18–21]. Of the diagnostic tools available for BMFs, the chromosome fragility test for Fanconi anemia (FA) and telomere length analysis for dyskeratosis congenita (DKC) should be included in screening studies. Further, the identification of specific gene mutations is essential for a definitive diagnosis of inherited BMFs.

At present, genetic tests for inherited BMFs are performed gene-by-gene using Sanger sequencing. However, this method has limitations in the diagnosis of inherited BMFs [22,23]. Inherited BMFs comprise more than 25 disease entities. More than 80 causative genes have been identified to date and this number continues to increase. As clinical and laboratory findings may overlap between inherited BMFs, the selection of target genes is challenging. Moreover, sequencing of multiple genes using Sanger methods is expensive and time-consuming [23]. To overcome these limitations of conventional genetic testing, several groups have proposed the clinical application of next generation sequencing (NGS) for the diagnosis of inherited BMFs [24–30]. NGS encompasses the following three approaches: target gene sequencing, whole exome sequencing (WES), and whole genome sequencing. Compared with WES and whole genome sequencing, target gene sequencing is relatively inexpensive and associated with a lower risk of incidental genetic discovery [31].

The Seattle group was the first to publish the result of clinical application of target gene sequencing for the diagnosis of inherited BMFs [30]. Its gene panel included 85 genes responsible for inherited BMFs and MDS, and had a median coverage of 549-fold, with 97.8% of bases having over 50-fold coverage and 98.2% of bases having over 10-fold coverage. Their study included 71 patients with previously unclassified BMFs and MDS after diagnostic work-up comprising chromosomal breakage testing for FA, telomere length testing for DKC, and pancreatic enzyme testing for Shwachman–Diamond syndrome. Further, patients were tested for gene mutations associated with a range of inherited BMFs, with all remaining unclassified after genetic work-up. Of the 71 patients, 58 were younger than 18 years and 32 had a positive family history. Eight of the 71 patients were found to harbor germline mutations in *GATA2*, *RUNX1*, *DKC1*, or *LIG4*. All patients lacked the characteristic physical anomalies and laboratory findings of these syndromes, with only four found to have a family history of inherited BMFs [30].

The Toronto group also developed an NGS panel comprising 72 known genes of inherited BMFs [22]. The average gene coverage was 99.1%, with an average read depth of 680-fold. A total of 158 patients with unknown gene mutations were included in this study. Of the 75 patients with clinically classified inherited BMF, 44 (59%) patients were found to have causal mutations by NGS. The diagnostic rates for each disease were 16/23 (70%) for Diamond–Blackfan anemia, 9/12 (75%) for FA, and 2/5 (40%) for DKC. The diagnoses of the four clinically-classified patients were amended according to the results of the NGS assay. Patients who fulfilled the diagnostic criteria of inherited BMFs and had no genetic diagnosis were recruited into the study. All included patients had chronic BMF in addition to a family history, physical malformations, or initial presentation earlier than one year of age. Patients with severe aplastic anemia (SAA) who did not respond to IST were also recruited. Of the 83 patients with unclassified inherited BMFs, 15 patients (18%) had mutations in *GATA2*, *WAS*, *G6PC3*, *TERT*, *TERC*, *TINF2*, *CXCR4*, *RPL5*, *MYH9*, *RTEL1*, or *MASTL* [22].

As a first-line diagnostic test, we analyzed 122 patients with clinical diagnoses of inherited BMFs using a target sequencing approach that covered 184 inherited BMFs and MDS-associated genes [32]. Our capture-based target sequencing covered 99.4% of the target region in 184 genes with more than 20 reads. Detected variants were considered to be causative variants reported to be pathogenic in previous literature or otherwise highly expected to be causative, such as nonsense, frameshift, and splice-site variants. In total, we identified 76 variants, which we considered to be pathogenic. In addition, we were able to identify pathognomonic copy number aberrations in eight patients. Patients with a clinical diagnosis of FA (15/22, 68%) or Shwachman–Diamond syndrome (4/6, 67%) achieved relatively high genetic diagnostic rates. A genetic diagnosis was achieved in approximately half of patients with Diamond–Blackfan anemia, (12/26, 46%) or DKC (5/13, 38%). In the majority of patients (45/55, 82%), the genetic diagnosis matched the clinical diagnosis. However, the clinical diagnosis was incompatible with the detected variant in the remaining 10 patients. We identified *RUNX1* mutation/deletion in three patients who developed thrombocytopenia during infancy and had been diagnosed with chronic thrombocytopenia of unknown cause. Accordingly, the diagnoses of these three patients were amended to familial platelet disorder with propensity to acute myelogenous leukemia. These results indicate the utility of genetic screening by NGS to elucidate genetic heterogeneity as a replacement for conventional Sanger sequencing in the diagnosis of inherited BMFs [32].

CHILDHOOD APLASTIC ANEMIA AND REFRACTORY CYTOPENIA OF CHILDHOOD

Childhood MDS is rare and differs from adult MDS in several aspects, thereby necessitating a pediatric approach to the diagnosis and management of MDS. The 2008 WHO classification proposed a new entity in childhood MDS, RCC [2]. Originally, this entity was proposed by the pathologists of EWOG–MDS [2]. Baumann and colleagues found that a diagnosis of RCC predicted an inferior response to IST compared with AA and an increased risk of clonal evolution, indicating BMT as a treatment of choice for RCC patients [33]. These data indicate that clinicians may be able to distinguish between AA, considered as an immune-mediated process, and RCC, caused by an underlying clonal stem-cell defect.

The WHO classification defines RCC as persistent cytopenia without an increase in blast count and bone marrow (BM)-aspirate smears demonstrating dysplastic changes in more than two cell lineages or >10% of cells within a single lineage. The spectrum of patients with RCC is wide, ranging from patients with severe hypocellular BM with mild dysplasia to those with normocellular BM with distinct dysplasia, meeting the criteria for refractory cytopenia with multilineage dysplasia (RCMD) in adults [2]. Currently, the WHO classification recommends that children who meet the criteria for RCMD should be considered as having RCC until the numbers of lineages involved are fully evaluated in order to provide an important prognostic discriminator [2]. Fig. 21.1 represents BM findings in AA, RCC, and RCMD.

RCC represents the most common subtype of childhood MDS, accounting for approximately half of all MDS cases. Mean corpuscular volume and hemoglobin F levels are typically elevated in RCC patients [34,35]. In contrast to adult refractory anemia, the majority of patients with RCC demonstrate a marked decrease in BM cellularity. Further, approximately 80% of RCC patients have a normal karyotype [2,35]. Accordingly, the histologic distinction between AA and hypocellular RCC is clinically challenging and thus requires a close collaboration between experienced hematologists and hematopathologists.

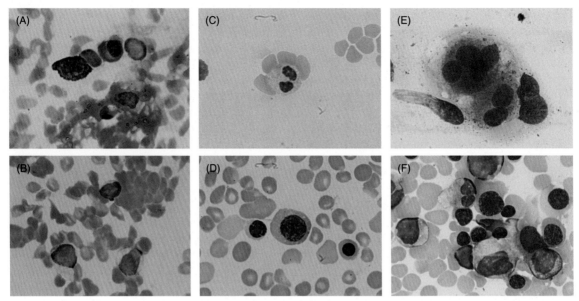

FIGURE 21.1   Representative bone marrow (BM) findings in aplastic anemia (AA), refractory cytopenia of childhood (RCC), and refractory cytopenia with multilineage dysplasia (RCMD). Representative BM findings demonstrating (A,B) severe hypocellular marrow without hematopoietic cells in AA, (C,D) hypocellular marrow with mild dysplasia in RCC, and (E,F) normocellular marrow with distinct dysplasia in RCMD. (A) Lymphocytes with a mast cell and a plasma cell. (B) Lymphocytes without hematopoietic cells. (C) Neutrophil with pseudo-Pelger-Huët anomaly. (D) Erythroblast with megaloblastic change. (E) Binucleated megakaryocyte and a megakaryocyte with separated nucleus. (F) Erythroblasts with megaloblastic changes and a neutrophil with pseudo-Pelger-Huët anomaly.

The evaluation of BM histomorphologic features is crucial for differentiation between AA and RCC, particularly in the absence of an identified chromosomal abnormality. A patchy biopsy pattern comprising erythroid islands of at least 20 progenitor cells with sparse maturation or clonal hematopoiesis is mandatory for the diagnosis of RCC. A higher mean corpuscular volume and marked left shift in myeloid differentiation are considered as confirmatory findings. Further, immunohistochemistry for megakaryocytes may have utility in distinguishing AA from RCC. The detection of micromegakaryocytes by immunohistochemistry is suggestive of RCC. The identification of MDS-related cytogenetic abnormalities, if present, may also facilitate differentiation between RCC and AA. Conversely, criteria for the diagnosis of severe AA include complete aplasia or severe hypoplasia of hematopoiesis

(>95% fatty marrow) with scarcely distributed BM cells without patchy area, particularly of erythropoietic precursors, and the absence of micromegakaryocytes [36].

Bauman et al. [36] reported a high concordance in the diagnosis of RCC among pathologists within EWOG–MDS. They performed a double blind interobserver study of 100 patients with SAA and RCC among 7 hematopathologists of EWOG–MDS and the German SAA study group. The reference pathologists failed to reach an agreement in only 4 of the 100 included patients [36]. Conversely, Forester et al. [37] recently demonstrated that even with the use of the methodology proposed by Bauman et al. [36], pathologists had a low concordance in terms of differentiating between AA and hypocellular RCC among patients in the United States. Japanese and Chinese pediatric hematologists and pathologists conducted a joint review of BM smears and trephine biopsy in 100 children with acquired BMFs according to WHO criteria [38]. In this study, hematologists and pathologists failed to reach an agreement regarding the diagnosis in 9 out of the 100 included patients. Although four patients were diagnosed as having AA by hematologists but as having RCC by pathologists, the final diagnoses was determined to be RCC in all four patients. In contrast, two patients were diagnosed as having RCC by hematologists but as AA by pathologists, with final diagnoses of RCC. No significant differences in clinical and laboratory findings and the distribution of diagnoses were observed between Japanese and Chinese children [38].

Historically, patients with hypocellular RCC have been given the diagnosis of AA in the majority of countries. The proposal of RCC remains controversial, has yet to be universally accepted, and requires further validation, particularly as the therapeutic and prognostic implications of a diagnosis of RCC have yet to be fully elucidated. Further, there is a lack of literature on the correlation between the morphologic classification of BMFs and clinical outcomes. In a previous report of EWOG–MDS, according to EWOG–MDS–1998 criteria, IST was reported to be less effective in the diagnosis of RCC than of SAA [33]. Further, higher rates of nonresponse, partial response, and treatment failure, including the emergence of clonal disease in children, were observed in patients diagnosed with RCC who received therapy according to the SAA protocol. These data indicate patients treated with IST according to the German SAA 94 protocol or EWOG–MDS–1998 protocol have different responses and outcomes. The risk of developing MDS or acute myelogenous leukemia post-IST was reportedly 3% for the period of reference pathology from 1998 to 2004, with only 3 of the 123 patients developing clonal disease. In contrast, 13 of 88 patients in the German SAA–1994 study developed clonal disease after IST during 1994–97 when reference pathology was not used [33].

Recently, Yoshimi et al. [39] reported the response rate to IST in patients with RCC who were enrolled in the prospective EWOG–MDS–1998 and EWOG–MDS–2006 studies. Horse ATG, which was used for IST in patients with AA and RCC, was withdrawn from the market and replaced by rabbit ATG in Europe and Asian countries in 2007. Horse ATG was used in the EWOG–MDS–1998 study ($n = 46$) and rabbit ATG ($n = 49$) was used in the EWOG–MDS–2006 study. The response rates with horse ATG and rabbit ATG were 59 and 47% at 4 months and 74 and 53% at 6 months, respectively. Clonal evolution was observed in eight patients at a median time of 14 months (6–64 months) after IST initiation (horse ATG, $n = 6$; rabbit ATG, $n = 2$). The cumulative incidence (CI) of clonal evolution at 4 years was 14% (7–30%) in the horse ATG group and 4% (1–17%) in the rabbit ATG group [39].

In Japan, we conducted the AA–97 study of horse ATG and cyclosporine for childhood AA between 1997 and 2008, when RCC had yet to be proposed by WHO, which provided a unique opportunity to compare long-term outcomes between patients with RCC and AA [40]. In order to assess the association between morphologic classification and clinical outcomes, we retrospectively classified stored BM samples according to the 2008 WHO criteria and evaluated the correlation between the diagnosis and clinical outcomes. We reviewed BM morphology in 186 children (median age, 8 years; range, 1–16 years) who were enrolled in the AA–97 study, with a median follow-up period of 87 months (range, 1–146 months). Morphologically, 62 (33%) patients were classified with AA, 94 (49%) with RCC, and 34 (18%) with RCMD. AA patients received granulocyte colony-stimulating factor more frequently and for longer durations than other patients. After 3 and 6 months, the respective response rates to IST did not significantly differ among the three groups (AA, 38% and 52%; RCC, 51% and 59%; and RCMD, 41% and 56%, respectively). The acquisition of chromosomal abnormalities was observed in five patients with AA, four patients with RCC, and three patients with RCMD. Although CI for total clonal evolution at 10 years did not significantly differ among the three groups, CI for the development of monosomy 7 was significantly higher in the AA group than in other groups. Multivariate analysis demonstrated only duration of granulocyte colony-stimulating factor administration of >40 days as a significant risk factor for the development of monosomy 7. Although the failure-free survival rate at 10 years did not significantly differ among the three groups, the overall survival rate at 10 years in the AA group (85 ± 5.1%) was significantly lower than those in the RCC (97 ± 1.9%) and RCMD (100%) groups. The findings of both EWOG and Japan indicate clinical outcomes do not differ between patients with AA and those with RCC who received IST. Further, strict morphologic differentiation of AA from RCC was not able to accurately predict clonal evolution in children with AA [40].

The majority of cases of AA are considered to be caused by immune-mediated suppression of hematopoiesis, whereas RCC results from defective hematopoiesis. There is a lack of reported studies that have evaluated the biologic differences between AA and RCC. T-cell receptor (TCR)-Vβ CDR3 region skewing and minor paroxysmal nocturnal hemoglobinuria (PNH) clones have been posited to contribute to immune-mediated myelosuppression [41–48]. The EWOG–MDS study demonstrated TCR-Vβ CDR3 skewing of BM in 40% of patients with RCC [41]. The same group found minor PNH clones in erythrocytes and/or granulocytes isolated from 36 of 87 (41%) patients with RCC [42]. The frequencies of these abnormalities among patients with RCC were comparable to those among patients with AA, implicating an immune-mediated pathogenesis in both diseases.

Further studies characterizing the molecular background of AA and RCC may facilitate differentiation between the two diseases and inform the validity of future classification systems. By using NGS, somatic mutations were frequently detected in myeloid malignancy-related genes, such as, *DNMT3A*, *BCOR*, and *ASXL1* in patients with adult MDS. To characterize the genetic background of childhood AA, RCC, and RCMD; we performed WES in 37 children with BMFs [49]. Furthermore, target sequencing for 88 IBMFs-associated genes and 96 myeloid malignancy-related genes were combined with WES. Blood samples obtained from 160 children with BMFs were examined. Of the patients analyzed by WES, somatic mutations were identified in 17 patients (46%), with no novel mutations detected using WES. In target sequencing, we observed somatic mutations in 33 children (21%). *BCOR* (*n* = 11) and *PIGA* (*n* = 4), previously proposed to be immune-escaped mutations in an adult study [50], were also recurrently mutated in children. However, the mutational frequency of myeloid malignancy-related genes, including *DNMT3A* and *ASXL1*, was substantially lower than in adult patients. We further compared the frequency of somatic mutations among AA, RCC, and RCMD groups and observed no statistically significant differences. The frequencies of somatic mutations were 7% in AA, 10% in RCC, and 23% in RCMD, which were much lower than adult MDS, wherein one or more somatic mutations are identified in over 80% of patients. The distribution of identified gene mutations was similar between AA and RCC, which predominately included *BCOR* [49]. Conversely, the prevalence of MDS-related gene mutations in patients with RCMD indicates a different molecular pathogenesis from AA/RCC. The results of our genetic analysis indicate that the distinction of RCC from AA with certainty is difficult, even with the use of molecular studies implementing NGS [49].

THE TREATMENT ALGORITHM FOR SEVERE AA IN CHILDREN

As the RCC proposal remains controversial, has not been universally accepted, and treatments are the same as those for AA, we will only

TABLE 21.1 Bone Marrow Transplantation (BMT) From an HLA Matched Family Donor (MFD) in Children With Severe Aplastic Anemia (SAA)

Study	N. of patients	Median age, y (range)	GR (%)	aGVHD (%)	cGVHD (%)	OS (%)
Kojima et al. (2000)[a]	37	9 (1–17)	0	6	3	97 at 10 years
Yagasaki et al. (2010)[b]	30	10 (0.8–18)	3	0	3	100 at 10 years
Szpecht et al. (2012)[c]	48	12.1 (2–18)	4.9	NA	NA	91.7 at 5 years
Kikuchi et al. (2013)[d]	329	10 (0–15)	8.8	13.4	17.3	89.7 at 10 years
Chen et al. (2013)[e]	53	10.5 (0.8–18)	15.1	14.3	19.6	90.6 at 5 years
Yoshida et al. (2014)[f]	213	9 (0–16)	2	12	6	92 at 10 years
Dufour et al. (2014)[g]	394	15 (12–17.5)	8	12	8	86 at 3 years
Dufour et al. (2015)[h]	396	2.7 (0–12)	2	8	6	91 at 3 years
Dufour et al. (2015)[i]	87	8.9 (0.5–18.6)	1	NA	NA	91 at 2 years

aGVHD, Acute graft-versus-host disease; cGVHD, chronic graft-versus-host disease; GR, graft rejection; HLA, human leukocyte antigen; N, number; NA, not available; OS, overall survival.

[a] Kojima S, Horibe K, Inaba J, et al. Long-term outcome of acquired aplastic anaemia in children: comparison between immunosuppressive therapy and bone marrow transplantation. Br J Haematol 2000;111:321–8.

[b] Yagasaki H, Takahashi Y, Hama A, et al. Comparison of matched-sibling donor BMT and unrelated donor BMT in children and adolescent with acquired severe aplastic anemia. Bone Marrow Transplant 2010;45:1508–13.

[c] Szpecht D, Gorczynska E, Kalwak K, et al. Matched sibling versus matched unrelated allogeneic hematopoietic stem cell transplantation in children with severe acquired aplastic anemia: experience of the Polish pediatric group for hematopoietic stem cell transplantation. Arch Immunol Ther Exp 2012;60:225–33.

[d] Kikuchi A, Yabe H, Kato K, et al. Long-term outcome of childhood aplastic anemia patients who underwent allogeneic hematopoietic SCT from an HLA-matched sibling donor in Japan. Bone Marrow Transplant 2013;48:657–60.

[e] Chen J, Lee V, Luo CJ, et al. Allogeneic stem cell transplantation for children with acquired severe aplastic anaemia: a retrospective study by the Viva-Asia Blood and Marrow Transplantation Group. Br J Haematol 2013;162:383–91.

[f] Yoshida N, Kobayashi R, Yabe H, et al. First-line treatment for severe aplastic anemia in children: bone marrow transplantation from a matched family donor versus immunosuppressive therapy. Haematologica 2014;99:1784–91.

[g] Dufour C, Pillon M, Passweg J, et al. Outcome of aplastic anemia in adolescence: a survey of the Severe Aplastic Anemia Working Party of the European Group for Blood and Marrow Transplantation. Haematologica 2014;99:1574–81.

[h] Dufour C, Pillon M, Socie G, et al. Outcome of aplastic anaemia in children. A study by the severe aplastic anaemia and paediatric disease working parties of the European Group Blood and Marrow Transplant. Br J Haematol 2015;169:565–73.

[i] Dufour C, Veys P, Carraro E, et al. Similar outcome of upfront-unrelated and matched sibling stem cell transplantation in idiopathic paediatric aplastic anaemia. A study on behalf of the UK Paediatric BMT Working Party, Paediatric Diseases Working Party and Severe Aplastic Anaemia Working Party of EBMT. Br J Haematol 2015;171:585–94.

discuss a treatment algorithm for severe AA in the present review. For children with SAA, BMT and IST have become accepted as the standard treatments during the past three decades. The current guideline recommends BMT from a MFD as the treatment of choice for childhood SAA. Table 21.1 summarizes the recently reported results of allogeneic BMT from a MFD in childhood SAA. The probability of overall survival exceeded 90% in the majority of studies [5,6,9,16,51].

Currently, the response rate to IST in children with AA is reportedly 30–70%, with an overall long-term survival rate of 90% [10,12,13]. Approximately one-third of patients are not expected to respond to IST, and 10–30% of responders relapse after initial therapy (Table 21.2) [11,12,52].

Recently, the Japanese group compared the outcome of children with SAA who received a BMT from a MFD ($n = 213$) or IST ($n = 386$) as first-line treatment [9]. While the 10-year

TABLE 21.2   Immune Suppressive Therapy With ATG and Cyclosporine in Children With AA and/or RCC

Study	Median age, y (range)	Disease	ATG type	N. of patients	6m response (%)	Relapse (%)	CE (%)	OS (%)
Kojima et al. (2000)[a]	9 (1–18)	AA	Horse	119	68	22	6	88 at 3 years
Saracco et al. (2008)[b]	9 (0.6–19)	AA	Horse	42	71	16	15	83 at 10 years
Scheinberg et al. (2008)[c]	<18	AA	Horse	77	77	33	8.5	80 at 10 years
Kamio et al. (2011)[d]	8.1 (0–17)	AA	Horse	441	60	11.9	4	87 at 10 years
Takahashi et al. (2013)[e]	9 (1–15)	AA	Rabbit	40	47.5	10.5	NA	93.8 at 2 years
Yoshimi et al. (2013)[f]	<18	AA	Horse	96	65	NA	NA	92 at 3 years
			Rabbit	32	34			92 at 3 years
Yoshida et al. (2014)[g]	9 (0–16)	AA	Horse	386	63	9.7	7	88 at 10 years
Hasegawa et al. (2014)[h]	7.8 (0.3–16.3)	RCC	Horse	29	45	15.4	7.6	87 at 5 years
Jeong et al. (2014)[i]	8 (0–17)	AA	Horse	297	60	7	6	92 at 10 years
			Rabbit	158	55	15	5	87 at 10 years
Garanito et al. (2014)[j]	8.1 (1.6–15)	AA	Rabbit	26	34.6	26.5	8.3	73.6 at 5 years
Dufour et al. (2014)[k]	15 (12–17.5)	AA	Horse	118	45	NA	NA	82 at 3 years
			Rabbit	25				
Yoshimi et al. (2014)[l]	10 (1.4–18.5)	RCC	Horse	46	74	15	12	91 at 4 years
			Rabbit	49	53	9	4	85 at 4 years
Hama et al. (2015)[m]	8 (1–16)	AA	Horse	61	52	NA	8	85 at 10 years
		RCC	Horse	91	59	NA	5	97 at 10 years
Dufour et al. (2015)[n]	2.7 (0–12)	AA	Horse	133	49.6	NA	NA	87 at 3 years
			Rabbit	34	29.4			
Pawelec et al. (2015)[o]	10.5 (0.5–17.5)	AA	Rabbit	63	44.4	14.3	3.6	67 at 10 years

AA, Aplastic anemia; ATG, antithymocyte globulin; CE, clonal evolution; m, months; N, number; NA, not available; OS, overall survival; RCC, refractory cytopenia of childhood.

[a] Kojima S, Hibi S, Kosaka Y, et al. Immunosuppressive therapy using antithymocyte globulin, cyclosporine, and danazol with or without human granulocyte colony-stimulating factor in children with acquired aplastic anemia. Blood 2000;96:2049–54.

[b] Saracco P, Quarello P, Iori AP, et al. Cyclosporin A response and dependence in children with acquired aplastic anaemia: a multicentre retrospective study with long-term observation follow-up. Br J Haematol 2008;140:197–205.

[c] Scheinberg P, Wu CO, Nunez O, Young NS. Long-term outcome of pediatric patients with severe aplastic anemia treated with antithymocyte globulin and cyclosporine. J Pediatr 2008;153:814–9.

[d] Kamio T, Ito E, Ohara A, et al. Relapse of aplastic anemia in children after immunosuppressive therapy: a report from the Japan Childhood Aplastic Anemia Study Group. Haematologica 2011;96:814–9.

[e] Takahashi Y, Muramatsu H, Sakata N, et al. Rabbit antithymocyte globulin and cyclosporine as first-line therapy for children with acquired aplastic anemia. Blood 2013;121:862–3.

[f] Yoshimi A, Niemeyer CM, Fuhrer MM, Strahm B. Comparison of the efficacy of rabbit and horse antithymocyte globulin for the treatment of severe aplastic anemia in children. Blood 2013;121:860–1.

[g] Yoshida N, Kobayashi R, Yabe H, et al. First-line treatment for severe aplastic anemia in children: bone marrow transplantation from a matched family donor versus immunosuppressive therapy. Haematologica 2014;99:1784–91.

[h] Hasegawa D, Chen X, Hirabayashi S, et al. Clinical characteristics and treatment outcome in 65 cases with refractory cytopenia of childhood defined according to the WHO 2008 classification. Br J Haematol 2014;166:758–66.

[i] Jeong DC, Chung NG, Cho B, et al. Long-term outcome after immunosuppressive therapy with horse or rabbit antithymocyte globulin and cyclosporine for severe aplastic anemia in children. Haematologica 2014;99:664–71.

[j] Garanito MP, Carneiro JD, Odone Filho V, Scheinberg P. Outcome of children with severe acquired aplastic anemia treated with rabbit antithymocyte globulin and cyclosporine A. J Pediatr 2014;90:523–7.

[k] Dufour C, Pillon M, Passweg J, et al. Outcome of aplastic anemia in adolescence: a survey of the Severe Aplastic Anemia Working Party of the European Group for Blood and Marrow Transplantation. Haematologica 2014;99:1574–81.

[l] Yoshimi A, van den Heuvel-Eibrink MM, Baumann I, et al. Comparison of horse and rabbit antithymocyte globulin in immunosuppressive therapy for refractory cytopenia of childhood. Haematologica 2014;99:656–63.

[m] Hama A, Takahashi Y, Muramatsu H, et al. Comparison of long-term outcomes between children with aplastic anemia and refractory cytopenia of childhood who received immunosuppressive therapy with antithymocyte globulin and cyclosporine. Haematologica 2015;100:1426–33.

[n] Dufour C, Pillon M, Socie G, et al. Outcome of aplastic anaemia in children. A study by the severe aplastic anaemia and paediatric disease working parties of the European Group Blood and Marrow Transplant. Br J Haematol 2015;169:565–73.

[o] Pawelec K, Salamonowicz M, Panasiuk A, et al. First-line immunosuppressive treatment in children with aplastic anemia: rabbit antithymocyte globulin. Adv Exp Med Biol 2015;836:55–62.

overall survival did not differ between the patients treated with IST and BMT (88% vs. 92%), the 10-year failure-free survival in patients receiving IST was significantly inferior to that in patients undergoing BMT (56% vs. 87%) [9]. The European group further analyzed the outcomes of 563 children aged 0–12 years with AA [5]. The 3-year overall survival rates after upfront BMT ($n = 396$) from a MFD or IST ($n = 167$) were reported to be 91 and 87%, respectively. However, the 3-year failure-free survival in the IST group was significantly worse than in the BMT group (33% vs. 87%) [5]. The findings of both large studies support the current algorithm for treatment decisions, which recommends BMT when a MFD is available in cases of childhood SAA.

Unrelated BMT is currently indicated for salvage therapy in children who fail to respond to IST. However, as summarized in Table 21.3, the results of BMT from a MUD have improved dramatically over the past two decades, with a proportion of studies reporting results comparable to BMT from a MFD [16,51,53,54]. Several investigators have advocated upfront MUD BMT if a suitable MUD can be found quickly [8,16]. Interestingly, our study found no difference in overall survival between MFD–BMT and MUD–BMT [51]. However, the incidence of posttransplant complications was substantially higher in MUD–BMT recipients than in MFD–BMT recipients. The incidence of acute and chronic graft-versus-host disease (GVHD) remains a major obstacle in MUD–BMT recipients. In addition, the incidence of viral reactivation was higher in MUD–BMT recipients than in MFD–BMT recipients [51].

As discussed, IST combining ATG and cyclosporine is currently administered as initial therapy in the majority of cases of childhood SAA as BMT is limited by the availability of MFD. Approximately 30–50% of patients are expected to not respond to IST. In fact, IST may be harmful as it increases the risk of serious infections and requires blood transfusions.

Therefore, biologic predictors of patient response to IST may prevent adverse events and decrease the cost of unnecessary treatments. The prediction of response to IST remains clinically challenging and novel biomarkers are highly anticipated. Several parameters and pretreatment blood counts have been proposed as markers of favorable IST responses, including shorter interval between diagnosis and treatment, higher absolute lymphocyte count, higher absolute reticulocyte count, and the presence of HLA–DR15 [47,55–58]. However, the predictive power of these markers is insufficient to inform clinical decisions, and patients with AA who lack an HLA identical sibling donor are still placed on IST without stratification. Minor PNH populations and short telomere length have been reported as predictive biomarkers of IST responsiveness in AA [59–64]. We evaluated the predictive power of two parameters and identified the absence of a PNH population and shorter telomere length as independent unfavorable predictors of IST [65]. The cohort was stratified into a poor prognosis group (PNH negative and shorter telomere length) and good prognosis group (PNH positive and/or longer telomere length). The response rates in the poor prognosis and good prognosis groups at 6 months were 19 and 70%, respectively. The combined absence of a minor PNH population and a short telomere length was determined to be an efficient predictor of poor response to IST (Fig. 21.2) [65].

Further, we recently demonstrated markedly increased thrombopoietin plasma levels to be associated with a poor IST response and lower failure-free survival in children with severe AA [66].

The findings of these studies may eventually be combined with blood count criteria to individualize therapy in AA children and may facilitate decision making regarding the choice between IST and upfront BMT as initial therapy based on the likelihood of IST efficacy.

TABLE 21.3 BMT From an HLA Matched Unrelated Donor (MUD) in Children With SAA

Study	N. of patients	Median age, y (range)	Conditioning	GR (%)	aGVHD (%)	cGVHD (%)	OS (%)
Kojima et al. (2001)[a]	15	11 (3–19)	CY/ATG/TBI	0	33	13	100 at 4 years
Vassiliou et al. (2001)[b]	8	7 (0–10)	CY/CP/TBI	0	25	0	100 at 3 years
Benesch et al. (2004)[c]	9	11 (6–16)	High CD34+ cell dose; TCD; CY/ATG/TLI; or TBI ± TT	NA	0	0	89 at 4 years
Kang et al. (2004)[d]	5	13 (7–18)	CY/FLU/ATG	0	0	0	80 at 2 years
Bunin et al. (2005)[e]	12	9 (1–20)	Partial TCD; TBI + CY/ Ara-C or Cy/TT or ATG	0	33	25	75 at 4 years
Perez-Albuerne et al. (2008)[f]	195	10	Various	15	43	35	51 at 5 years
Kosaka et al. (2008)[g]	31	8 (0–17)	CY/ATG/TBI; FLU/CY/ ATG/TBI	16	13	13	93 at 3 years
Yagasaki et al. (2010)[h]	31	9 (3–17)	Various	6	37	27	93.8 at 10 years
Szpecht et al. (2012)[i]	38	10.5 (3–23)	Various	19.5	NA	NA	64 at 5 years
Dufour et al. (2015)[j]	29	8.9 (0.5–18.6)	Various	4	NA	NA	96 at 2 years

aGVHD, acute graft-versus-host disease; ATG, antithymocyte globulin; cGVHD, chronic graft-versus-host disease; CP, alemtuzumab; CY, cyclophosphamide; Flu, fludarabine; GR, graft rejection; N, number; NA, not available; OS, overall survival, TBI, total body irradiation; TCD, T-cell depletion; TLI, total lymphoid irradiation; TT, thiotepa; Various, indicates 4 or more conditioning regimens.

[a] Kojima S, Inaba J, Yoshimi A, et al. Unrelated donor marrow transplantation in children with severe aplastic anaemia using cyclophosphamide, anti-thymocyte globulin and total body irradiation. Br J Haematol 2001;114:706–11.

[b] Vassiliou GS, Webb DK, Pamphilon D, Knapper S, Veys PA. Improved outcome of alternative donor bone marrow transplantation in children with severe aplastic anaemia using a conditioning regimen containing low-dose total body irradiation, cyclophosphamide and Campath. Br J Haematol 2001;114:701–5.

[c] Benesch M, Urban C, Sykora KW, et al. Transplantation of highly purified CD34+ progenitor cells from alternative donors in children with refractory severe aplastic anaemia. Br J Haematol 2004;125:58–63.

[d] Kang HJ, Shin HY, Choi HS, Ahn HS. Fludarabine, cyclophosphamide plus thymoglobulin conditioning regimen for unrelated bone marrow transplantation in severe aplastic anemia. Bone Marrow Transplant 2004;34:939–43.

[e] Bunin N, Aplenc R, Iannone R, et al. Unrelated donor bone marrow transplantation for children with severe aplastic anemia: minimal GVHD and durable engraftment with partial T cell depletion. Bone Marrow Transplant 2005;35:369–73.

[f] Perez-Albuerne ED, Eapen M, Klein J, et al. Outcome of unrelated donor stem cell transplantation for children with severe aplastic anemia. Br J Haematol 2008;141:216–23.

[g] Kosaka Y, Yagasaki H, Sano K, et al. Prospective multicenter trial comparing repeated immunosuppressive therapy with stem-cell transplantation from an alternative donor as second-line treatment for children with severe and very severe aplastic anemia. Blood 2008;111:1054–9.

[h] Yagasaki H, Takahashi Y, Hama A, et al. Comparison of matched-sibling donor BMT and unrelated donor BMT in children and adolescent with acquired severe aplastic anemia. Bone Marrow Transplant 2010;45:1508–13.

[i] Szpecht D, Gorczynska E, Kalwak K, et al. Matched sibling versus matched unrelated allogeneic hematopoietic stem cell transplantation in children with severe acquired aplastic anemia: experience of the Polish pediatric group for hematopoietic stem cell transplantation. Arch Immunol Ther Exp 2012;60:225–33.

[j] Dufour C, Veys P, Carraro E, et al. Similar outcome of upfront-unrelated and matched sibling stem cell transplantation in idiopathic paediatric aplastic anaemia. A study on behalf of the UK Paediatric BMT Working Party, Paediatric Diseases Working Party and Severe Aplastic Anaemia Working Party of EBMT. Br J Haematol 2015;171:585–94.

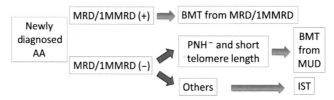

FIGURE 21.2 The treatment algorithm for children with SAA. Patients with available MRD/HLA1 locus mismatched related donor (1MMRD), first-line treatment is BMT. For patients without MRD/1MMRD, upfront BMT from a matched unrelated donor MUD could be considered as a potential option in children with the combination of paroxysmal nocturnal hemoglobinuria PNH⁻ clone and short telomere length. IST is considered first-line therapy for the others.

References

[1] Barrett J, Saunthararajah Y, Molldrem J. Myelodysplastic syndrome and aplastic anemia: distinct entities or diseases linked by a common pathophysiology? Semin Hematol 2000;37:15–29.

[2] Baumann I, Niemeyer CM, Bennett JM, Shannon K. Childhood myelodysplastic syndrome. In: Swerdlow SH, Campo E, Harris NL, editors. WHO classification of tumors of haematopoietic and lymphoid tissues. 4th ed. Lyon, France: IARC Press; 2008.

[3] Bacigalupo A, Brand R, Oneto R, et al. Treatment of acquired severe aplastic anemia: bone marrow transplantation compared with immunosuppressive therapy—The European Group for Blood and Marrow Transplantation experience. Semin Hematol 2000;37:69–80.

[4] Doney K, Leisenring W, Storb R, Appelbaum FR. Primary treatment of acquired aplastic anemia: outcomes with bone marrow transplantation and immunosuppressive therapy. Seattle Bone Marrow Transplant Team. Ann Intern Med 1997;126:107–15.

[5] Dufour C, Pillon M, Socie G, et al. Outcome of aplastic anaemia in children. A study by the severe aplastic anaemia and paediatric disease working parties of the European Group Blood and Bone Marrow Transplant. Br J Haematol 2015;169:565–73.

[6] Kojima S, Horibe K, Inaba J, et al. Long-term outcome of acquired aplastic anaemia in children: comparison between immunosuppressive therapy and bone marrow transplantation. Br J Haematol 2000;111:321–8.

[7] Locasciulli A, van't Veer L, Bacigalupo A, et al. Treatment with marrow transplantation or immunosuppression of childhood acquired severe aplastic anemia: a report from the EBMT SAA Working Party. Bone Marrow Transplant 1990;6:211–7.

[8] Samarasinghe S, Webb DK. How I manage aplastic anaemia in children. Br J Haematol 2012;157:26–40.

[9] Yoshida N, Kobayashi R, Yabe H, et al. First-line treatment for severe aplastic anemia in children: bone marrow transplantation from a matched family donor versus immunosuppressive therapy. Haematologica 2014;99:1784–91.

[10] Jeong DC, Chung NG, Cho B, et al. Long-term outcome after immunosuppressive therapy with horse or rabbit antithymocyte globulin and cyclosporine for severe aplastic anemia in children. Haematologica 2014;99:664–71.

[11] Kojima S, Hibi S, Kosaka Y, et al. Immunosuppressive therapy using antithymocyte globulin, cyclosporine, and danazol with or without human granulocyte colony-stimulating factor in children with acquired aplastic anemia. Blood 2000;96:2049–54.

[12] Locasciulli A, Oneto R, Bacigalupo A, et al. Outcome of patients with acquired aplastic anemia given first line bone marrow transplantation or immunosuppressive treatment in the last decade: a report from the European Group for Blood and Marrow Transplantation (EBMT). Haematologica 2007;92:11–8.

[13] Scheinberg P, Wu CO, Nunez O, Young NS. Long-term outcome of pediatric patients with severe aplastic anemia treated with antithymocyte globulin and cyclosporine. Journal Pediatr 2008;153:814–9.

[14] Bacigalupo A, Marsh JC. Unrelated donor search and unrelated donor transplantation in the adult aplastic anaemia patient aged 18-40 years without an HLA-identical sibling and failing immunosuppression. Bone Marrow Transplant 2013;48:198–200.

[15] Bacigalupo A, Oneto R, Bruno B, et al. Current results of bone marrow transplantation in patients with acquired severe aplastic anemia. Report of the European Group for Blood and Marrow Transplantation. On behalf of the Working Party on Severe Aplastic Anemia of the European Group for Blood and Marrow Transplantation. Acta Haematol 2000;103:19–25.

[16] Dufour C, Veys P, Carraro E, et al. Similar outcome of upfront-unrelated and matched sibling stem cell transplantation in idiopathic paediatric aplastic anaemia. A study on behalf of the UK Paediatric BMT Working Party, Paediatric Diseases Working Party and Severe Aplastic Anaemia Working Party of EBMT. Br J Haematol 2015;171:585–94.

[17] Viollier R, Socie G, Tichelli A, et al. Recent improvement in outcome of unrelated donor transplantation for aplastic anemia. Bone Marrow Transplant 2008;41:45–50.

[18] Fogarty PF, Yamaguchi H, Wiestner A, et al. Late presentation of dyskeratosis congenita as apparently acquired aplastic anaemia due to mutations in telomerase RNA. Lancet 2003;362:1628–30.

[19] Liang J, Yagasaki H, Kamachi Y, et al. Mutations in telomerase catalytic protein in Japanese children with aplastic anemia. Haematologica 2006;91:656–8.

[20] Vulliamy TJ, Walne A, Baskaradas A, Mason PJ, Marrone A, Dokal I. Mutations in the reverse transcriptase component of telomerase (TERT) in patients with bone marrow failure. Blood Cells Mol Dis 2005;34:257–63.

[21] Yamaguchi H, Baerlocher GM, Lansdorp PM, et al. Mutations of the human telomerase RNA gene (TERC) in aplastic anemia and myelodysplastic syndrome. Blood 2003;102:916–8.

[22] Ghemlas I, Li H, Zlateska B, et al. Improving diagnostic precision, care and syndrome definitions using comprehensive next-generation sequencing for the inherited bone marrow failure syndromes. J Med Genet 2015;52:575–84.

[23] Teo JT, Klaassen R, Fernandez CV, et al. Clinical and genetic analysis of unclassifiable inherited bone marrow failure syndromes. Pediatrics 2008;122:e139–48.

[24] Amstutz U, Andrey-Zurcher G, Suciu D, Jaggi R, Haberle J, Largiader CR. Sequence capture and next-generation resequencing of multiple tagged nucleic acid samples for mutation screening of urea cycle disorders. Clin Chem 2011;57:102–11.

[25] Berg JS, Evans JP, Leigh MW, et al. Next generation massively parallel sequencing of targeted exomes to identify genetic mutations in primary ciliary dyskinesia: implications for application to clinical testing. Genet Med 2011;13:218–29.

[26] DaRe JT, Vasta V, Penn J, Tran NT, Hahn SH. Targeted exome sequencing for mitochondrial disorders reveals high genetic heterogeneity. BMC Med Genet 2013;14:118.

[27] Meder B, Haas J, Keller A, et al. Targeted next-generation sequencing for the molecular genetic diagnostics of cardiomyopathies. Circ Cardiovasc Genet 2011;4:110–22.

[28] Rizzo JM, Buck MJ. Key principles and clinical applications of "next-generation" DNA sequencing. Cancer Prev Res 2012;5:887–900.

[29] Vasta V, Ng SB, Turner EH, Shendure J, Hahn SH. Next generation sequence analysis for mitochondrial disorders. Genome Med 2009;1:100.

[30] Zhang MY, Keel SB, Walsh T, et al. Genomic analysis of bone marrow failure and myelodysplastic syndromes reveals phenotypic and diagnostic complexity. Haematologica 2015;100:42–8.

[31] Lipson D, Capelletti M, Yelensky R, et al. Identification of new ALK and RET gene fusions from colorectal and lung cancer biopsies. Nat Med 2012;18:382–4.

[32] Okuno Y, Muramatsu H, Yoshida k, et al. Target gene sequencing for genetic diagnosis of congenital bone marroe failure syndromes. Japan J Clin Hematol 2015;56:243.

[33] Yoshimi A, Baumann I, Fuhrer M, et al. Immunosuppressive therapy with anti-thymocyte globulin and cyclosporine A in selected children with hypoplastic refractory cytopenia. Haematologica 2007;92:397–400.

[34] Kardos G, Baumann I, Passmore SJ, et al. Refractory anemia in childhood: a retrospective analysis of 67 patients with particular reference to monosomy 7. Blood 2003;102:1997–2003.

[35] Niemeyer CM, Baumann I. Classification of childhood aplastic anemia and myelodysplastic syndrome. Hematol Am Soc Hematol Educ Program 2011;2011:84–9.

[36] Baumann I, Fuhrer M, Behrendt S, et al. Morphological differentiation of severe aplastic anaemia from hypocellular refractory cytopenia of childhood: reproducibility of histopathological diagnostic criteria. Histopathology 2012;61:10–7.

[37] Forester CM, Sartain SE, Guo D, et al. Pediatric aplastic anemia and refractory cytopenia: a retrospective analysis assessing outcomes and histomorphologic predictors. Am J Hematol 2015;90:320–6.

[38] Yang W, Zhang P, Hama A, Ito M, Kojima S, Zhu X. Diagnosis of acquired bone marrow failure syndrome during childhood using the 2008 World Health Organization classification system. Int J Hematol 2012;96:34–8.

[39] Yoshimi A, van den Heuvel-Eibrink MM, Baumann I, et al. Comparison of horse and rabbit antithymocyte globulin in immunosuppressive therapy for refractory cytopenia of childhood. Haematologica 2014;99:656–63.

[40] Hama A, Takahashi Y, Muramatsu H, et al. Comparison of long-term outcomes between children with aplastic anemia and refractory cytopenia of childhood who received immunosuppressive therapy with antithymocyte globulin and cyclosporine. Haematologica 2015;100:1426–33.

[41] Aalbers AM, van den Heuvel-Eibrink MM, Baumann I, et al. T-cell receptor Vbeta skewing frequently occurs in refractory cytopenia of childhood and is associated with an expansion of effector cytotoxic T cells: a prospective study by EWOG-MDS. Blood Cancer J 2014;4:e209.

[42] Aalbers AM, van der Velden VH, Yoshimi A, et al. The clinical relevance of minor paroxysmal nocturnal hemoglobinuria clones in refractory cytopenia of childhood: a prospective study by EWOG-MDS. Leukemia 2014;28:189–92.

[43] Epling-Burnette PK, Painter JS, Rollison DE, et al. Prevalence and clinical association of clonal T-cell expansions in myelodysplastic syndrome. Leukemia 2007;21:659–67.

[44] Epperson DE, Nakamura R, Saunthararajah Y, Melenhorst J, Barrett AJ. Oligoclonal T cell expansion in myelodysplastic syndrome: evidence for an autoimmune process. Leukemia Res 2001;25:1075–83.

[45] Fozza C, Contini S, Galleu A, et al. Patients with myelodysplastic syndromes display several T-cell expansions, which are mostly polyclonal in the CD4(+) subset and oligoclonal in the CD8(+) subset. Exp Hematol 2009;37:947–55.

[46] Kochenderfer JN, Kobayashi S, Wieder ED, Su C, Molldrem JJ. Loss of T-lymphocyte clonal dominance in patients with myelodysplastic syndrome responsive to immunosuppression. Blood 2002;100:3639–45.

[47] Maciejewski JP, Follmann D, Nakamura R, et al. Increased frequency of HLA-DR2 in patients with paroxysmal nocturnal hemoglobinuria and the PNH/aplastic anemia syndrome. Blood 2001;98:3513–9.

[48] Wang H, Chuhjo T, Yasue S, Omine M, Nakao S. Clinical significance of a minor population of paroxysmal nocturnal hemoglobinuria-type cells in bone marrow failure syndrome. Blood 2002;100:3897–902.

[49] Narita A, Okuno Y, Muramatsu H, et al. Genetic background of idiopathic bone marrow failure syndromes in children. Blood 2015;126:3610.

[50] Yoshizato T, Dumitriu B, Hosokawa K, et al. Somatic mutations and clonal hematopoiesis in aplastic anemia. N Engl J Med 2015;373:35–47.

[51] Yagasaki H, Takahashi Y, Hama A, et al. Comparison of matched-sibling donor BMT and unrelated donor BMT in children and adolescent with acquired severe aplastic anemia. Bone Marrow Transplant 2010;45:1508–13.

[52] Fuhrer M, Rampf U, Baumann I, et al. Immunosuppressive therapy for aplastic anemia in children: a more severe disease predicts better survival. Blood 2005;106:2102–4.

[53] Kojima S, Inaba J, Yoshimi A, et al. Unrelated donor marrow transplantation in children with severe aplastic anaemia using cyclophosphamide, anti-thymocyte globulin and total body irradiation. Br J Haematol 2001;114:706–11.

[54] Vassiliou GS, Webb DK, Pamphilon D, Knapper S, Veys PA. Improved outcome of alternative donor bone marrow transplantation in children with severe aplastic anaemia using a conditioning regimen containing low-dose total body irradiation, cyclophosphamide and Campath. Br J Haematol 2001;114:701–5.

[55] Chang MH, Kim KH, Kim HS, et al. Predictors of response to immunosuppressive therapy with antithymocyte globulin and cyclosporine and prognostic factors for survival in patients with severe aplastic anemia. Eur J Haematol 2010;84:154–9.

[56] Oguz FS, Yalman N, Diler AS, Oguz R, Anak S, Dorak MT. HLA-DRB1*15 and pediatric aplastic anemia. Haematologica 2002;87:772–4.

[57] Scheinberg P, Wu CO, Nunez O, Young NS. Predicting response to immunosuppressive therapy and survival in severe aplastic anaemia. Br J Haematol 2009;144:206–16.

[58] Yoshida N, Yagasaki H, Hama A, et al. Predicting response to immunosuppressive therapy in childhood aplastic anemia. Haematologica 2011;96:771–4.

[59] Afable MG 2nd, Shaik M, Sugimoto Y, et al. Efficacy of rabbit anti-thymocyte globulin in severe aplastic anemia. Haematologica 2011;96(9):1269–75.

[60] Maciejewski JP, Rivera C, Kook H, Dunn D, Young NS. Relationship between bone marrow failure syndromes and the presence of glycophosphatidyl inositol-anchored protein-deficient clones. Br J Haematol 2001;115:1015–22.

[61] Sakaguchi H, Nishio N, Hama A, et al. Peripheral blood lymphocyte telomere length as a predictor of response to immunosuppressive therapy in childhood aplastic anemia. Haematologica 2014;99:1312–6.

[62] Sugimori C, Chuhjo T, Feng X, et al. Minor population of CD55– CD59– blood cells predicts response to immunosuppressive therapy and prognosis in patients with aplastic anemia. Blood 2006;107:1308–14.

[63] Tutelman PR, Aubert G, Milner RA, Dalal BI, Schultz KR, Deyell RJ. Paroxysmal nocturnal haemoglobinuria phenotype cells and leucocyte subset telomere length in childhood acquired aplastic anaemia. Br J Haematol 2014;164:717–21.

[64] Wang H, Chuhjo T, Yamazaki H, et al. Relative increase of granulocytes with a paroxysmal nocturnal haemoglobinuria phenotype in aplastic anaemia patients: the high prevalence at diagnosis. Eur J Haematol 2001;66:200–5.

[65] Narita A, Muramatsu H, Sekiya Y, et al. Paroxysmal nocturnal hemoglobinuria and telomere length predicts response to immunosuppressive therapy in pediatric aplastic anemia. Haematologica 2015;100:1546–52.

[66] Elmahdi S, Muramatsu H, Narita A, et al. Markedly high plasma thrombopoietin (TPO) level is a predictor of poor response to immunosuppressive therapy in children with acquired severe aplastic anemia. Pediatr Blood Cancer 2016;63:659–64.

Index

Printed in the United States
By Bookmasters